Peace Ethology

Peace Ethology

Behavioral Processes and Systems of Peace

Edited by

Peter Verbeek
University of Alabama at Birmingham
Alabama, USA

Benjamin A. Peters
University of Michigan
Michigan, USA

Registered Office(s)
John Wiley & Sons, Inc., 111 River Street, Hoboken, NJ 07030, USA
John Wiley & Sons Ltd, The Atrium, Southern Gate, Chichester, West Sussex, PO19 8SQ, UK

Editorial Office
9600 Garsington Road, Oxford, OX4 2DQ, UK

For details of our global editorial offices, customer services, and more information about Wiley products visit us at www.wiley.com.

Wiley also publishes its books in a variety of electronic formats and by print-on-demand. Some content that appears in standard print versions of this book may not be available in other formats.

Library of Congress Cataloging-in-Publication Data

Names: Verbeek, Peter, 1952– editor. | Peters, Benjamin A., 1971– editor.
Title: Peace ethology : behavioral processes and systems of peace / edited by Peter Verbeek,
    Benjamin A. Peters.
Description: Hoboken, NJ : John Wiley & Sons, 2018. | Includes bibliographical references and index. |
Identifiers: LCCN 2018008628 (print) | LCCN 2018014557 (ebook) | ISBN 9781118922538 (pdf) |
    ISBN 9781118922521 (epub) | ISBN 9781118922514 (cloth)
Subjects: LCSH: Peace-building–Research. | Peace–Research. | Human behavior.
Classification: LCC HM1126 (ebook) | LCC HM1126 .P413 2018 (print) | DDC 327.1/72–dc23
LC record available at https://lccn.loc.gov/2018008628

Cover Design: Wiley
Cover Image: Courtesy of Peter Verbeek

Set in 10/12pt Warnock by SPi Global, Pondicherry, India
Printed in Singapore by C.O.S. Printers Pte Ltd

10 9 8 7 6 5 4 3 2 1

Dedicated on behalf of Peter to Pieter and Christina, Teruo and Toshiko, and Mamiko. And dedicated on behalf of Benjamin to Sayaka, Noah, Kai, and Sola. Peace to you all.

Contents

List of Contributors

Otto Adang
Department of Psychology, Faculty
of Behavioral and Social Sciences,
University of Groningen, The Netherlands
and
Police Academy of The Netherlands,
Research Department, The Netherlands

Saleem H. Ali
Department of Geography & Center
for Energy and Environmental Policy,
University of Delaware, Newark, USA

Gabriel van den Brink
Centrum Èthos, Faculty of Philosophy,
Free University, Amsterdam,
The Netherlands

Douglas P. Fry
Department of Anthropology,
University of Alabama at Birmingham
(UAB), Birmingham, USA

Ellen Furnari
Webster University, Pleasant Hill, USA

Daniel Hyslop
Institute for Economics and Peace,
St. Leonards, Australia

Wiebren S. Jansen
Department of Social, Health &
Organizational Psychology, Utrecht
University, Utrecht, The Netherlands

Misja van de Klomp
Applied Safety & Security
Studies, HU University of Applied
Sciences Utrecht, Utrecht,
The Netherlands

Kathleen Kostelny
Columbia Group for Children in
Adversity, Beaverdam, USA

Harry Kunneman
University of Humanistic Studies
Utrecht, The Netherlands

Thomas Morgan
Institute for Economics and Peace,
St. Leonards, Australia

Darcia Narvaez
Department of Psychology,
University of Notre Dame,
Notre Dame, USA

Sabine Otten
Department of Psychology, Faculty
of Behavioral and Social Sciences,
Groningen University,
The Netherlands

Benjamin A. Peters
Global Scholars Program, College
of Literature, Science, and the Arts,
University of Michigan, USA

Joám Evans Pim
Center for Global Nonkilling,
Honolulu, USA
and
Åbo Akademi University, Finland

Teresa Romero
Joseph Banks Laboratories, School of
Life Sciences, University of Lincoln,
Lincoln, UK

Cary J. Roseth
Educational Psychology and Educational
Technology Program, Michigan State
University, East Lansing, USA

Robert M. Sapolsky
Departments of Biological Sciences,
Neurology, and Neurosurgery,
Stanford University, Stanford, USA

Juliette Schaafsma
Tilburg School of Humanities,
Tilburg University, The Netherlands

Nurit Shnabel
The School of Psychological Sciences,
Tel-Aviv University, Israel

Sara Stronks
University of Applied Sciences, School of
Governance, Law & Urban Development,
The Netherlands

Peter Verbeek
Department of Anthropology,
University of Alabama at Birmingham
(UAB), Birmingham, USA

Todd Walters
International Peace Park Expeditions,
Takoma Park, USA

Michael Wessells
Columbia University, Beaverdam, USA

Foreword

Robert M. Sapolsky

It can be awe-inspiring, if deeply puzzling at times, to contemplate the human capacity for obsessive specialization, to consider the range of things that humans can devote their lives to in study and scholarship. You can be a coniologist or a caliologist – experts in the sciences of dust and of birds' nests, respectively – and spend years in monastic solitude, becoming the definitive expert on some subspecialty of each. There's batologists and brontologists, studying brambles and thunder, doing their research with manic focus that is at the cost of vacations, hobbies, or personal relationships. Or there's vexillologists and zygologists, with their hard-earned, dazzling knowledge of flags and of methods for fastening things together. It just goes on and on – odontology and odonatology, phenology and phonology, parapsychology and parasitology. A rhinologist and a nosologist can meet, fall in love, and perhaps have a child who becomes a rhinological nosologist, studying the classification of diseases of the nose.

In recent decades, there has been the emergence of what must seem like one of the most unlikely "-ology"'s of all, peace ethology, an emerging behavioral science of peace that is producing robust findings. And the notion of there being such a realm of scholarship must seem quixotic to many. This is the case for at least three reasons.

The first one is mammoth, and is obvious to anyone who has noted what humans have been up to in recent millennia. The capacity of humans for violence, and for victimization of the weak by the strong, is so great that devoting one's scholarly life to the scientific study of peace must feel like trying to document the beauty of snowflakes in the Sahara. When it comes to peace, we have a pretty dismal track record as a species, with our occasional capacity for living peaceably being barely maintained by a thin veneer of rules, laws, ethics, and morality.

But despite that, there is room for optimism. This is because, while it is initially hard to believe, we have been becoming more peaceful in recent centuries, have shown an extraordinary increase in empathy and for feeling moral imperatives to protect those in need. For the first time in recorded human history, the majority of Earth's people vote in electoral democracies; most leaders are opposed to the likes of slavery, child labor, and domestic violence; nearly all nations are signatories to international agreements regarding the treatment of prisoners and of civilians in warfare, the banning of certain weapons, and the international criminalization of certain acts of war; and most such nations are willing to support apolitical multinational peacekeeping forces that can be sent anywhere on the globe. Sure, all of this is rife with hypocrisy, lip service, and corruption. But it is still a stunningly different world than it was a few centuries ago.

One of the main points of this volume is that there is little reason anymore to think that human prosociality, when it does flourish, is solely or even mostly the outcome of that thin veneer of culture, of each society's equivalent of fire and brimstone. This conclusion is based on a trio of fields of study that have challenged our views of the roots of human goodness:

1) Rather than being the outcome of features of culture specific to our species, some of the best of human behaviors and our core of prosociality are shared with numerous other primates. Yes, yes, other primates kill avidly, carrying out competitive infanticide, having organized intergroup violence, and systematically eradicating all the members of another group. But humans are not alone in having the capacities for empathy, altruism, and cooperation among nonrelatives, reconciliation, a sense of justice, and third-party peacekeeping. Humans may do all those in remarkably abstract ways – for example, we can be galvanized into prosocial activism by the plight of a fictional character in a novel ("So you're the little woman who wrote the book that started this great war," Abraham Lincoln reportedly said to Harriet Beecher Stowe). But when we do so, the roots of those impulses are not confined to our own species.

2) Developmental psychologists have shown how the rudiments of empathy and a sense of justice are there in kids, in toddlers, even in preverbal infants. Humans of astonishingly young ages can detect instances of unequal treatment, have a preference for pro- over antisocial individuals, and choose to mete out punishment accordingly. For example, have toddlers observe puppets interacting, some being mean to others, some being kind; afterward, given a choice, they would rather hold and play with the kind puppets, and will advocate giving a treat to a good puppet over one who is a jerk – and all before such children are old enough to comprehend their first sermon.

3) Finally, there is little reason to think that the long arc of hominid history has been filled with warfare. Instead, the behavior of the few remaining contemporary hunter-gatherers, and the archeological and paleontological records, suggest that the vast majority of our time as a species has been spent in small hunter-gatherer bands that are fairly egalitarian in nature, and that have various means (e.g., a fusion/fission structure) to deal with conflict without escalated violence.

Collectively, these three bodies of work suggest that the salutary trends of the last few centuries do not represent humans breaking new grounds of prosociality, but rather something resembling a recovery to our pre-agricultural past.

Despite that, many might still view peace ethology skeptically for a second reason. This is because of a commonplace and simplistic view that "peace" merely equals the absence of conflict. Or, perhaps worse, that peace equals a level of conflict that people collectively deem to be tolerable and inevitable. When viewed this way, studying peace is somewhat akin to, say, biomedical scientists studying the absence of fever. Yet, as will be shown throughout the volume, the making and maintaining of peace is an intensely active process.

But despite that, the prospects of being a peace ethologist might still seem inauspicious, for a third reason that is closely related to the second one – the "-ology" part of *peace ethology* suggests a topic that is subject to scientific exploration, that has underlying rules and patterns. And for many, the notion that there are systematic ways in which peace can be fostered, that its facilitation can be a subject of scholarship, seems foolish. Yet, the scholarship is there and is quickly growing, in all sorts of areas. At the reductive end of things, neuroscientists are learning, for example, the circumstances in which the neuromodulator oxytocin promotes prosocial behavior and when it does the opposite; brain-imaging studies show that while the brain has an implicit, automatic tendency to make Us/Them dichotomies, it is incredibly easy to manipulate the dichotomizing process, turning Them's into Us's. Meanwhile, psychologists fruitfully explore how much our moral acts are the outcome of moral reasoning versus moral intuition, when one dominates the other, and with what sorts of outcomes. Game theorists and evolutionary biologists elucidate the circumstances where cooperation can be jumpstarted amid a sea of noncooperators. Sociologists, demographers, and geographers explore the time-honored *contact theory*, demonstrating rules for when contact between groups worsens conflict and when it lessens it. Anthropologists identify commonalities across cultures in means of conflict resolution. And people of heroic devotion, who might be classified (to borrow a term from molecular medicine) as "translational" scientists, learn the best ways to do some of the hardest tasks on Earth – the likes of setting up Truth and Reconciliation Commissions, or reintegrating child soldiers back into their communities.

Making peace and preserving it will never have anything akin to the laws of thermodynamics. Nonetheless, as this volume demonstrates, peace ethology is indeed now a rigorous intellectual and scientific venture, one with more consequences than those of nearly all of the other -ologists combined.

Acknowledgments

We are grateful to the editors and staff at Wiley for their unwavering support for this project. We are indebted to the chapter contributors who entrusted us with their work. Their work is an inspiration to us, and we are privileged to be able to share it with the world through this volume. The idea for the volume goes back to the 2013 Lorentz Center workshop entitled "Obstacles and Catalysts of Peaceful Behavior" at Leiden University. One of us co-organized the workshop with Douglas Fry, and several of the chapter contributors participated in it. Special thanks go to the Lorentz Center's Mieke Schutte, Henriette Jensenius, and Ikram Cakir for their kind support, and to the workshop sponsors for their generous financial backing. We also gratefully acknowledge Miyazaki International College (MIC) and its founder, Hisayasu Otsubo. We began our collaboration at MIC's pioneering School of International Liberal Arts, where we benefited from teaching and discussing peace ethology research with our intellectually curious and critically perspicacious students. In addition, MIC generously supported us through funding and time allocations for conference participation and research. We also thank the peace ethology students at the University of Alabama at Birmingham (UAB) for their critical thoughts and for keeping us honest and focused on peace ethology's future. Kacey Keith merits special mention for her graduate work at UAB on sharing the methods and findings of peace ethology with the community at large and for coining the fortuitous label of "prosocial learning for a prosocial species" for that effort. Last but not least, we recognize the scholars who came before us and blazed a trail for peace ethology. Without the vision and exemplary science of Theodore Lentz, Niko Tinbergen, Frans de Waal, and others, our own work on peace ethology would never have seen the light of day.

Peter Verbeek & Benjamin A. Peters

1

The Nature of Peace

Peter Verbeek and Benjamin A. Peters

At the time that we are writing the introductory chapter to this volume, 100 years after the start of a "war to end all wars" and 70 years after the end of World War II, the world is not at peace. While we are writing this chapter in relative comfort, an untold number of our fellow human beings of all ages are suffering the effects of direct or structural violence. Even here, in one of the most peaceful countries in the world, people suffer these effects when there is bullying, domestic violence, assault, rape, and homicide, and these ill effects extend to those who are victims of discrimination, labor exploitation, and poverty, to name only a few examples. And yet, we believe that this is a promising time for peace. We see new opportunities for peace in behavioral science, in the global policy arena, and in everyday life. And we propose that these basic and applied opportunities for peace are intertwined.

This book develops and advances the behavioral science of peace. It offers new concepts for integrating knowledge systems concerning peace across disciplines, and it provides examples of recent research on behavioral processes and systems of peace that illustrate the integrative framework that we propose. The book grew out of a weeklong interdisciplinary workshop at the Lorentz Center of Leiden University in the Netherlands entitled "Obstacles and Catalysts of Peaceful Behavior" (OCPB). Fifty-three scientists from three continents and a range of disciplines, including anthropology, ethology, evolutionary biology, neuroscience, political science, and psychology, attended the workshop in March 2013. Of the 23 authors in this book, 13 attended the workshop. While previous interdisciplinary gatherings at the Lorentz Center addressed behavioral aspects of peace ("Aggression and Peacemaking in an Evolutionary Context" in 2010 – see Fry 2013; and "Context, Causes and Consequences of Conflict" – see Kruk & Kruk de Bruin 2010), OCPB stood out due to its exclusive focus on peaceful behavior. One of the participants captured the synergistic mixture of topics addressed during the workshop and the promise this holds for the study of peace as follows: "It was very interesting to see how apparently disconnected realities, such as molecular biology, canine ethology, cooperation in primates, oxytocin, and Japan's Article 9, came together and made sense in developing an alternative insight on peaceful behavior." The aim of this book is to channel this synergy further by presenting a peace ethology approach to the behavioral processes and systems of peace.

Peace Ethology: Behavioral Processes and Systems of Peace, First Edition.
Edited by Peter Verbeek and Benjamin A. Peters.
© 2018 John Wiley & Sons Ltd. Published 2018 by John Wiley & Sons Ltd.

Operationalizing Peace Concepts

A traditional perspective on peace links it to the absence of direct violence, in particular organized mass killing in war (Galtung 1996, 2012). Other forms of direct violence implied in this *negative* notion of peace include the examples mentioned in this chapter such as physical bullying, assault, and homicide, and extend to torture and the intentional destruction of homes and communities of targeted victims (cf. Opotow 2012). The more recent *positive* notion of peace is based on the absence of structural violence (Galtung 1996, 2012). Structural violence in this context refers to harm caused to people through, for example, social injustice, discrimination, prejudice, social or moral exclusion, and poverty linked to these conditions, and their intended or unintended cultural justifications (cf. Galtung 2012). Christie (2012) interprets these two complementary perspectives on peace as "direct peace" and "structural peace," with the former achieved through peacemaking and the latter through peacebuilding (Table 1.1).

Conceptualizing peace as the absence of violence tends to concentrate intellectual and practical energy on the study of obstacles to peace at the relative expense of the study of catalysts to peace. Moreover, implicit in this approach is the notion of peace as a *state*, specifically a state that occurs with the absence of direct and structural violence. In this volume, we present a dynamic approach to peace. We investigate and discuss peace as *process*, more specifically a complex of *behavioral processes* and the *behavioral systems* that may ensue as a function of these processes. Our treatment of peace as process reflects a contemporary perspective of peace, both in practice and in science, as evidenced, for example, in Nobel Peace Prize Laureate Óscar Arias Sánchez' suggestion, "Peace is a never-ending process, the work of many decisions by many people in many countries. It is an attitude, a way of life, a way of solving problems and resolving conflicts" (Sánchez 1995 cited in Verbeek 2008). This is mirrored by psychologists Morton Deutsch and Peter Coleman, who propose, "Peace is never achieved, but rather is a process that is fostered by a variety of cognitive, affective, behavioral, structural, institutional, spiritual, and cultural components" (Deutsch & Coleman 2012). Going by these two quotes alone, we can identity multiple levels and domains at which the processes of peace can be measured, including "decision-making, attitudes, life-styles, and conflict resolution" (from Sánchez 1995), and "cognitive and emotional functioning,

Table 1.1 Three dimensions of peace: direct, structural, and sociative peace.

	Direct	Structural	Sociative
Violence	Direct violence[1]	Structural violence[1]	None/aggression[3]
Peace	**Direct peace**[2]	**Structural peace**[2]	**Sociative peace**[3]
	Negative[1]	Positive[1]	
	(Peacemaking)[2]	(Peacebuilding)[2]	(Peacekeeping)[4]

[1] Galtung (1996, 2012).
[2] Christie (2012).
[3] Gregor (1996), cited in Verbeek (2008).
[4] Verbeek (2013).
Note: Peace terms adopted in this chapter are in boldface font.
Source: Adapted from Christie (2012).

behavior, (social) structures, institutional functioning, spirituality, and culture" (from Deutsch & Coleman 2012).

The process-based concept of peace that we propose here transcends peace as a response to direct or structural violence (*direct peace* and *structural peace*) to include peace concerned with the preservation of harmony in relations, for example through the pursuit, establishment, or deepening of mutual or reciprocal interests, tolerance, helping and sharing, and the active avoidance of aggressive confrontations (*sociative peace*; Verbeek 2008; cf. Gregor 1996). Table 1.1 shows our three-dimensional concept of peace in comparison to previous conceptualizations.

Our approach to peace is comparative and transcends the human condition as we consider the natural origins and behavioral manifestations of peace across species (de Waal 2000; Verbeek this volume, Chapter 16) in conjunction with the evolved human potential for peace (Fry 2006, 2012; de Waal 2012). In nature, aggression and peace are not antithetical but, rather, linked in recurring relationships that express themselves in flexible phenotypes and evolving genotypes (Verbeek this volume, Chapter 16; 2013; Kunneman this volume, Chapter 15). Until about four decades ago, and similar to work on peace in humans, science focused almost exclusively on the aggressive dimension of natural relationships and virtually ignored nature's peaceful solutions to the propagation of life (Verbeek this volume, Chapter 16; 2013). However, the paradigm in behavioral science is shifting toward a new look at the interplay of aggression and peace in nature, and this allows for a fresh perspective on peace in human nature and how to draw on it (Verbeek this volume, Chapter 16; 2013; Fry this volume, Chapter 14; Kunneman this volume, Chapter 15).

We operationally define the natural phenomenon of aggression as behavior through which species, individuals, families, groups, and communities pursue active control of resources and the social environment at the expense of others (cf. de Boer in Kruk & Kruk-de Bruin 2010). In our view, aggression can be *species-typical* or *species-atypical*. The former is context-dependent aggressive behavior that is commonly shown by members of the species, while the latter is context-dependent aggressive behavior that is infrequently shown by members of the species (cf. Haller & Kruk 2006; Verbeek *et al.* 2007; Verbeek 2013). Violence, in our conceptual framework, is escalated aggressive behavior that is out of inhibitory control (de Boer *et al.* 2009). An important question in the context of the study of direct peace is whether war, as an organized form of direct violence, is species-typical or species-atypical for humans. Fry and Verbeek address this question in their respective chapters in this volume (see also Wrangham 1999; Sussman 2013; Verbeek 2013; and Wilson *et al.* 2014 for a range of comparative perspectives on this issue).

Like aggression, we view peace as a natural phenomenon that culture may modify. We operationally define peace as

> Behavioral processes and systems through which species, individuals, families, groups, and communities negate direct and structural violence (*direct peace*; *structural peace*), keep aggression in check or restore tolerance in its aftermath (*sociative peace*), maintain just institutions and equity (*structural peace*), and engage in reciprocally beneficial and harmonious interactions (*sociative peace*). (Table 1.1; Verbeek 2008, 2013; cf. Coleman & Deutsch 2012 and definitions contained therein)

Peace processes, in our conceptual framework, are sequential and interrelated behaviors that enable peaceful relations within and across social domains. Flourishing peace processes can give rise to and arise from *peace systems*, which we define as institutions or arrangements that pattern their members' interactions toward peace. Fry (2012) introduced the concept of peace systems at the level of nations and cultures, and we extend it herewith across species and social domains. Peace systems, thus defined, are patterns of social behavior that promote or sustain peace.

Observing Peace

Considering that peace transcends individual species and social domains, as the chapters in this volume demonstrate, we need a multidisciplinary or even transdisciplinary (Kunneman this volume, Chapter 15) approach to study and understand peace. This raises the issue of how to integrate different systems of knowledge (Galtung 2010). We deal with this by following up on ethologist Niko Tinbergen's call to apply the "aims and methods" of ethology to the study of war and peace (Tinbergen 1968). We scaffold our conceptual framework with an ethology of peace that applies ethology's four principal questions about the *proximate causation, development, function*, and *evolution* of behavior to the study of peace processes and systems (Tinbergen 1963; Verbeek 2008). Figure 1.1 models our approach.

Our peace ethology model shown in Figure 1.1 visualizes the flow of evolutionary (biological and cultural) and developmental inputs on behavioral peace processes and their proximate causes and consequences on individuals (selves), relationships, and institutions. It is in this interplay between and among individuals, relationships, and institutions that the functions of peace processes and systems come to the fore. The model is a feedback model, as initial effects of peace processes on individuals,

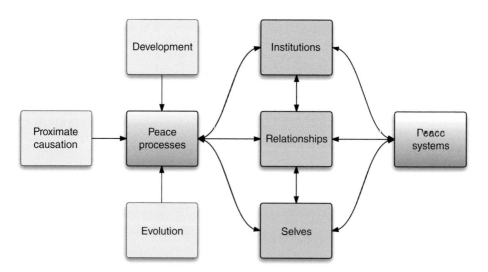

Figure 1.1 A peace ethology model of behavioral processes and systems of peace. *Note*: Peace processes as between and among individuals, families, groups, and communities.

relationships, and institutions are expected to generate and give form to subsequent peace processes. Positive-feedback loop reiterations involving individuals, relationships, and institutions can give rise to peace systems, which, in turn, feed back to pattern peace processes in space and time.

Selves, Relationships, and Institutions

Selves, relationships, and *institutions* are in themselves seen as processes in our model. For example, *selves,* in our conceptual framework, can develop as peaceful selves, in part as a function of peace process behaviors and experiences. We define the peaceful self as characterized by virtuous dispositions for benevolence and justice and efficacious in nonviolent conflict transformation and peacemaking (Verbeek *et al.* 2015). We suggest that the peaceful self is *enabled* by our evolved dispositions for peace as expressed, for example, in social behavioral dispositions (e.g., Jaeggi *et al.* 2010; Fry this volume, Chapter 14; Kunneman this volume, Chapter 15; Verbeek this volume, Chapter 16), emotional functioning (e.g., empathy: Preston & de Waal 2002; Decety & Jackson 2006), and associated brain mechanisms (e.g., Immordino-Yang *et al.* 2009; Krill & Platek 2012; Piper *et al.* 2015), *nurtured* in our evolved developmental niche (Narvaez this volume, Chapter 6; cf. Leckman *et al.* 2014), and *shaped* by narratives of solicitude and justice (e.g., Peters this volume, Chapter 11; Kunneman personal communication; cf. Ricoeur 1992; Howell & Larsen 2015). As the bidirectional arrows in the model indicate, selves, including peaceful selves, result from – and continue to be reciprocally affected by – relationships with others and the institutions in which these relationships may be embedded.

Relationships. Hinde (1979, 1987) proposed a useful distinction between social "interactions" and social "relationships." According to Hinde, an interaction (or relation) involves a series of interchanges over a limited span of time, and the behavior can be described in terms of the content of the interchanges (fighting, talking, kissing, etc.). Hinde proposed that if two individuals (and, by extension, families, groups, and communities) who know each other have a series of interactions over time, the course of each interaction might be influenced by experience in the preceding ones. In this case, we speak of those interacting as having a relationship. Inherent to Hinde's definition is the notion that relationships are behavioral processes, and we apply this notion to our model.

Institutions. As the number and frequency of interactions increase over time, and as relationships become routinized, implicit or explicit rules of behavior may emerge that pattern behavioral processes in the individuals embedded in those relationships. When these rules develop and persist to the point that individuals who did not participate in the original set of interactions that gave rise to them learn and follow the rules, we can say that an institution is emergent. As the bidirectional arrows of the model indicate, as changes in individual behaviors and relationships evolve, the institutions evolve as well. In this way, institutions can temporally transcend the lifetime of any one individual or relationship while remaining in a process of emergent flow. They persist to the degree that they underpin and shape the behavioral processes of individuals new to the institution. Thus, institutions structure subsequent behavioral processes including peace processes, and through ongoing changes, whether subtle or punctuated, they themselves change (Thelen 2003).

Research Questions

Our peace ethology process model affords and scaffolds a multilevel investigation of the behavioral processes and systems of peace by addressing ethological questions along and across its conceptual links. In terms of *proximate causation*, it allows us to ask: what biological, psychological, political, cultural, and environmental factors make peace processes happen at any given time, and how can learning and experience modify them? Regarding *development*, we can ask: when and how do peace processes first emerge in the behavioral repertoire of species, individuals, groups, communities, and cultures? And what is the capacity for change or transformation of peace processes within these developmental domains in response to different environmental conditions? Concerning *function*, we can ask: what are the immediate and delayed benefits of peace processes, and how do they affect the survival, well-being, and lifetime success of individuals, groups, communities, and cultures? And finally, with regard to *evolution*, we can ask: why and how did the ability to engage in peace processes evolve over generations and evolutionary time in species, individuals, groups, communities, and cultures? And how do peace processes compare across extant species, communities, and cultures? In the four subsections that follow, we review how the contributors to this volume address a number of these questions in their accounts of behavioral processes and systems of peace.

Answers from Research

Proximate Causation

In Part One of the volume, our contributors seek to identify and analyze biological, psychological, cultural, or environmental factors that make peace processes happen at a given time, and how learning and experience can modify them. In Chapter 2, Nurit Shnabel approaches these questions through social-psychological research on interpersonal and intergroup reconciliation by testing the Needs-Based Model. Shnabel shows how restoring victims' sense of agency and perpetrators' sense of moral-social standing through the apology–forgiveness cycle increases the willingness of both to reconcile. With important implications for restorative justice interventions, Shnabel's work demonstrates how restoring parties' positive identities is a proximate cause of peace after conflict.

In a related vein, but informed by theories from social, organizational, and evolutionary psychology, Sabine Otten, Juliette Schaafsma, and Wiebren Jansen present findings on inclusion in culturally diverse settings as a pathway to peace in Chapter 3. Exclusion has negative costs related to well-being and group functioning and increases the probability of conflict and aggression, whereas behavioral processes of inclusion enhance peace. As their work shows, the successful promotion of all-inclusive multiculturalism and cognitive processes like self-anchoring act as proximate causes for peace between minority and majority members of culturally diverse groups. These findings have obvious importance in an age when diversity has become increasingly common in social organizations.

Turning to the proximate causes of peacekeeping and peacemaking in chimpanzees (*Pan troglodytes*), in Chapter 4 Teresa Romero presents findings on causal factors that lead uninvolved bystanders to initiate friendly contact with recent recipients of aggression. Specifically, she presents three hypotheses about the function of bystander affiliation and examines the possible underlying causes of each. These include: consolation, which may begin with some level of emotional perspective-taking; mediated reconciliation, which would follow from knowledge of third-party relationships; and self-protection, the underlying mechanisms of which may be individual recognition, associative learning, and responses to aversive stimuli.

Chapter 5 concludes the unit and presents the findings of scholar-practitioners Saleem Ali and Todd Walters on the Experiential Peacebuilding Cycle. Focusing on the Balkans, Iraq, Indonesia, and the United States, they show how a problem–solution proposition focused on a common environmental concern can act as a proximate cause of peaceful behaviors. In addition, they explain how learning and experiential peacebuilding modify behavioral processes toward resilient relationships and sustainable peace.

Development

The contributors in Part Two present findings on two related developmental questions. First, they identify when and how peace processes first emerge. Second, they present findings on the relationship between environmental conditions and the capacity for change or transformation of peace processes. Darcia Narvaez addresses these questions in Chapter 6 by analyzing how *Homo sapiens'* cultural and childbearing heritages provide the evolved developmental niche through which peaceful behaviors and relations emerge. Starting with anthropological data on small-band hunter-gatherer societies, she identifies core social elements that affect the development of humans' optimal peaceful behaviors and follows this by analyzing the adverse effects of more historically recent childbearing and childrearing practices.

In Chapter 7, Cary Roseth approaches the questions that structure this section through a review of the literature on children's social development as it pertains to experiences of conflict. Such experiences likely promote the development of peaceful behavioral processes, and studies of these processes provide evidence for a natural tendency among children to resolve conflicts through peacemaking and to maintain peaceful relations.

Chapter 8 and Chapter 9 concern the development of peaceful behavioral processes in the context of postconflict societies. Ellen Furnari provides evidence in Chapter 8 that the development of robust relationships characterized by trust, cooperation, and acceptance enhances effective peacekeeping after conflict. Focusing the analysis at the community level, Furnari's research highlights the development of such robust relationships as a core strategy and practice of peacekeeping. Her study also offers insights into the comparative benefit of cooperative, unarmed civilian peacekeeping versus coercive, military peacekeeping.

In a related vein, Mike Wessells and Kathleen Kostelny examine the role that communities play in reintegrating former child soldiers in postconflict environments. In Chapter 9, they take a community resilience approach to show how this specific peace process, the sustained reintegration of child soldiers, develops through peacebuilding, restorative justice, education, child protection, and mental and psychosocial support.

Function

The contributors in Part Three assess the function of behavioral processes and systems of peace in order to identify and analyze their immediate and delayed benefits. Additionally, they investigate how these affect the survival, well-being, and success of communities. In Chapter 10, Otto Adang, Sarah Stronks, Misja van de Klomp, and Gerard van den Brink use the Relational Model (de Waal 1996) to assess the function of particular behavioral processes in peacemaking after police–citizen group confrontations. In particular, they emphasize the function of face-to-face meetings between the police and citizens following conflict. Such "critical moments" function to promote reconciliation by altering the meaning of events and redefining the relations of the parties involved. Furthermore, they show how the behavioral processes of assessments of value, compatibility, and security (cf. Cords & Aureli 2000) function to enhance community relationships in postconflict peacemaking.

In Chapter 11, Benjamin Peters assesses the function of constitutions as systems of peace and of peace constitutions in particular. Specifically, he shows how liberal democratic constitutions function to limit the species-atypical behaviors of state- and warmaking, and how peace constitutions do so with optimal effectiveness by prohibiting war and the maintenance of military forces, protecting the right to live in peace, and promoting the development of cultures of peace. Using the cases of Costa Rica and Japan, he demonstrates how peace constitutions have benefited national communities by preventing their participation in war and by eliminating organizations of violence at the disposal of the state for use against civil society.

In an analysis that reaches both prior to and beyond the state, in Chapter 12 Joám Evans Pim examines how decentralized peace systems function to reduce violence and killing and enhance peaceful coexistence with neighboring societies. Using empirical evidence from our nomadic forager past and historically recent and contemporary cases, he shows how decentralized, self-governing communities function to achieve peaceful societies akin to what Gandhi termed "Oceanic Circles."

In Chapter 13, Daniel Hyslop and Thomas Morgan follow with an examination of how investing in eight key areas of social and institutional development that are related to structural peace can increase a country's overall resilience and level of peace. They term these eight areas the Pillars of Peace and estimate the benefit of "perfect peacefulness" to the global economy at $9.8 trillion US dollars.

Evolution

In Part Four, our contributors ask why and how the ability to engage in peace processes evolved over generations or evolutionary time in species, individuals, groups, communities, and cultures. Furthermore, they demonstrate how peace processes compare across extant species, communities, and cultures. In Chapter 14, Douglas Fry reviews evidence that confirms *Homo sapiens'* evolved capacity to cooperate, manage peaceful relationships, and resolve disputes without violence. Reviewing findings from nonhuman animal behavior as well as archeological and nomadic forager data, he demonstrates that humans share these evolved capacities for peaceful behavior with other animals, and he connects them to the very real possibilities of abolishing war and handling disputes justly and nonviolently.

Further broadening the peace horizon in Chapter 15, Harry Kunneman recognizes *Homo sapiens'* place within a wider, evolutionary transspecies peace heritage. He does so by distinguishing three evolved social patterns into which all life forms fall and identifies one, ergopoietic relations, as the most promising route to the future evolution of transspecies peace.

Research conducted during the past decades suggests that peaceful behavior is ubiquitous in nature. In the final chapter of this section, Chapter 16, Peter Verbeek reviews and discusses peaceful behavior in a wide range of nonhuman animals. He discusses how explaining the evolution of peaceful behavior has become a chief challenge for behavioral science. Psychiatrist and environmentalist Ian McCallum points out that, "strictly speaking, there is no such thing as human nature. There is only nature and the very human expression of it" (McCallum 2012). Verbeek follows this line of reasoning and makes the case that studying the role of peaceful behavior in the survival and propagation of nonhuman animal life has direct significance for improving our understanding of the evolved abilities for peace in humans.

Shifting Paradigms: Three Dimensions of Peace and Global Issues

The work of the 23 authors united in this volume sheds new light on how species (Chapters 4, 14, 15, and 16), individuals (Chapters 2, 7, and 14), families (Chapter 6), groups (Chapters 3 and 5), and communities (Chapters 8, 9, 10, 11, 12, and 13) can, and do, make, build, and keep peace. The basic and applied work in the volume reflects a paradigm shift in behavioral science: away from a singular focus on direct peace and toward an integration of direct, structural, and sociative peace. As Fry comments on this paradigm shift, "the point is not to deny the obvious human capacity to engage in war and acts of violence, but rather to balance the traditional overemphasis on competition and violence with a brighter view of human nature that is consistent with the evidence from anthropology to zoology" (Fry this volume, Chapter 14). We add that the paradigm shift shows that *scientifically* we are finally getting serious about finding out how peace works.

Paradigm shifts in science do not come about in a social vacuum, and recent developments in the global policy arena mirror the new thinking about peace in behavioral science. Traditionally, (direct) peace has been seen as a necessary condition for policy work on global issues to succeed. For example, in a recent report from the Sustainable Development Solutions Network (SDSN) for the Secretary-General of the United Nations (SDSN 2013), the authors state, "The most important public good is *peace*." They add that "personal security, ending conflict, and *consolidating peace* are all necessary components of good governance for sustainable development" (our italics).

This one-dimensional focus on direct peace as a condition for policy work is changing to a multidimensional view of peace as part and parcel of policy work, as the case of global health policy illustrates. Like peace, health is more and more seen as a process, specifically as "a process leading to physical, mental, social, and spiritual well being" as well as a "resource for the full realization of the human potential" (Simonelli *et al.* 2014). Health is also increasingly seen as the product of respect for universal rights, including the rights to food, housing, work, education, human dignity, life, nondiscrimination, privacy, access to information, and the freedoms of association, assembly, and movement, among others (cf. CESCR 2000, cited in Cotter *et al.* 2009). As the implementation of universal rights is meant to negate structural violence, implementing universal rights to health is an obvious aspect of structural peace. This is perhaps nowhere as apparent as in efforts to tackle climate change, which a panel of medical and health experts recently described as the greatest global health opportunity of the twenty-first century (Watts *et al.* 2015). Simply put, then, working for health is working for peace.

Like global health, sustainable development is also linked to universal rights. In a recent letter to all permanent UN missions, for example, the UN High Commissioner for Human Rights emphasized the need to make all sustainable development policies and goals consistent with international human rights law and called for efforts "to chart a fresh course, and to embrace a new paradigm of development built on a foundation of human rights, equality and sustainability" (Pillay 2015; see also Office of the High Commissioner for Human Rights 2012). The UN Secretary-General (2014) mirrors this position in a synthesis report on the post-2015 sustainable development agenda. It follows that, like working for health, working for sustainable development is working for peace.

As we mentioned at the start of this chapter, we believe that this is a promising time for peace. Paradigm shifts in behavioral science and the public policy arena are changing traditional one-dimensional views of peace into multidimensional conceptual perspectives. To move from the conceptual to the practical, we now need to work on a better understanding of the behavioral processes that foster peace through universal rights and create conditions for sustained health, sustainable development, and human flourishing. We believe that our peace ethology model can be instrumental in these efforts.

References

CESCR. (2000). General Comment 14 to Article 12 of the International Covenant on Economic, Social and Cultural Rights. E/C.12/2000/4. Geneva: United Nations.

Christie, D.J. (2012). Peace psychology: Definitions, scope, and impact. In D.J. Christie (ed.), *The encyclopedia of peace psychology*. Chichester: Wiley-Blackwell.

Coleman, P.T., & Deutsch, M. (2012). *Psychological components of sustainable peace* New York: Springer.

Cords, M., & Aureli, F. (2000). Reconciliation and relationship qualities. In F. Aureli & F.B.M. de Waal (Eds.), *Natural conflict resolution*. Berkeley: University of California Press.

Cotter, L.E., Chevrier, J., El-Nachef, W.N., Radhakrishna, R., Rahangdale, L., Weiser, S.D., & Lacopino, V. (2009). Health and human rights education in U.S. schools of medicine and public health: Current status and future challenges. *PLoS ONE, 4*(3): e4916. doi:10.1371/journal.pone.0004916

de Boer, S.F., Caramaschi, D., Natarajan, D., & Koolhaas, J.M. (2009). The vicious cycle towards violence: Focus on the negative feedback mechanisms of brain serotonin neurotransmission. *Frontiers in Behavioral Neuroscience, 3*(52), 1–6.

Decety, J., & Jackson, P.L. (2006). A social-neuroscience perspective on empathy. *Current Directions in Psychological Science, 15*(2), 54–58.

Deutsch, M., & Coleman, P.T. (2012). Psychological components of sustainable peace: An introduction. In P.T. Coleman & M. Deutsch (Eds.), *Psychological components of sustainable peace*. New York: Springer.

de Waal, F.B.M. (1996). Conflict as negotiation. In W.C. McGrew, L.F. Marchant, & T. Nishida (Eds.), *Great ape societies*. Cambridge: Cambridge University Press.

de Waal, F.B.M. (2000). Primates: A natural heritage of conflict resolution. *Science, 289*, 586–590.

de Waal, F.B.M. (2012). The antiquity of empathy. *Science, 336*, 874–876.

Fry, D.P. (2006). *The human potential for peace: An anthropological challenge to assumptions about war and violence*. New York: Oxford University Press.

Fry, D.P. (2012). Life without war. *Science, 336*, 879–884.

Fry, D.P. (Ed.). (2013). *War, peace, and human nature*. New York: Oxford University Press.

Galtung, J. (1996). *Peace by peaceful means*. London: Sage Publications.

Galtung, J. (2010). Peace studies and conflict resolution: The need for transdisciplinarity. *Transcultural Psychiatry, 47*, 20–32.

Galtung, J. (2012). Peace, positive and negative. In D.J. Christie (Ed.), *The encyclopedia of peace psychology*. Chichester: Wiley-Blackwell.

Gregor, T. (1996). *A natural history of peace*. Nashville, TN: Vanderbilt University Press.

Haller, J., & Kruk, M.R. (2006). Normal and abnormal aggression: Human disorders and novel laboratory models. *Neuroscience and Biobehavioral Reviews, 30*, 292–303.

Hinde, R.A. (1979). *Towards understanding relationships*. New York: Academic.

Hinde, R.A. (1987). *Individuals, relationships and culture: Links between ethology and the social sciences*. Cambridge: Cambridge University Press.

Howell, A.J., & Larsen, D.J. (2015). Other-oriented hope reflects an orientation toward others. In A.J. Howell & D.J. Larsen (Eds.), *Understanding other-oriented hope: An integral concept within hope studies*. New York: Springer.

Immordino-Yang, M.H., McColl, A., Damasio, A., & Damasio, D. (2009). Neural correlates of admiration and compassion. *Proceedings of the National Academy of Sciences, 106*(19), 8021–8026.

Jaeggi, A.V., Burkart, J.M., & Van Schaik, C.P. (2010). On the psychology of cooperation in humans and other primates: Combining the natural history and experimental evidence of prosociality. *Philosophical Transactions of the Royal Society B, 365*, 2723–2735.

Krill, A.L., & Platek, S.M. (2012). Working together may be better: Activation of reward centers during a cooperative maze task. *PLoS ONE, 7*(2), e30613.

Kruk, M.R., & Kruk-de Bruin, M. (2010). *Discussions on context, causes and consequences of conflict*. Leiden: The Lorentz Center, Leiden University.

Leckman, J.F., Panter-Brick, C., & Salah, R. (Eds.). (2014). *Pathways to Peace: The transformative power of children and families*. Cambridge, MA: The MIT Press.

McCallum, I. (2012). A wild psychology. In P.H. Kahn Jr. & P.H. Hasbach (Eds.), *Ecopsychology, science, totems, and the technological species*. Cambridge, MA: The MIT Press.

Office of the High Commissioner for Human Rights. (2012). *Human rights indicators: A guide to measurement and implementation*. Geneva: United Nations.

Opotow, S. (2012). Moral exclusion. In D.J. Christie (Ed.), *The encyclopedia of peace psychology*. Chichester: Wiley-Blackwell.

Pillay, N. (2015). *Human rights in the post-2015 agenda* [Letter to all permanent missions in New York and Geneva]. Geneva: UN High Commissioner for Human Rights.

Piper, W.T., Saslow, L.R., & Saturn, S.R. (2015). Autonomic and prefrontal events during moral elevation. *Biological Psychology, 108*, 51–55.

Preston, S.D., & de Waal, F.B.M. (2002). Empathy: Its ultimate and proximate bases. *Behavioral and Brain Sciences, 25*(1), 1–20.

Ricoeur, P. (1992). *Oneself as another.* Chicago: Chicago University Press.

Sánchez, Ó.A. (1995). Understanding, tolerance, freedom and democracy. In M. Thee (Ed.), *Peace!* Paris: UNESCO.

Simonelli, I., Mercer, R., Bennett, S., Clarke, A., Fernandes, G.A.I., Fløtten, K., Maggi, S., Robinson, J.E., Simonelli, F., Vaghri, Z., Webb, E., & Goldhagen, J. (2014). A rights and equity-based "Platform and Action Cycle" to advocate child health and well being by fulfilling the rights of children. *The Canadian Journal of Children's Rights, 1*(1), 199–218.

Sussman, R.W. (2013). Why the legend of the killer ape never dies: The enduring power of cultural beliefs to distort our view of human nature. In D.P. Fry (Ed.), *War, peace, and human nature: The convergence of evolutionary and cultural views.* New York: Oxford University Press.

Sustainable Development Solutions Network (SDSN). (2013, June). *An action agenda for sustainable development* [Report for the UN Secretary-General]. Geneva: United Nations. www.unsdsn.org

Thelen, K. (2003). How institutions evolve. In J. Mahoney & D. Rueschemeyer (Eds.), *Comparative historical analysis in the social sciences.* Cambridge: Cambridge University Press.

Tinbergen, N. (1963). On aims and methods of ethology. *Zeitschrift für Tierpsychologie, 20*, 410–433.

Tinbergen, N. (1968). On war and peace in animals and man: An ethologist's approach to the biology of aggression. *Science, 160*, 1411–1418.

UN Secretary-General. (2014). *The road to dignity by 2030: Ending poverty, transforming all lives and protecting the planet* [Synthesis report of the Secretary-General on the post-2015 sustainable development agenda]. New York: United Nations.

Verbeek, P. (2008). Peace ethology. *Behaviour, 145*, 1497–1524.

Verbeek, P. (2013). An ethological perspective on war and peace. In D.P. Fry (Ed.), *War, peace, and human nature: The convergence of evolutionary and cultural views.* New York: Oxford University Press.

Verbeek, P., Iwamoto, T., & Murakami, N. (2007). Differences in aggression among wild type and domesticated fighting fish are context dependent. *Animal Behaviour, 73*, 75–83.

Verbeek, P., Kunneman, H., & Peters, B.A. (2015, March 12–14). *The peaceful self: An interdisciplinary window on motivation of virtue.* Milwaukee, WI: Interdisciplinary Moral Forum, Marquette University. http://epublications.marquette.edu/smv_imf/IMF/Friday/3/

Watts, N., Adger, W.N., Agnolucci, P., Blackstock, J., Byass, P., Cai, W., Chaytor, S., Colbourn, T., Collins, M., Cooper, A., Cox, P.M., Depledge, J., Drummond, P., Ekins, P., Galaz, V., Grace, D., Graham, H., Grubb, M., Haines, A., Hamilton, I., Hunter, A., Jiang, X., Li, M., Kelman, I., Liang, L., Lott, M., Lowe, R., Luo, Y., Mace, G., Maslin, M., Nilsson, M., Oreszczyn, T., Pye, S., Quinn, T., Svensdotter, M., Venevsky, S., Warner, K., Xu, B., Yang, Y., Yin, Y., Yu, C., Zhang, Q., Gong, P., Montgomery, H., & Costello, A. (2015).

Health and climate change: Policy responses to protect public health. *The Lancet.* 10.1016/S0140-6736(15)60854-6

Wilson, M., Boesch, C., Fruth, B., Furuichi, T., Gilby, I.C., Hashimoto, C., Hobaiter, C.L., Hohmann, G., Itoh, N., Koops, K., Lloyd, J.N., Matsuzawa, T., Mitani, J.C., Mjungu, D.C., Morgan, D., Mullar, M.N., Mundry, R., Nakamura, M., Pruetz, J., Pusey, A.E., Riedel, J., Sanz, C., Schel, A.M., Simmons, N., Waller, M., Watts, D.P., White, F., Wittig, R.M., Zuberbühler, K., & Wrangham, R.W. (2014). Lethal aggression in *Pan* is better explained by adaptive strategies than human impacts. *Nature, 513*, 414–417.

Wrangham, R. (1999). Evolution of coalitionary killing. *Yearbook of Physical Anthropology, 42*, 1–30.

Part One

Proximate Causation

2

A Social-Psychological Perspective on the Proximate Causation of Peaceful Behavior: The Needs-Based Model of Reconciliation

Nurit Shnabel

A main message of the present volume, arising from Verbeek and Peters' introductory chapter as well as from Parts Three and Four about the function and evolution of peace systems, is that both human and nonhuman societies need mechanisms that enable conflicting parties to reconcile and thus maintain valuable relationships and prevent (at least some of) the negative consequences of conflict, aggression, and lack of cooperation. Among humans, a primary social mechanism that facilitates reconciliation following transgressions is the *apology forgiveness cycle*, in which the perpetrator takes responsibility and expresses remorse for the harm caused to the victim, who, in turn, reciprocates by granting forgiveness to the perpetrator despite the wrongdoing (Tavuchis 1991). Tavuchis' (1991) seminal work on the sociology of this cycle suggests that it has the power to dramatically, almost "magically," transform the relations between former adversaries and replace the downward spiral of alienation and aggression with an upward spiral of goodwill and generosity. The Needs-Based Model of reconciliation (Nadler & Shnabel 2008; Shnabel & Nadler 2008), the theoretical framework presented in this chapter, was developed in an attempt to understand, from a social-psychological perspective, how this "magic" works.

Anchored in the theoretical tradition of Social Identity Theory (Tajfel & Turner 1986), the main tenet of the Needs-Based Model is that transgressions, at both the interpersonal and intergroup levels, threaten specific dimensions in the identities of victims and perpetrators. As long as these threats are not removed, they serve as barriers to reconciliation and might even lead to the conflict's escalation. However, restoring victims' and perpetrators' positive identities, which can be done through the apology forgiveness cycle, should serve as a catalyst for reconciliation, increasing victims' and perpetrators' readiness to show goodwill toward each other. As this brief description implies, in terms of the four principal questions that guide ethological research (Tinbergen 1963), the Needs-Based Model concerns the immediate causation of conciliatory behavior. That is, it aims to identify factors within the organism (e.g., the motivation to restore positive identity) and outside of it (e.g., one's social role, of victim or perpetrator, within a given social context) that facilitate or hinder conciliatory behavior.

I open the present chapter by defining reconciliation and distinguishing it from the related concepts of conflict settlement and resolution. I then introduce the theoretical

Peace Ethology: Behavioral Processes and Systems of Peace, First Edition.
Edited by Peter Verbeek and Benjamin A. Peters.

perspective of the Needs-Based Model of reconciliation and present empirical findings that support its hypotheses regarding the dynamics between victims and perpetrators in contexts of both interpersonal and intergroup transgressions. I then move on to extend the model to "dual conflicts" in which there are no consensual, clear-cut roles of victims and perpetrators because both adversaries transgress against each other. I conclude by summarizing the theoretical insights provided by the model to the scientific understanding of peaceful behavior. I also point to the practical implications of the model for the planning of interventions intended to promote such behavior among conflicting individuals and groups.

Reconciliation and Conflict Resolution

The concept of reconciliation was introduced into the scientific discourse by primatologists de Waal and van Roosmalen (1979) in their reports of friendly reunions between former chimpanzee opponents soon after aggressive confrontations. However, it took almost two more decades before mainstream social psychology began to devote greater attention to the study of reconciliation (Nadler 2012). Thus, ironically, research on how nonhuman animals reconcile and make up following conflicts was ahead of the corresponding research on humans (de Waal 2000).

One reason for this relatively late introduction of the concept and study of reconciliation into mainstream social psychology may be that during the second half of the twentieth century, the study of conflict in the social sciences (including social psychology) was dominated by a realist approach to conflict and its resolution. According to this view, "disputes between persons and between groups are grounded in conflicts of material interests" (Scheff 1994, p. 3). These interests may range from natural resources such as land or water in contexts of international conflicts to pocket knives in contexts of competing groups of children within a summer camp setting (Sherif *et al.* 1961). The realist viewpoint has generated much interest and insight over past decades, for example through the introduction of concepts and theorizing based on game theory into social psychology (Jones 1998). However, it also encouraged a somewhat limited view of conflict settlement as reaching an agreed-upon formula (e.g., a contract or peace treaty) for distributing these contested resources, be it land or pocket knives, between the adversarial parties (see Kelman 2008). As such, the realist approach may have contributed to drawing attention away from the emotional, non-instrumental processes involved in conflicts and their resolution. Yet, such non-instrumental, socio-emotional processes have critical influence, even when the ultimate goal is of instrumental nature (e.g., maintaining ceasefire at an agreed-upon border; see Furnari, this volume, Chapter 8). These socio-emotional processes are exactly the core of reconciliation.

While reconciliation is broadly viewed as the process of "removing conflict-related emotional barriers that block the way to healing a discordant relationship" (Shnabel *et al.* 2008, p. 162), its exact definition remains somewhat elusive and controversial. Kelman (2008) has proposed a distinction between reconciliation on one hand, and conflict settlement and resolution on the other hand. Conflict settlement consists of finding the formula for the division of contested resources between the adversarial parties, in line with the realist approach discussed here. Conflict resolution involves building a pragmatic, trustful partnership wherein both sides view cooperation as best serving their interests (i.e., establishing constructive "working relations"; Rouhana 2004). Finally, Kelman views reconciliation as a process of identity change in which each party strengthens the core elements of its own identity while accommodating the other. This process involves the removal of the negation of the other as an element of one's own identity and the ability to acknowledge the other's account of the conflict without having to agree with it fully. To illustrate, in the reconciliation process, each party learns to stop blaming the other party for being the sole party responsible for the conflict.

The view of reconciliation as a process of identity change, involving accepting rather than negating the adversary's identity, is also reflected in the definition of reconciliation by Staub *et al.* (2005) as "a process that must include a changed psychological orientation toward the other" (p. 301), which eventually leads to mutual acceptance. The King Center, an institution dedicated to the advancement of the legacy of Dr. Martin Luther King, further emphasizes mutual acceptance as a core element of reconciliation by defining it as the "bringing together of adversaries in a spirit of community after a conflict has been resolved" (King Center n.d.). Other definitions, however, emphasize the restoration of trustworthy relations as the essence of reconciliation. For example, Exline and Baumeister (2000) define reconciliation as "a willingness to come together to work, play, or live in an atmosphere of trust" (p. 136).

The lack of a widely accepted definition of reconciliation may stem from the novelty of this concept within the conflict resolution literature and political discourse (Nadler 2012), as well as from the fact that it is an abstract concept that can denote both a process and an outcome (i.e., a state or condition), and might be hard to distinguish from related concepts such as peacemaking (Verbeek 2008). Another source of this lack of clarity regarding the concept of reconciliation may be that it is overloaded with multiple meanings. In particular, Rouhana (2004) argues that in contexts characterized by structural violence (i.e., relatively permanent unequal social arrangements that privilege some groups at the expense of others; Galtung 1969), reconciliation should mean the promotion of *positive peace*, that is, reaching a just, equal social arrangement (see Christie *et al.* [2008] for the distinction between positive and negative peace, the latter denoting the mere cessation of violence). Reconciliation in the sense of promoting mutually positive emotional orientations among conflicting groups, yet without the promotion of an equal, just social arrangement, is fake and superficial, because it serves the continuation of the oppression of the underprivileged group.

While the importance of reaching structural equality between advantaged and disadvantaged groups to the reconciliation process is beyond doubt, equating the two concepts may be viewed as a conceptual stretching of the latter (see Meierhenrich 2008). Therefore, in the present chapter, I refer to reconciliation as the process of mending and "healing" broken relationships (see also Staub *et al.* 2005). I have two reasons for conceptualizing reconciliation in this way. First, the view of reconciliation as a *relationship-oriented process* is a thread common to the various definitions of this concept. Second, according to Social Dominance Theory (Sidanius & Pratto 1999), structural violence (i.e., group-based inequality) is restricted to societies that produce sustainable economic surplus (e.g., there is generally no structural violence among hunter-gatherer societies, which lack sufficient economic surplus; see also Narvaez, this volume, Chapter 6). For this reason, including the promotion of an equal, just social arrangement in the definition of reconciliation reduces the parsimony of this concept, as it makes it applicable to a narrower range of contexts. For the purposes of the present volume, which focuses on the evolution of peaceful behavior and systems at both the individual and intergroup levels and within various human and nonhuman societies, it is important to use a parsimonious definition of reconciliation, which captures its core essence – the restoration of trust between and positive identities of former adversaries (see also Verbeek [2008] for the distinction between restorative peace, which involves the restoration of harmony to relations following conflict, and sociative peace, which involves the negation of structural violence through social justice).

The Needs-Based Model of Reconciliation

As mentioned in the opening of this chapter, the apology forgiveness cycle is a major social mechanism for facilitating reconciliation, namely, for restoring victims' and perpetrators' positive identities and mutual trust. According to Tavuchis (1991), the language people use when referring to the apology forgiveness cycle (e.g., expressions such as "owing" or "accepting" an apology) implies that "something almost tangible is being bartered" (p. 33). This observation led us to conceptualize reconciliation as an *act of social exchange* in which the conflicting parties exchange symbolic, psychological "commodities"; these symbolic commodities restore victims' and perpetrators' positive identities (Shnabel & Nadler 2008) and mutual trust (Shnabel *et al.* 2014).

To understand the nature of the psychological "commodities" that victims and perpetrators exchange, we turned to theorizing about the "Big Two" (Abele *et al.* 2008), according to which there are two fundamental content dimensions along which people perceive and judge themselves and others: the *agency* dimension, representing traits such as "strong," "competent," "influential," and "self-determined"; and the *moral-social* dimension (aka the communion dimension), representing traits such as "moral," "warm," and "trustworthy."

Based on this theorizing, the Needs-Based Model argues that transgressions cause asymmetric threats to victims' and perpetrators' identities (SimanTov-Nachlieli *et al.* 2013). Victims who feel inferior regarding their power (Foster & Rusbult 1999), honor (Scheff 1994), and perceived control (Baumeister *et al.* 1994) experience threat to the agency dimension of their identity. The lack of agency makes them particularly vulnerable in light of the possibility that the perpetrator would repeat the transgression, for which victims are especially vigilant (McCullough *et al.* 2013). In contrast, perpetrators suffer from moral inferiority (Exline & Baumeister 2000). Because rejection is the social sanction imposed upon those who violate the norms or moral standards of their community (Tavuchis 1991), perpetrators are said to experience anxiety over social exclusion, which is sometimes reflected in feelings such as guilt (Baumeister *et al.* 1994), shame (Exline & Baumeister 2000), or repentance (North 1998). In addition, perpetrators may be concerned about the victim's possible intention to take revenge or avoid them, two common responses among victims, who wish to deter the perpetrator and reduce the probability of future harm to the self (see McCullough *et al.* 2013 for a discussion of acceptance, revenge, avoidance, and forgiveness as four possible responses by victims following transgressions).

The experience of these different identity threats brings about different motivational states among victims and perpetrators. Victims, who feel weak and humiliated, experience the need to restore their sense of agency (i.e., ability to determine their own and others' outcomes). For this reason, victims often show heightened power-seeking behavior (Foster & Rusbult 1999). Victims' attempt to (re)gain power may also lead to heightened aggressive, vengeful behavior (Frijda 1994). In contrast, perpetrators, who feel that their moral image is impaired, experience the need to restore their positive moral identity and regain acceptance in the community from which they feel potentially excluded (for the fundamental need to belong and feel included, see Otten *et al.*, this volume, Chapter 3). Sometimes, perpetrators cope with their culpability through moral disengagement (e.g., minimizing the severity of the harm or blaming the victim for bringing it upon herself; Bandura 1990). Yet perpetrators' need to gain moral-social acceptance may also lead them to seek forgiveness and show heightened helping behavior toward others, including their victims (Shnabel *et al.* 2008).

The Needs-Based Model further posits that as long as victims' and perpetrators' needs remain unsatisfied, they serve as barriers to reconciliation. However, an exchange interaction through which victims and perpetrators satisfy each other's needs for *empowerment* and *acceptance* (respectively) should increase their willingness to reconcile with each other. One practice through which perpetrators can empower the victims is by apologizing to them. Perpetrators' acknowledgment of responsibility for causing the victims injustice constitutes an admission of owing them a moral debt, which only the victim can cancel (Minow 1998). This returns control to the hands of the victim. Other strategies through which perpetrators can empower their victims include pointing out the victims' achievements and capabilities (because a sense of high competence is a critical component of empowerment; e.g., Brookings & Bolton 2000), or, in the case of a victimized group, by appealing to national pride or expressing respect for the group's culture and values. This array of strategies can empower victims and restore their impaired sense of agency.

A key strategy through which victims may restore perpetrators' moral image includes expressing forgiveness, understanding for the circumstances that compelled the perpetrators' actions, and sympathy for their emotional distress. Enright *et al.* (1998) view these expressions of empathy toward the perpetrator as "gifts" that victims can offer to those who have wronged them. Such "gifts" mitigate the moral inferiority engendered by the perpetrator role (Exline & Baumeister 2000) and provide reassurance that they belong to, rather than remain excluded from, their designated moral community. Victims can also satisfy perpetrators' need for social acceptance by expressing willingness to form friendships with them or engage in economic or cultural cooperation. For example, the willingness of many Israeli Jews to buy German-made products may be interpreted as an expression of social acceptance following an era in which most Israeli Jews boycotted German-made products (Shnabel *et al.* 2008).

A successful social exchange of the symbolic "commodities" of empowerment and acceptance can promote victims' and perpetrators' willingness to reconcile with each other through two routes (Shnabel *et al.* 2014). First, such an exchange can symbolically erase the roles of "powerless victim" and "morally inferior perpetrator"; that is, it can restore the conflicting parties' positive identities and place them on an equal footing (North 1998). In addition, an empowering message conveyed by perpetrators (e.g., expressions of apology and respect toward the victims) may imply an assurance that the transgression would not reoccur and even a promise for reparation (Blatz *et al.* 2009). Similarly, an accepting message conveyed by the victims (e.g., expressions of forgiveness

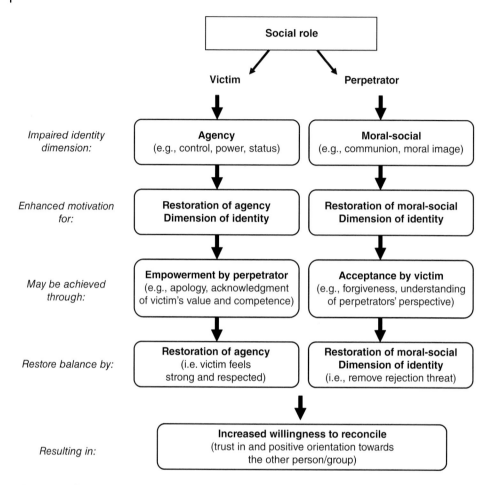

Impaired identity dimension:

Enhanced motivation for:

May be achieved through:

Restore balance by:

Resulting in:

Figure 2.1 The Needs-Based Model of reconciliation.

and empathy) toward the perpetrators implies that they do not intend to hold a grudge against them (Shnabel *et al.* 2014). Due to these implications, empowering and accepting messages conveyed by perpetrators to victims and vice versa can restore their trust in each other's good intentions and, thus, further increase the conflicting parties' readiness to reconcile. Figure 2.1 summarizes the process proposed by the Needs-Based Model.

Empirical Evidence in Support of the Needs-Based Model

A series of experiments by Shnabel and Nadler (2008) provide direct empirical support for the model's hypotheses in contexts of interpersonal transgressions. The first study used the "creativity-test" experimental paradigm in which participants were randomly assigned to be either "writers," who composed marketing slogans for a list of products, or "judges," who evaluated these slogans. There were two types of writer–judge dyads: control dyads and experimental dyads. In both types of dyads, writers had allegedly failed their "test," whereas judges had allegedly passed it. However, the instructions provided to participants were different in the experimental and control dyads. In the experimental dyads, judges were instructed to be strict in their evaluations. They were also told that being too nice could harm their own chances of passing the test. By the end of the session, participants (i.e., judges and writers) in the experimental dyads were told that the judges passed the test, whereas writers failed it due to the harsh evaluations they received from their judges. In the control dyads, judges were instructed to be relatively lenient. Participants were later informed that the judges passed the test, whereas writers failed it due to the decision of an external committee. Thus, in both the experimental and control dyads, judges passed the test and writers failed it, which allowed us to control for information about success or failure. However, only in the experimental dyads did judges deliberately fail their partners to improve their own chances of passing the test; hence, their success was gained at the expense of the writers. Self-report questionnaires that measured participants' self-perceptions and psychological needs revealed that, as predicted, writers in the experimental dyads (i.e., victims) had the lowest sense of agency and the highest need for power (i.e., compared to writers in the control dyads and to judges in either the experimental or control dyads). Correspondingly, judges in the experimental dyads (i.e., perpetrators) had the lowest moral image and the highest need for acceptance (i.e., compared to judges in the control dyads and to writers in the experimental or control dyads). The same pattern of results occurred in an experimental paradigm that used real-life transgressions by asking participants to recall a personal episode in which they had either hurt or been hurt by a significant other.

An additional set of studies supported the model's prediction that addressing victims' and perpetrators' needs should increase their willingness to reconcile with each other. The first study again used the creativity-test experimental paradigm to randomly assign participants to the role of victim or perpetrator. Following the transgression (i.e., after participants learned that the writer failed the test due to the harsh evaluations of the judge, who herself passed the test), participants received a message from their counterpart that expressed, depending on the experimental condition, empowerment (i.e., acknowledgment of their high competence), social acceptance (i.e., acknowledgment of their high social skills), or neither. As expected, victims' readiness to reconcile (measured through self-report questionnaires) was highest in the empowerment condition (compared to the acceptance or control conditions), whereas perpetrators' readiness to reconcile was highest in the acceptance condition (compared to the empowerment or control conditions). The same pattern of results occurred in two experiments that used role-playing scenarios of transgressions – one describing a supervisor's refusal of a seemingly legitimate request by an employee, and the other describing a situation in which an employee returning from maternity leave discovered that her attractive job in an organization had been taken over by a fellow worker.

The experiments conducted by Shnabel and Nadler (2008) pointed to restoration of positive identity as the mechanism responsible for the increase in victims' and perpetrators' willingness to reconcile. That is, empowering messages from the perpetrators increased victims' sense of agency, and this increase, in turn, led to heightened willingness to reconcile. Correspondingly, accepting messages from the victims improved perpetrators' moral image, and this improvement, in turn, led to heightened willingness to reconcile. Subsequent studies by Shnabel *et al.* (2014) revealed that besides the restoration of positive identities, the exchange of empowering and accepting messages between victims and perpetrators also restored their trust in each other's positive intentions. Specifically, empowering messages from the perpetrators restored the victims' belief that the perpetrators would not repeat the transgression, and accepting messages from the victims restored the perpetrators' belief that the victims would not hold a grudge or try to take revenge. The restoration of trust, in turn, mediated victims' and perpetrators' willingness to reconcile. Thus, there are two different routes, identity restoration and trust building, through which empowering and accepting messages can promote reconciliation.

It is interesting to note, in this regard, that empowering and accepting messages from non-involved third parties successfully restored victims' and perpetrators' positive identities and thus contributed indirectly to reconciliation. However, these messages failed to restore victims' and perpetrators' mutual trust and, thus, were relatively ineffective in bringing about reconciliation, compared to identical messages whose source was the other conflict party (Shnabel *et al.* 2014). These findings, which stand in stark contrast to third parties' high effectiveness in promoting conflict settlement (Carnevale & Pruitt 1992), highlight the importance of direct communication and dialogue between victims and perpetrators for the socio-emotional process of reconciliation.

Applying the Needs-Based Model to Intergroup Contexts

Self-Categorization Theory (Turner *et al.* 1987), a prominent perspective within social psychology, argues that when a given ingroup–outgroup distinction is salient, people define themselves less in terms of their unique characteristics as individuals and more in terms of prototypical attributes of their ingroup. Therefore, feelings of victimization or guilt can be experienced "by association" (Doosje *et al.* 1998), that is, individuals can feel victimized or guilty due to historical or contemporary events in which their ingroup was involved, regardless of their own personal involvement. For example, US Americans may feel victimized when reminded of the September 11th events and as perpetrators when reminded of the nuclear bombing of Hiroshima and Nagasaki, even if they did not personally participate in these events.

Based on the logic of Self-Categorization Theory, we theorized that the psychological dynamics between members of victimized and perpetrating groups should correspond to the dynamics between individual victims and perpetrators (Shnabel *et al.* 2009). That is, we predicted that members of victimized and perpetrating groups should be motivated to restore their ingroup's respective sense of agency or positive moral image and show greater willingness to reconcile with the outgroup when its representatives convey empowering or accepting messages to the ingroup. To test this possibility empirically, we conducted two experiments.

The first experiment (Shnabel *et al.* 2009) focused on relations between Germans and (Israeli) Jews. We reasoned that when reminding subjects of the Holocaust, Jews would perceive themselves as victims and Germans would perceive themselves as perpetrators. To examine the Needs-Based Model's predictions regarding the psychological needs associated with these social roles, we exposed Jewish and German participants to two speeches, allegedly made by the outgroup's representative at the opening ceremony of the Holocaust Memorial in Berlin. The speeches were identically phrased for Jewish and German participants, and their main message was either acceptance or empowerment of the recipient group. The accepting message highlighted that "[w]e, the [Germans/Jews], should accept the [Jews/Germans] and remember that we are all human beings," whereas the empowering message highlighted that "[i]t is the [Germans'/ Jews'] right to be strong and proud of their country and to have the power to determine their own fate." In line with predictions, Jews reported feeling less powerful than Germans and showed greater readiness to reconcile with the Germans following an empowering compared to an accepting message from a German representative. Correspondingly, Germans reported having a more negative moral image than Jews and showed greater readiness to reconcile with the Jews following an accepting compared to an empowering message from a Jewish representative.

Although the pattern of results supported the model, the different pattern of responses to messages obtained among Jews and Germans could be alternatively attributed to cultural differences between the two groups. To rule out this alternative explanation, we replicated this experiment in the context of relations between Israeli Jews and Israeli Arabs. Because, as discussed in greater detail here, both Jews and Arabs often perceive themselves as the real victims of the Jewish-Arab conflict (Shnabel & Noor 2012), we focused on a particular historical event for which there is a consensus as to the victimization of Arabs by the Jewish side: the Kefar Kasem killings. In this event, which took place in October 1956, the Jewish-Israeli border patrol killed 43 unarmed Arab civilians for violating a curfew that had recently been imposed. Using the same experimental procedure as in the German-Jewish study, we exposed Israeli Arabs and Jews to speeches conveying messages of empowerment or acceptance. The speeches were ostensibly made by representatives of their outgroup on the 50th anniversary of the killings. Consistent with predictions, while Arabs were more willing to reconcile following a message of empowerment than of acceptance from the Jews, Jews were more willing to reconcile following a message of acceptance than of empowerment from the Arabs.

Taken together, the findings of the two studies suggest that it is the social role within a specific context, rather than preexisting cultural values, which determined Jews' preference for a particular type of message: in a context where they were placed in the social role of victims, an empowering message was found to be more effective in promoting reconciliation than an accepting message, whereas in a context in which they were placed in the social role of perpetrators, an accepting message was found to be more effective.

In addition, the same pattern of results was replicated in two experiments (Shnabel, Ulrich *et al.* 2013) that examined the Needs-Based Model in contexts of relations between groups of unequal status (i.e., structural disparity). For example, in one study, participants were BA students of a university that was presented, depending on the experimental condition, as either advantaged or disadvantaged compared to a competing university in terms of access to scarce spots in a master's program. Participants also learned that because the advantaged university favors its own students, the chances of being accepted to this MA program are lower for students who graduated from the disadvantaged, compared to the advantaged, university. Following the exposure to the information about group inequality, participants were exposed to a message from a representative of their outgroup that expressed either acceptance, through reassuring their ingroup's warmth, or empowerment, through reassuring their ingroup's competence.

Consistent with the Needs-Based Model's logic, advantaged group members showed more positive attitudes (e.g., greater readiness to share resources such as labs or libraries) toward the disadvantaged group following a warmth-reassuring message, whereas disadvantaged group members showed more positive attitudes toward the advantaged group following a competence-reassuring message. In addition, following the "right" type of message, both advantaged and disadvantaged group members revealed greater readiness to act for equality (e.g., sign a petition aiming to change the admission regulations for master's programs). Thus, even though acting for equality might come at the expense of their privileged position, advantaged group members showed generosity in response to a message that removed the threat posed to their identity as warm and moral. As for disadvantaged group members, apparently the reassurance of their high competence and capabilities restored their sense of collective efficacy, that is, their belief that their ingroup can improve its situation through unified efforts. These studies suggest that messages that restore advantaged and disadvantaged groups' positive identities and mutual trust can promote reconciliation, even in its broadest meaning of reaching a more just, equal social arrangement (i.e., the meaning proposed by Rouhana [2004], as discussed in this chapter).

Applying the Needs-Based Model to Conflict of Dual Social Roles

After gaining support for the Needs-Based Model in contexts in which the social roles of victims and perpetrators (or disadvantaged and advantaged) were consensual and clear-cut, we turned to extend the model into contexts that are marked by duality of social roles, that is, contexts in which both parties have transgressed against each other. Our predictions regarding the experience of identity threats, emotional needs, and responses to messages among *duals* (i.e., individuals or group members who serve as both victims and perpetrators at the same time) were straightforward: in line with the model's logic, we predicted that duals would experience threats to both their agency and their moral image, be motivated to restore both identity dimensions, and show increased willingness to reconcile following either empowering or accepting messages from their adversaries (SimanTov-Nachlieli & Shnabel 2014).

In terms of behavior, however, the prediction was less straightforward because the experience of victimization versus perpetration could potentially influence behavior in opposite directions (i.e., antisocially vs. prosocially, respectively). Specifically, on one hand, the experience of victimization often leads to heightened aggressive, antisocial behavior: victims were found to feel entitled to behave antisocially (Zitek *et al.* 2010), and their frustration may lead them to behave aggressively (Dollard *et al.* 1939), such as by taking revenge (Frijda 1994), in an attempt to restore their impaired sense of agency. The experience of perpetration, on the other hand, may lead to prosocial behavior. As discussed earlier, although perpetrators sometimes attempt to deny their responsibility for causing injustice (Bandura 1990; Schönbach 1990), when faced with the immorality of their acts, they may also try to restore their positive identity through reconciling and compensating their victims (Estrada-Hollenbeck & Heatherton 1998), including offering them help.

Although duals experience both victimization and perpetration simultaneously, we expected their need for agency, which should lead to heightened antisocial tendencies, to exert greater influence on their behavior than their need for morality, which should potentially lead to prosocial tendencies. This expectation was based on research in the person perception domain, which found that individuals' agency-related self-perceptions (e.g., perceived competence) had greater influence on their emotional responses than their morality-related self-perceptions (Wojciszke 2005). A similar pattern was found at the intergroup level, where the perceived desirability of ingroup attributes was primarily agency- or competence-based rather than morality-based (Phalet & Poppe 1997).

We conducted two experiments to test our "primacy of agency" hypothesis (SimanTov-Nachlieli & Shnabel 2014). The first experiment induced participants with "duality" in the lab. For this purpose, we used a modified version of the Dictator Game (Kahneman *et al.* 1986), in which one player, the "proposer," allocates valuable resources (e.g., material payoffs) between herself and the second player, who serves as the "recipient." Specifically, each participant in the experiment was asked to divide valuable extra credit points between herself and another player, knowing that the other player was asked to do the same. Participants then received bogus feedback about their own and the other player's allocations. This feedback constituted the experimental manipulation: participants assigned to the victim condition learned that the other player allocated the extra credit points unfairly, participants in the perpetrator condition learned that they allocated the extra credit points unfairly, participants in the dual condition learned that both they and the other player allocated the extra credit points unfairly, and control participants learned that neither side allocated the extra credit points unfairly.

Following the assignment to social roles, participants filled out self-report questionnaires, which measured their experience of identity threats, motivations, and responses to messages. Replicating previous findings, these measures revealed that compared to the control participants, victims felt less agentic and were motivated to gain more power, whereas perpetrators felt less moral and were motivated to restore their moral image. Also, whereas control participants responded equally positively to empowering and accepting messages from the other player, victims responded more positively to an empowering message, whereas perpetrators responded more positively to an accepting message. Most importantly, we found that, compared to control participants, duals experienced threats to both their agency and their moral image and were motivated to restore both identity dimensions. Moreover, duals responded equally positively to accepting and empowering messages from the other player.

In terms of behavior, however, duals resembled victims. Specifically, in the last phase of the experiment, participants were given the opportunity to either deny or donate credit points to the other player, indicating vengeful (antisocial) or generous (prosocial) behavior, respectively. Victims and duals (but not perpetrators) denied significantly more credit points from the other player compared to participants in the control condition. Perpetrators (but not victims and duals) donated significantly more credit points to the other player compared to participants in the control condition. These findings support our theorizing regarding the "primacy of agency" effect of duality: like victims, duals' heightened need for restoration of agency translated into greater vengeful, antisocial behavior; unlike perpetrators, duals' heightened need to restore moral image failed to translate into greater prosocial behavior.

The second experiment replicated the same pattern of results in a context of a dual conflict between groups. Participants in this experiment were Israeli Jews who were randomly assigned into three different roles – "pure" victims, "pure" perpetrators, and duals. Depending on the experimental condition, participants were asked to recall and write about two incidents in which Palestinians victimized their ingroup (e.g., the Passover massacre of 2002 in which a suicide bomber killed 30 unarmed Israeli civilians), two incidents in which their ingroup victimized Palestinians (e.g., the 1994 Cave of the Patriarchs massacre in which an Israeli settler opened fire inside a mosque and killed 29 unarmed Palestinian civilians), or one victimization and one perpetration incident. Replicating previous findings, victims reported an impaired sense of agency, a wish to restore power, and heightened aggressive tendencies (e.g., increased support for using unrestricted force against any act of Palestinian terrorism), whereas perpetrators reported an impaired moral image, a wish to restore a positive moral identity, and increased helping tendencies (e.g., greater willingness to provide humanitarian aid to Gaza). As for duals, they experienced impairment for, and wished to restore, both agency and moral image. In terms of behavior, however, duals showed only heightened aggressiveness, not heightened helpfulness, toward Palestinians.

The results of these two experiments are consistent with previous theorizing that the experience of victimization is more profound psychologically than the experience of perpetration (Baumeister 1997). For example, in contexts of protracted intergroup conflicts characterized by mutual violence, both conflicting parties develop a deep sense of victimhood but not of perpetration (Noor *et al.* 2012). These two experiments also shed light on how conflict may escalate, suggesting that when conflicting individuals or groups transgress against each other, they are likely to respond with further aggression. Tragically, what is seen as a just retribution by one party is likely to be seen by the other party as unjustified aggression that needs to be further avenged, thus fueling the continued cycle of aggression (Newberg *et al.* 2000).

Agency Affirmation: Reassuring Dual Conflicting Parties' Strength

Although the "primacy of agency" effect pessimistically revealed that duals show heightened levels of aggressiveness and vengefulness, we theorized that the fact that duals did show heightened need for restoration of their moral image leaves room for optimism. Specifically, we reasoned that addressing duals' primary need for agency may allow their need for restoration of positive moral identity to come to the fore and exert its positive effect on their behavior toward their adversaries. To examine this possibility in the context of a dual intergroup conflict, we developed an "agency affirmation" intervention, in which members of groups involved in a dual conflict were exposed to a text that reassured their ingroup's strength, competence, and resiliency. We hypothesized that once group members would be reminded of their ingroup's agency, they would be more willing to relinquish some power for the sake of moral considerations and that this greater willingness, in turn, would lead to their lower vengefulness and greater helpfulness toward each other. A series of four experiments tested the effectiveness of this agency affirmation intervention among Palestinians and Israeli Jews (Shnabel *et al.* 2016 and SimanTov-Nachlieli *et al.* 2017).

In the first experiment, which focused on Israeli Jewish participants, we pitted the agency affirmation intervention against a moral threat manipulation, which was found to increase prosocial tendencies in various interpersonal and intergroup contexts (e.g., Hopkins *et al.* 2007). Participants assigned to the agency affirmation condition were exposed to a text that affirmed Israel's strength by reminding participants that Israel is a strong nation that has proved its power, self-determination, and resilience in many domains such as the economy, technical achievements, and military might. Participants assigned to the moral threat condition were exposed to a text that portrayed Israel in a way that undermined its positive moral identity. Participants assigned to the agency-affirmation-and-moral-threat condition were exposed to a text that combined both agency affirmation and moral threat, whereas control participants read no text. We found that the exposure to agency affirmation reduced Israeli Jews' aggressiveness against Palestinians while increasing their helpfulness toward them. By contrast, the exposure to a moral threat did not affect Israeli Jews' aggressive or helpful tendencies. Moreover, the positive effect of agency affirmation was mediated by participants' willingness to relinquish power for morality (e.g., their belief that Israel should give up its power superiority in order to be just and fair with the Palestinians). An additional experiment replicated these findings and further revealed that a morality affirmation intervention (i.e., the reassurance of the moral-social dimension of the Israeli Jews' ingroup's identity) failed to set in motion a process leading to greater prosocial tendencies.

The third study revealed that the positive effect of agency affirmation was replicated even under conditions of intense security threat. In particular, we tested the effectiveness of this intervention during the military operation "Pillar of Cloud," which took place in November 2012. During this operation, the Israeli Defense Force's (IDF) air force bombed more than 1500 sites in the Gaza Strip, while Hamas and other Palestinian militant groups fired over 1500 rockets into Israel. At the sixth day of the operation, we recruited Israeli Jewish participants to take part in an online experiment in which they were randomly assigned either to a control, no-text condition or to the agency affirmation condition, which exposed them to a text based on the previous study, yet adjusted to the wartime context. For example, the text referred to the effectiveness of the Israeli Iron-Dome anti-rocket defense system, which was used during the operation for the first time and proved to be highly efficient. Compared to the control condition, participants in the agency affirmation condition showed greater readiness to relinquish power for the sake of moral considerations (measured using self-report items such as "Israel should restrain its operations in Gaza to maintain its positive moral image in the world"). This greater readiness, in turn, led to less vengefulness and more helpfulness toward Palestinians. To illustrate, compared to control participants, participants in the agency affirmation condition were less supportive of military operations that would harm Palestinian citizens and more supportive of providing them with humanitarian aid.

The fourth experiment examined the effectiveness of agency affirmation among Palestinians. We were concerned that Palestinians might be less susceptible than Israelis to an affirmation of their ingroup's agency (i.e., it might be harder to effectively affirm the Palestinians' ingroup's agency), due to their ingroup's relative inferiority in terms of military force, economic conditions, and so on. Notably, despite the substantial power asymmetry, the obtained pattern of results generally corresponded to the one obtained among Israeli Jews. In line with expectations, exposing Palestinian participants to a text that affirmed the Palestinians' strength and resilience (i.e., reminding participants that the Palestinian nation is strong, cohesive, and known worldwide for its inner strength and resiliency) increased their willingness to relinquish power for morality (e.g., readiness to give up the use of violence in order to be just and moral in the conflict against the Jews). This, in turn, led to greater helping tendencies toward Israelis (e.g., providing Israelis with humanitarian aid in case of a natural disaster). Once again, the exposure to moral threat did not affect participants' prosocial tendencies.

In summary, a series of experiments revealed that an affirmation of their ingroup's agency increased mutual prosocial tendencies among members of groups involved in a dual conflict. Interestingly, experiments conducted by Shnabel, Halabi *et al.* (2013) imply that even an indirect affirmation of agency can carry such positive effects. Specifically, Shnabel and colleagues exposed Israeli Jews and Palestinians to a text that induced them with a "common perpetrator identity" by highlighting that both parties are equipped with lethal weapons and have actively inflicted substantial harm upon each other. The exposure to this text increased Israeli Palestinians' and Jews' sense of their ingroup's agency (e.g., their belief that their ingroup has the power and resources to solve the conflict) compared to a control, neutral-text condition. The increase in agency, in turn, translated into reduced engagement in competition over the victim status and greater forgiveness tendencies. Thus, drawing group members' attention to their ingroup's strength, even through a reminder of how this strength was misused against the outgroup, can address their pressing need for agency and allow their need for restoration of positive moral image to come to the fore. While this work focused on the group level, we are currently conducting additional research to extend our understanding of the effects of agency affirmation at the individual level.

Implications for the Science of Peaceful Systems and Behavior

In the present chapter, I have argued that conflicting individuals and groups experience threat to the two fundamental dimensions of their identity, namely, the agency and communion (i.e., moral-social) dimensions; consequently, they feel enhanced needs for empowerment and acceptance. These basic needs can manifest in various ways. For example, in the context of a conflict between competing universities, the need for empowerment or acceptance may be manifested as students' wish for reassurance of their competence or warmth, whereas in the context of a prolonged violent conflict between national groups, the need for empowerment or acceptance may manifest as the wish for military superiority or recognition of the ingroup's morality by other nations. While the Needs-Based Model focuses on the link between addressing these needs and reconciliation among humans, it may be interesting to examine whether such a link exists among nonhuman animals as well (e.g., other primates). What may be unique to humans, however, is the translation of these basic needs for empowerment and acceptance into more abstract identity-related motivations, such as the motivation to establish one's victim status in an attempt to gain support from third parties (Noor *et al.* 2012) or to protect one's moral image (Sullivan *et al.* 2012).

The empirical findings presented in this chapter suggest that addressing the identity-related motivations of conflicting parties can lead to their mutual expressions of generosity, even at the expense of giving up power and privilege for the sake of moral, prosocial considerations. These findings are theoretically important in light of the fact that many social-psychological theories highlight individuals' and groups' motivation to maximize their outcomes (e.g., interdependence theory; see Rusbult & Van Lange 1996) and gain power and dominance (e.g., social dominance theory; Sidanius & Pratto 1999). These motivations are often associated with competitive, aggressive, and antisocial behavior. However, the research presented in this chapter reveals that conflicting individuals and groups also have strong moral-social motivations, which can be harnessed to promote mutual prosocial behavior. As such, this research may be viewed as part of the general trend within social and biological sciences of paying greater attention to humans' and other species' capacity for cooperation, empathy, and prosocial behavior. This capacity has been overlooked in earlier literature, which was predominated by the idea that "animal life, and by extension human nature, is based on unmitigated competition" and therefore "a hallmark of humanity is aggression" (de Waal 2012, p. 874).

Drawing attention toward these "better angels of our nature" is important not only for theoretical but also for practical reasons. The Needs-Based Model tells us that restoring adversarial parties' positive identities can not only prevent further conflict escalation but also set in motion an upward spiral of goodwill (see also Ali & Walters [this volume, Chapter 5] for a discussion of how peacebuilding programs may initiate such an upward spiral). The purpose of restorative justice interventions, which focus on addressing the emotional needs of adversaries rather than on merely punishing perpetrators (Wachtel & McCold 2001), is to initiate such an upward spiral. Such conciliatory interventions range from international peacemaking tribunals such as the South African Truth and Reconciliation Commission (TRC) to innovations within schools, social services, communities, and the criminal justice system (e.g., in the form of structured encounters between criminals and their victims) (Boyes-Watson 2008).

Importantly, the insights gained from work on the Needs-Based Model suggest that these interventions should be planned in a way that allows victims to feel empowered and perpetrators to feel accepted. For example, in encounters between criminals and their victims, the victim should be able to determine, or at least influence, the appropriate punishment (Shnabel *et al.* 2008); in dialogue groups between former victims and perpetrators (e.g., Jews and Germans), emphasis should be put on developing mutual empathy, such that members of the perpetrating group would not feel morally condemned due to their group affiliation (Maoz & Bar-On 2002); and in the reintegration of former child soldiers, these children should make restitution to their communities (e.g., through rebuilding a school or a health post), while community members need to accept the returnees as "our children" and learn to see them not only as perpetrators but also as those who had been subjected to considerable suffering themselves (see Wessells & Kostelny, this volume, Chapter 9). I hope that the present chapter and volume will encourage future research and development of such theory-informed interventions to promote peaceful systems and behavior.

References

Abele, A.E., Cuddy, A.J.C., Judd, C.M., & Yzerbyt, V.Y. (2008). Fundamental dimensions of social judgment. *European Journal of Social Psychology*, 38, 1063–1065.

Bandura, A. (1990). Selective activation and disengagement of moral control. *Journal of Social Issues*, 46, 27–46.

Baumeister, R.F. (1997). *Evil: Inside human violence and cruelty*. New York: Henry Holt.

Baumeister, R.F., Stillwell, A.M., & Heatherton, T.F. (1994). Guilt: An interpersonal approach. *Psychological Bulletin, 115*, 243–267.

Blatz, C.W., Schumann, K., & Ross, M. (2009). Government apologies for historical injustices. *Political Psychology, 30*, 219–241.

Boyes-Watson, C. (2008). *Peacemaking circles and urban youth: Bringing justice home*. St. Paul, MN: Living Justice Press.

Brookings, J.B., & Bolton, B. (2000). Confirmatory factor analysis of a measure of intrapersonal empowerment. *Rehabilitation Psychology, 45*, 292–298.

Carnevale, P.J., & Pruitt, D.G. (1992). Negotiation and mediation. *Annual Review of Psychology, 43*, 531–582.

Christie, D.J., Tint, B.S., Wagner, R.V., & Winter, D.D. (2008). Peace psychology for a peaceful world. *American Psychologist, 63*, 540–552. doi:10.1037/0003-066x.63.6.540

de Waal, F.B.M. (2000). Primates: A natural heritage of conflict resolution. *Science, 289*, 586–600.

de Waal, F.B.M. (2012). The antiquity of empathy. *Science, 336*, 874–876. doi:10.1126/science.1220999

de Waal, F.B.M., & van Roosmalen, A. (1979). Reconciliation and consolation among chimpanzees. *Behavioral Ecology and Sociobiology, 5*, 55–60.

Dollard, J., Doob, L., Miller, N., Mowrer, O., & Sears, R. (1939). *Frustration and aggression*. New Haven, CT: Yale University.

Doosje, B., Branscombe, N.R., Spears, R., & Manstead, A.S.R. (1998). Guilty by association: When one's group has a negative history. *Journal of Personality and Social Psychology, 75*, 872–886.

Enright, R.D., Freedman, S., & Rique, J. (1998). The psychology of interpersonal forgiveness. In R.D. Enright & J. North (Eds.), *Exploring forgiveness* (pp. 46–62). Madison, WI: University of Wisconsin Press.

Estrada-Hollenbeck, M., & Heatherton, T. F. (1998). Avoiding and alleviating guilt through prosocial behavior. In J. Bybee (Ed.), *Guilt and children* (pp. 215–231). San Diego, CA: Academic Press.

Exline, J.J., & Baumeister, R.F. (2000). Expressing forgiveness and repentance: Benefits and barriers. In M.E. McCullough, K.I. Pargament, & C.E. Thoresen (Eds.), *Forgiveness: Theory, research and practice* (pp. 133–155). New York: Guilford Press.

Foster, C.A., & Rusbult, C.E. (1999). Injustice and powerseeking. *Personality and Social Psychology Bulletin, 25*, 834–849.

Frijda, N.H. (1994). The lex talionis: On vengeance. In S.H.M.V. Goozen, N.E.V.d. Poll, & J.A. Sergeant (Eds.), *Emotions: Essays on emotion theory* (pp. 263–289). Hillsdale, NJ: Lawrence Erlbaum.

Galtung, J. (1969). Violence, peace and peace research. *Journal of Peace Research, 3*, 176–191.

Hopkins, N., Reicher, S., Harrison, K., Cassidy, C., Bull, R., & Levine, M. (2007). Helping to improve the group stereotype: On the strategic dimension of prosocial behavior. *Personality and Social Psychology Bulletin, 333*, 776–788.

Jones, E. (1998). Major developments in five decades of social psychology. In D.T. Gilbert, S.T. Fiske, & G. Lindzey (Eds.), *The handbook of social psychology* (4th ed., Vol. *1*, pp. 3–57). New York: McGraw-Hill.

Kahneman, D., Knetsch, J., & Thaler, R. H. (1986). Fairness as a constraint on profit seeking: Entitlements in the market. *American Economic Review, 76*, 728–741.

Kelman, H.C. (2008). Reconciliation from a social-psychological perspective. In A. Nadler, T. Malloy, & J.D. Fisher (Eds.), *The social psychology of intergroup reconciliation* (pp. 15–32). New York: Oxford University Press.

King Center. (N.d.). Glossary of nonviolence Retrieved from http://www.thekingcenter.org/glossary-nonviolence0000

Maoz, I., & Bar-On, D. (2002). From working through the Holocaust to current ethnic conflicts: Evaluating the TRT group workshop in Hamburg. *Group, 26*, 29–48.

McCullough, M.E., Kurzban, R., & Tabak, B.A. (2013). Cognitive systems for revenge and forgiveness. *Behavioral and Brain Sciences, 36*, 1–58.

Meierhenrich, J. (2008). Varieties of reconciliation. *Law and Social Inquiry, 33*, 195–231.

Minow, M. (1998). *Between vengeance and forgiveness: Facing history after genocide and mass violence.* Boston: Beacon Press.

Nadler, A. (2012). Reconciliation: Instrumental and socioemotional aspects. In D.J. Christie & C. Montiel (Eds.), *Encyclopedia of peace psychology.* Hoboken, NJ: Wiley-Blackwell.

Nadler, A., & Shnabel, N. (2008). Instrumental and socioemotional paths to intergroup reconciliation and the Needs-Based Model of Socioemotional Reconciliation. In *The social psychology of intergroup reconciliation* (pp. 37–56). New York: Oxford University Press.

Newberg, A.B., d'Aquili, E.G., Newberg, S.K., & deMarici, V. (2000). The neuropsychological correlates of forgiveness. In M.E. McCullough, I. Pargament, & C.E. Thoresen (Eds.), *Forgiveness: Theory, research and practice* (pp. 91–108). New York: Guilford.

Noor, M., Shnabel, N., Halabi, S., & Nadler, A. (2012). When suffering begets suffering: The psychology of competitive victimhood between adversarial groups in violent conflicts. *Personality and Social Psychology Review, 16*, 351–374.

North, J. (1998). The "ideal" of forgiveness: A philosopher's exploration. In R.D. Enright & J. North (Eds.), *Exploring forgiveness* (pp. 15–34). Madison, WI: University of Wisconsin.

Phalet, K., & Poppe, E. (1997). Competence and morality dimensions in national and ethnic stereotypes: A study in six eastern-European countries. *European Journal of Social Psychology, 27*, 703–723.

Rouhana, N.N. (2004). Group identity and power asymmetry in reconciliation processes: The Israeli-Palestinian case. *Peace and Conflict: Journal of Peace Psychology, 10*, 33–52.

Rusbult, C.E., & Van Lange, P.A.M. (1996). Interdependence processes. In E.T. Higgins & A.W. Kruglanski (Eds.), *Social psychology: Handbook of basic principles* (pp. 564–596). New York: Guilford Press.

Scheff, T.J. (1994). *Bloody revenge: Emotions, nationalism, and war.* Boulder, CO: Westview Press.

Schönbach, P. (1990). *Account episodes: The management or escalation of conflict.* New York: Cambridge University Press.

Sherif, M., Harvey, O.J., White, B.J., Hood, W.R., & Sherif, C.W. (1961). *Intergroup cooperation and competition: The Robbers Cave experiment.* Norman, OK: University Book Exchange.

Shnabel, N., Halabi, S., & Noor, M. (2013). Overcoming competitive victimhood and facilitating forgiveness through re-categorization into a common victim or perpetrator identity. *Journal of Experimental Social Psychology, 49*, 867–877.

Shnabel, N., & Nadler, A. (2008). A needs-based model of reconciliation: Satisfying the differential emotional needs of victim and perpetrator as a key to promoting reconciliation. *Journal of Personality and Social Psychology, 94*, 116–132.

Shnabel, N., Nadler, A., Canetti-Nisim, D., & Ullrich, J. (2008). The role of acceptance and empowerment from the perspective of the Needs-Based Model. *Social Issues and Policy Review, 2*, 159–186.

Shnabel, N., Nadler, A., & Dovidio, J.F. (2014). Beyond need satisfaction: Empowering and accepting messages from third parties ineffectively restore trust and consequent reconciliation. *European Journal of Social Psychology, 44*, 126–140.

Shnabel, N., Nadler, A., Ullrich, J., Dovidio, J.F., & Carmi, D. (2009). Promoting reconciliation through the satisfaction of the emotional needs of victimized and perpetrating group members: The needs-based model of reconciliation. *Personality and Social Psychology Bulletin, 35,* 1021–1030.

Shnabel, N., & Noor, M. (2012). Competitive victimhood among Jewish and Palestinian Israelis reflects differential threats to their identities: The perspective of the Needs-Based Model. In K.J. Jonas & T. Morton (Eds.), *Restoring civil societies: The psychology of intervention and engagement following crisis* (pp. 192–207). Malden, MA: Wiley-Blackwell.

Shnabel, N., SimanTov-Nachlieli, I., & Halabi, S. (2016). The power to be moral: Affirming Israelis and Palestinians' agency promotes mutual pro-social tendencies across group boundaries. *Journal of Social Issues, 72,* 566–583.

Shnabel, N., Ullrich, J., Nadler, A., Dovidio, J.F., & Aydin, A.L. (2013). Warm or competent? Improving intergroup relations by addressing threatened identities of advantaged and disadvantaged groups. *European Journal of Social Psychology, 43,* 482–492.

Sidanius, J., & Pratto, F. (1999). *Social dominance: An intergroup theory of social hierarchy and oppression.* New York: Cambridge University Press.

SimanTov-Nachlieli, I., Shnabel, N., Aydin, A. L., & Ullrich, J. (2017). Agents of pro-sociality: Affirming conflicting groups' agency promotes mutual pro-social tendencies. *Political Psychology.* DOI: 10.1111/pops.12418

SimanTov-Nachlieli, I., & Shnabel, N. (2014). Feeling both victim and perpetrator: Investigating duality within the Needs-Based Model. *Personality and Social Psychology Bulletin, 40,* 301–314. doi:10.1177/0146167213510746

SimanTov-Nachlieli, I., Shnabel, N., & Nadler, A. (2013). Individuals' and groups' motivation to restore their impaired identity dimensions following conflicts: Evidence and implications. *Social Psychology, 44,* 129–137.

Staub, E., Pearlman, L.A., Gubin, A., & Hagengimana, A. (2005). Healing, reconciliation, forgiving and the prevention of violence after genocide or mass killing: An intervention and its experimental evaluation in Rwanda. *Journal of Social and Clinical Psychology, 24,* 297–334.

Sullivan, D., Landau, M.J., Branscombe, N.R., & Rothschild, Z.K. (2012). Competitive victimhood as a response to accusations of ingroup harm doing. *Journal of Personality and Social Psychology, 102,* 778–795. 10.1037/a0026573

Tajfel, H., & Turner, J.C. (1986). The social identity theory of inter-group behavior. In S. Worchel & L.W. Austin (Eds.), *Psychology of intergroup relations.* Chicago: Nelson-Hall.

Tavuchis, N. (1991). *Mea culpa: A sociology of apology and reconciliation.* Stanford, CA: Stanford University Press.

Tinbergen, N. (1963). On aims and methods of ethology. *Zeitschrift für Tierpsychologie, 20,* 410–433.

Turner, J.C., Hogg, M.A., Oakes, P.J., Reicher, S.D., & Wetherell, M.S. (1987). *Rediscovering the social group: A self-categorization theory.* Oxford: Blackwell.

Verbeek, P. (2008). Peace ethology. *Behaviour, 145,* 1497–1524.

Wachtel, T., & McCold, P. (2001). Restorative justice in everyday life. In H. Strang & J. Braithwaite (Eds.), *Restorative justice in civil society.* New York: Cambridge University Press.

Wojciszke, B. (2005). Morality and competence in person- and self-perception. *European Review of Social Psychology, 16,* 155–188.

Zitek, E.M., Jordan, A.H., Monin, B., & Leach, F.R. (2010). Victim entitlement to behave selfishly. *Journal of Personality and Social Psychology, 98,* 245–255.

3

Inclusion as a Pathway to Peace: The Psychological Experiences of Exclusion and Inclusion in Culturally Diverse Social Settings

Sabine Otten, Juliette Schaafsma, and Wiebren S. Jansen

In times of globalization, international mobility, and growing emancipation of minority groups, social systems such as companies, schools, and universities are typically characterized by diversity, for example in terms of ethnicity, gender, age, religion, or sexual orientation. Yet, even though diversity has become a typical feature of modern social groups, it is a feature that may not only be experienced as enriching, but also pose a challenge to tolerance and peaceful relations among group members. Therefore, the growing complexity of modern societies has triggered an increasing interest in understanding how social exclusion can be prevented and how social systems can be built that provide safe feelings of inclusion for members of all groups involved.

Advantages of diverse as compared to homogeneous groups have been shown especially for the domain of creativity and innovation (e.g., Nijstad & Paulus 2003; van der Zee & Paulus 2008). Yet, regarding social functioning of diverse groups, outcomes are generally less positive. Research – especially in the work context – has revealed that diversity in groups may enhance the probability of conflict and communication problems (e.g., Williams & O'Reilly 1998; Joshi & Roh 2009). Rather than a larger diverse group seeing itself as an entity, it is often subdivided into subgroups, and these subgroups will typically differ in terms of status and access to relevant resources (e.g., Homan *et al.* 2007). Accordingly, group members – especially those from minority groups – may experience disadvantages and exclusion. As we will outline in more detail in this chapter, such experiences are costly as they not only are prone to negatively affect individual well-being and group functioning, but also may enhance the probability of conflict and aggression. Yet, while exclusion may pose a threat to peaceful relations within and between groups, the reliable experience of inclusion can be assumed to enhance and secure positive relations among the members of today's complex social groups.

In the present chapter, we summarize relevant theories and empirical evidence regarding the psychological experience of exclusion and inclusion in social groups, especially in groups that are diverse. The studies we report on mostly refer to *individual* experiences of inclusion and exclusion, and they measure how these experiences translate into well-being and harmonious – as opposed to conflictual – relations with others. Hence, when we refer to implications of these findings for securing or increasing

Peace Ethology: Behavioral Processes and Systems of Peace, First Edition.
Edited by Peter Verbeek and Benjamin A. Peters.
© 2018 John Wiley & Sons Ltd. Published 2018 by John Wiley & Sons Ltd.

peaceful relations, we refer to this micro level. Our point is that through experiencing inclusion in diverse group settings, people will be less prone to have conflicts with members from other groups, and will be more prone to realize the added value of those who are different. However, we assume that the more social groups succeed in avoiding exclusion and fostering inclusion of their members from different subgroups, the more peace can be secured on a broader level (i.e., between subgroups as a whole).

In this chapter, we lean on theories from social, organizational, and evolutionary psychology and on research findings stemming from both experimental and field research, with a strong representation of research from the work and organizational contexts. In the first part of this chapter, we explain why inclusion is important but also why exclusion is highly likely, and we focus on the characteristics, underlying processes, and consequences of the experience of social *exclusion*. Next, we deal with the psychological experience of *inclusion*, its characterizing features, and its determinants and consequences.

Importantly, we argue that inclusion is more than just the absence of exclusion. In our view, inclusion not only implies that people can safely rely on being part of the group, but also means that both similarities and differences are appreciated and openly expressed, thereby supporting individual well-being, group performance, and peaceful, harmonious intragroup interactions (Otten & Jansen 2014).

The Importance of Inclusion and the Inevitability of Exclusion

Human beings are an exceptionally social species with a strong need to affiliate with others and to belong to social groups (e.g., Baumeister & Leary 1995; Williams 2001). It has been argued that this need is deeply rooted in our evolutionary past. Lacking the typical defenses of other large mammals (e.g., claws, fangs, speed, and physical strength) and with an extended period of offspring dependency on parental care, our ancestors were probably only able to survive harsh environments by living in cooperative groups. These groups could not only defend them against predators and rival outgroups, but also help them find nutritious food and take care of their offspring. Exclusion from the group most likely posed a serious threat to survival: excluded individuals were unlikely to reproduce themselves and probably also faced an early death (e.g., Leary & Cottrell 2013; see also Narvaez [this volume, Chapter 6] for a discussion on the importance of belonging in small-band hunter-gatherer societies).

There is reason to believe that, as a result of such evolutionary pressures, human beings have developed the ability to quickly notice even very subtle cues of exclusion and reflexively respond with pain once such signals are detected (see also Cacioppo *et al.* [2011], cited in Verbeek [2013], for similar effects of social isolation in other social animals). Research has found that even brief and seemingly innocuous episodes of exclusion are distressing and immediately threaten fundamental human needs such as the need to belong, the need for self-esteem, the need for control, and the need for a meaningful existence (e.g., Wirth *et al.* 2010; Cacioppo *et al.* 2011; Wesselmann *et al.* 2012). Furthermore, there is evidence that the dorsal anterior cingulate cortex – which is also involved in the experience of physical pain – is active when people are being excluded (e.g., Eisenberger *et al.* 2003). This initial pain response to exclusion tends to be subject to few if any individual or situational moderators, suggesting that exclusion hurts regardless of the circumstances (e.g., Williams 2007).

Despite the importance of being included, exclusion is an inevitable part of life in contemporary society. Already at an early age, people may have their first experiences with being ignored, excluded, or rejected. Case studies among young children suggest that they use or experience exclusion in a variety of ways during supervised play (e.g., Barner-Barry 1986). And in their adult lives, people are also likely to become either a victim or a perpetrator of some form of exclusion. For example, in an event-contingent diary study in Australia, Nezlek *et al.* (2012) found that people reported on average one episode a day in which they had felt ignored or excluded. To some extent, this is because people simply lack the time and resources to establish social relationships with everyone they meet, but there are also evolutionary reasons for why people have to be selective about whom they accept in their midst. Even though group living can be very beneficial, it also comes with risks as it may increase competition for resources (e.g., food and mates), attract free riders, and increase the likelihood of disease and parasite transmission. It has been argued that, in light of these potential threats, people should prefer groups of relatively moderate size and should not be willing to include everyone. More specifically, they should be motivated only to include individuals who provide fitness benefits and to exclude those who generate fitness costs (e.g., Brewer 1991; Leary 2001; Leary & Cottrell 2013).

Excluding People Who Are Different

So, groups tend to be critical about whom they include, and some people are more likely to be avoided or rejected than others. Members of other groups (i.e., outgroups), for instance, often tend to be less desirable interaction partners than those of one's own group (the ingroup). Evolutionary psychologists have argued that this is because outgroup members are unlikely to be close kin and also because – in our ancestral environment – humans had to compete (though not necessarily directly or violently) with other groups over valuable but sometimes scarce resources. Forming relationships with outsiders may not have been obvious for that reason, but it also may not have been desirable because it most likely increased the pressure on limited resources within a group. Furthermore, the game of reciprocity may have been more difficult with outgroup members, as they may have had different norms about mutual cooperation and sharing. Thus, establishing relationships with outgroup members may have created significant fitness costs in our evolutionary past, and by limiting their access to the group, these costs could be contained (e.g., Brewer 1999; Gil-White 2001; Leary & Cottrell 2013; but see Narvaez [this volume, Chapter 6] for examples of small-band hunter-gatherers with strong reciprocity norms, who easily share food with members of other groups).

The human tendency to create "us" versus "them" distinctions is likely to be a result of this evolutionary adaptation: group boundaries can be used as a proxy to determine who can be trusted, and who will probably cooperate (Brewer 1997, 1999; Brewer & Carporael 2006). Although people do not necessarily have hostile attitudes toward outgroup members, research suggests that they generally do see them in a less positive light as compared to their ingroup members and also tend to favor the ingroup over the outgroup when allocating positive resources (e.g., Tajfel *et al.* 1971; Tajfel 1981). This tendency to favor the ingroup has been found to extend across all forms of group membership – even when this is based on seemingly arbitrary criteria – and has also been found in cultures across the world. For example, in a study on reciprocal attitudes among 30 groups in East Africa, Brewer and Campbell (1976) found that almost all groups rated the ingroup more positively than the outgroup on a host of dimensions (e.g., friendliness and honesty).

Even though "ingroup love" rather than "outgroup hate" is typically the dominant motive in intergroup relations (Brewer 1999), the tendency to perceive one's own group as better typically also means that the ingroup and its members are seen as deserving more than those from other groups. Hence, as a result of the propensity to differentiate between ingroups and outgroups, many forms of direct or indirect discrimination may develop, leading to the avoidance, rejection, or exclusion of people who are different. In everyday society, this may particularly affect ethnic minority group members. They are often easily classified as outgroup members and are, as a result of this, likely to encounter discrimination and exclusion in many aspects of their everyday lives. For example, in a relatively recent large-scale study conducted in the European Union, a quarter of the ethnic minority group participants reported having recently felt discriminated against or excluded because of their ethnic background (Eurobarometer 2009). To some extent, this may occur relatively subtly during their daily interactions with majority group members. For instance, laboratory research has found that majority group members may indirectly express prejudice when they interact with ethnic minority group members, by expressing less nonverbal friendliness (e.g., Dovidio *et al.* 2002). In line with this, research on the quality of ethnic minority group members' everyday interactions shows that they tend to feel less liked, less respected, and less accepted during interactions that involve majority group members as compared to interactions that do not involve majority group members (Schaafsma *et al.* 2010).

But ethnic minority group members may also experience exclusion and discrimination at a broader or institutional level. For example, there is evidence that it is often more difficult for them to have access to or participate in a variety of domains such as schools, the housing market, and the labor market. In this regard, research demonstrates that they are more likely to be excluded from housing or rental opportunities (e.g., Turner & Ross 2003). Moreover, across various countries, employers have been found to be reluctant or unwilling to hire ethnic minority group members, even when they are equally qualified as compared to ethnic majority group members (e.g., Bertrand & Mullainathan 2004; Carlsson & Rooth 2006; EUMC 2006; Eurobarometer 2009; Andriessen *et al.* 2010, 2012; Brynin & Güveli 2012).

Research suggests that it may be relatively difficult for people to recover from such experiences with group-based exclusion or discrimination. For example, in a study by Goodwin *et al.* (2010), Caucasian and African American adults were either included or briefly excluded during a virtual ball-toss game (Cyberball). They found that participants responded more negatively to exclusion by outgroup members than by ingroup members and also had more difficulty recovering from outgroup exclusion because they were more likely to attribute it to prejudice or racism. In addition, survey data show that more chronic experiences with or perceptions of exclusion and discrimination are related to higher levels of depression and stress, lower levels of life satisfaction and happiness, and lower levels of self-esteem (e.g., Williams & Chung 1997; Branscombe *et al.* 1999; Pascoe & Smart Richman 2009). Thus, believing that one is the victim of group-based exclusion or discrimination can result in a variety of negative mental health outcomes.

Moreover, in the work context, a recent survey among Dutch employees revealed that not only actual experiences of exclusion or unfair treatment but also the mere expectation of possibly being a target of prejudice were negatively related to organizational identification, trust in the organization, and trust in society (Otten & van der Zee 2014). Not surprisingly, such expectations were significantly higher among ethnic minority employees. In addition, de Vroome and collaborators (2011) found in a large survey among immigrants in the Netherlands that structural integration (i.e., having a job in a Dutch organization) only enhanced identification with the Dutch host society if immigrants did not experience exclusion and discrimination at work.

Exclusion as a Potential Threat to Peace

Given the importance of being included by others and the potential risks and negative effects of being excluded, one might expect that excluded individuals should engage in behaviors that are likely to promote inclusion and acceptance by others. Paradoxically, however, rejected individuals have often been found to become aggressive and to engage in behaviors that are likely to reduce the likelihood of them being accepted again. For example, in a series of experiments, Twenge *et al.* (2007) manipulated social exclusion by telling participants that they would later end up alone in life, or by telling them that nobody wanted to work with them. They found that exclusion caused a substantial reduction in prosocial behavior: socially excluded participants were less helpful, less cooperative, and also less willing to volunteer for follow-up experiments.

In another set of studies, Twenge and colleagues (2001) found that excluded participants were more likely to aggress toward someone not directly involved in the rejection experience (and as such seemed to redirect their aggression), by blasting them with higher levels of aversive noise. In addition, studies on real-world crime and violence have shown a link between perceived rejection and aggression in daily life, such as domestic violence, school shootings, homicides, and gang violence (e.g., Walsh *et al.* 1987; Garbarino 1999; Leary *et al.* 2001). For example, in an analysis of news reports of US school shootings, Leary and colleagues (2001) discovered that in 13 of the 15 instances, the shooters experienced chronic or acute social rejection.

Obviously, these antisocial reactions to exclusion are unlikely to promote social inclusion and may, for that reason, seem maladaptive. It has been argued, however, that when the need for control is sufficiently thwarted as a result of the exclusion, this can outweigh the desire to be liked. This may also happen when people believe that they have no opportunities for re-inclusion. In such a situation, aggression may actually be a functional response as it may be a means through which people are able to restore a sense of personal power or control over others (e.g., Warburton *et al.* 2006). In line with this idea, Warburton *et al.* (2006) found that participants who had been given a task that was designed to increase their feelings of control were less likely to aggress following exclusion than participants who had completed a task that decreased their feelings of control.

Nevertheless, even though – from an individual perspective – antisocial responses to exclusion may in some ways be functional, such reactions can also pose a serious threat to peace, particularly when exclusion takes place in an intergroup context.

For example, there is evidence that exclusion by ethnic outgroup members results in more hostile or aggressive reactions than exclusion by ethnic ingroup members. In a study among African American and Caucasian students, for instance, Mendes *et al.* (2008) found that social rejection by different-race evaluators resulted in more anger than rejection by same-race evaluators. Research by Schaafsma and Williams (2012) suggests that such negative feelings are likely to extend beyond those involved in the contact situation. They examined how adolescents from different ethnic groups in the Netherlands (of Dutch, Moroccan, and Turkish descent) responded to being excluded by ethnic in- and outgroup members. For this purpose, they let participants play an online game of Cyberball with two fictitious players who either had the same ethnic background as them or who belonged to a different ethnic group. They found that participants who had been excluded by outgroup members during this game not only expressed more aggressive intentions toward the excluders than participants who had been excluded by ingroup members, but also expressed more hostile feelings toward the outgroup at large. Importantly, this was because people were more likely to attribute the exclusion by outgroup members to prejudice and racism than exclusion by ingroup members.

Moreover, exclusion – and chronic exclusion in particular – may contribute to prejudice and intolerance toward outgroup members and impede the successful integration of ethnic minority members. For example, a recent study among German participants found that socially excluded participants showed less tolerance toward Muslims openly practicing Islam in Germany and also supported restrictive naturalization policies more (Aydin *et al.* 2014). People who experience exclusion on a regular basis may also assert their religious identity more strongly or become more susceptible to radical or religious fundamentalist beliefs. Several surveys have found relationships between feelings of social exclusion on the one hand, and religious identification or fundamentalist orientations on the other (e.g., Gijsberts 2010; Phalet & Ter Wal 2004; Verkuyten & Yildiz 2010).

Finally, evidence from laboratory studies indicates that excluded individuals experience higher levels of religious affiliation and also have stronger intentions to engage in religious behaviors as compared to non-excluded individuals (Aydin *et al.* 2010). According to Schaafsma and Williams (2012), fundamentalist religious beliefs may become particularly attractive for excluded individuals because such beliefs are likely to provide them with certainty and meaning again and may help them regain lost needs of control, belonging, and self-esteem. In support of this idea, they found in their study that adolescents with Christian and Muslim beliefs endorsed fundamentalist religious beliefs more strongly after having been excluded, but that this was most likely to occur following exclusion by ingroup members (see also Ali & Walter [this volume, Chapter 5], who make the case that empowerment is an essential element of peacebuilding transformation processes).

Taken together, the findings from the studies described here suggest that exclusion may be a precursor to the polarization and radicalization of ethnic groups and as such constitutes an important obstacle to establishing or maintaining peaceful interethnic relations. Given these potentially disruptive effects, it is important to focus on the social integration and inclusion of different ethnic groups within society, and to create opportunities for contact and cooperation between them.

Defining Inclusion

Indeed, the concept of inclusion has received increasing attention in the public debate, on political agendas, and in the literature (e.g., Roberson 2006; Bilimoria *et al.* 2008; Lirio *et al.* 2008; Shore *et al.* 2011). But what exactly is inclusion? Or, more correctly, how do we define this concept in the context of the present chapter?

Level of Analysis

First of all, inclusion can be seen as both a group-level and individual-level characteristic. On the group level, an organization or other social setting can be described as being more or less inclusive. In the most descriptive way, it means that a larger group comprises members from various subgroups, but it may also refer to actual organizational practices that are implemented to facilitate the inclusion of members from various subgroups (such as making sure that hiring committees in organizations comprise members from various subgroups). On the individual level, inclusion refers to a psychological experience. That is, inclusion has been defined as the degree to which the group member feels part of its group (e.g., Janssens & Zanoni 2008), or as the degree to which (s)he feels part of critical processes in the group (e.g., being informed about changes, being asked for an opinion, etc.; Mor Barak & Cherin 1998). In this section, and in line with the theory and research reported in the first part of this chapter, we will focus on the individual group member as the relevant unit of analysis. We assume that characteristics of the individual group member, and especially minority versus majority status, play a relevant role in the psychological experiences of inclusion and exclusion.

Inclusion versus Identification

A second relevant aspect is that inclusion is experienced based on signals that group members receive from the social group. Thus, even though perceived inclusion is a subjective experience, it is determined by the group (think, for example, of an organization that explicitly promotes diversity in its mission statement). This makes inclusion different from group identification (Jansen *et al.* 2014; Otten & Jansen 2014). Like inclusion, identification describes how closely individual group members feel linked to the group. Yet, in the process of identification, it is the individual who signals whether (s)he likes or dislikes a certain group. Therefore, identification and inclusion are in most instances related yet distinct concepts. A group may safely include me, but I need not necessarily value this inclusion a lot. Conversely, I may – at least in the short run – strongly identify with a group that fails to signal that I fit in.

Inclusion as a Two-Dimensional Concept

In theories of and research on peace, there is often a distinction made between negative peace, defined as the end of violence, and positive peace, characterized by reliable social justice (e.g., Christie *et al.* 2001). Similarly, and in line with previous literature (Shore *et al.* 2011), we assume that inclusion is not simply the absence of exclusion. It is more than that. More specifically, we have defined inclusion as "the extent to which an individual perceives that the group provides him or her with a sense of belonging and authenticity" (Jansen *et al.* 2014). This implies that, while a state of *inclusion* is achieved if both belonging and room for authenticity are experienced as high, a state of *exclusion* entails that a group member neither gets signals that she safely belongs to and is appreciated by the group nor feels that she may exhibit her true self. Low scores on one dimension coupled with high scores on the other define two other psychological stages: *separation* implies that people dare to be authentic, but feel peripheral within the group, and *assimilation* means that a safe feeling to belong is coupled with the idea of having very little room for exhibiting the true self (Otten & Jansen 2014). Like the literature on social exclusion, the literature on inclusion leans on the need to belong as a fundamental human motive (Baumeister & Leary 1995). This need can plausibly account not only for the negative effects of exclusion but also for the beneficial effects of inclusion on human functioning and well-being. The authenticity dimension, however, is – at least at first glance – probably less obvious and self-evident. In situations where the group (e.g., the organization or a work team) is psychologically relevant, people's *social* rather than their *personal* self will typically be salient. According to Self-Categorization Theory (Turner *et al.* 1987), this implies that the focus of attention will shift from unique, individual characteristics to shared attributes within the group. Yet, the more groups become diverse and the more they comprise individuals from various subgroups, the higher the probability is that being different will become or stay psychologically relevant for the individual group members. And the more obvious such differences within the group are (e.g., because having a different ethnicity goes along with different looks and, often, different religious values), the more relevant the authenticity dimension should be for the well-being and well-functioning of group members (Homan *et al.* 2007).

We assume that group members' well-being is not only determined by the extent to which their need to belong is satisfied, but also depends on the degree to which they perceive to be valued for their idiosyncratic attributes. Such a two-dimensional approach already has some tradition in social psychology; first, according to Optimal Distinctiveness Theory (ODT; Brewer 1991), the need to belong has an antagonist, namely, the need to be distinct. Group memberships – but also interpersonal relations – are considered optimal if these two needs are in balance. Importantly, the idea is that the two needs are negatively interdependent: the more easily one can become and stay a member of a certain group, the more difficult it will be to feel unique and distinct as a member of this group. To resolve this conflict, ODT holds that people can also satisfy their need for uniqueness at the intergroup level. That is, they may feel more distinct by contrasting their own group from other groups (Brewer 1991). However, also within groups, there are possibilities to properly balance the two needs. A growing literature reveals that encouraging group members to use their individuality when becoming part of the group (i.e., by having a say in the creation of group norms) and acknowledging within-group differences need not weaken group members' commitment to and identification with their group (e.g., Jans *et al.* 2012; van Veelen *et al.* 2013a, 2013b; see also Otten & Jansen 2014).

A second theoretical concept that is closely associated with the present conceptualization of inclusion is Self-Determination Theory (SDT; Deci & Ryan 2000). Similar to ODT, SDT posits that group membership and people's well-being and functioning in social groups are determined by the satisfaction of two fundamental needs: relatedness and autonomy. Relatedness is defined as the need to be connected to others (Deci & Ryan 2000); hence, this dimension can be seen as closely related or even equivalent to what has been labeled the "need to belong" in other literature (Brewer

1991; Baumeister & Leary 1995). The need for autonomy involves the desire to experience choice and the wish to behave in accordance with one's integrated sense of self (Deci & Ryan 2000; cf. Jansen *et al.* 2014). While the need to be distinct refers to the wish to be different and distinguishable from others, the need for autonomy comes closer to the authenticity dimension in our definition of inclusion. Autonomy can be experienced if group members feel that they are allowed to do and allowed to be what is in line with their own representation of the self. However, such a "true" self may either resemble or be different from other known group members. As we will outline further in this chapter, this is relevant if we consider how perceived inclusion may differ for minority and majority members.

In a recent review of the organizational literature on inclusion, Shore *et al.* (2011) also suggested a two-dimensional conceptualization, wherein inclusion is determined by perceived belonging and perceived room to be unique. However, we consider it relevant to have authenticity rather than uniqueness as the defining second dimension of inclusion. Only then, as we will argue in more detail here, can a possible negative interdependence between facilitating inclusion for minority and majority members be avoided.

Diversity Ideologies and Inclusion in Diverse Groups

In diversifying social contexts, the starting point is typically a situation in which the majority group defines the relevant norms and has most access to relevant resources, thereby regulating the exclusion or inclusion of minority members. Scoring high on prototypicality normally implies having a higher status within the diverse group (e.g., Mummendey & Wenzel 1999; Verkuyten 2006; Dovidio *et al.* 2007). Against this background, attempts to facilitate the inclusion of minority members have often focused on convincing the majority of the added value in diversity (e.g., Cox 1993; Ely & Thomas 2001). Striving for diverse and heterogeneous, rather than homogeneous, groups has been associated with a higher chance to be innovative and creative, with a better chance to reach out to diversifying markets, but also with acting according to relevant societal standards of fairness (see van der Zee & Otten 2014).

It is plausible to assume that such possible assets of the inclusion of minority members should also appeal to majority members. However, a strong focus on appreciating uniqueness in attempts to promote diversity may also imply that those who are mainstream (i.e., majority group members) experience a loss of status and appreciation by the organization. This suggests a possible negative interdependence between promoting uniqueness and enhancing the inclusion of majority and minority members: while promoting uniqueness could be experienced as an inclusion signal by the minority members, it might be experienced as undermining the inclusion (or at least the status) of majority members. In fact, recent research has shown that a mere focus on promoting diversity and the inclusion of minority group members may be associated with feelings of exclusion in majority members (Plaut *et al.* 2011).

Colorblindness versus Multiculturalism

In line with the above argument, there is evidence showing that the type of diversity ideology (i.e., the goals and procedures that are associated with the implementation and management of diversity) are highly relevant for minority and majority members' well-being and functioning in diverse organizations (or other types of larger, complex social groups) (Stevens *et al.* 2008; Plaut *et al.* 2011). The two most prominent diversity ideologies are multiculturalism and colorblindness. In a nutshell, the multicultural ideology is pluralistic and sees diversity as an asset for the organization; group differences should be appreciated and used in order to maintain or improve the quality of the organization. In contrast, a colorblind ideology is assimilationist by ignoring or minimizing group differences and their possible role within the group (cf. Plaut *et al.* 2009). While the former ideology values being *a*typical and different, the latter values fitting in and being *prototypical*. Given these differences, it is not surprising that minority members often favor a multiculturalism ideology over a colorblind approach, while the reverse is true for majority members (e.g., Shelton *et al.* 2006; Verkuyten 2006; Wolsko *et al.* 2006). Moreover, recent research by Plaut and collaborators (2011) revealed that majority members (white Americans) implicitly associate multiculturalism with exclusion rather than inclusion.

Diversity ideologies not only may affect group members' feelings of inclusion or exclusion, but also are correlated with well-being and well-functioning. This was shown in a recent survey of Dutch employees by Vos and collaborators (Vos *et al.* 2014). Their data indicated that both minority and majority members benefit from high levels of perceived inclusion, in terms of not only their subjective well-being but also their (self-reported) functioning. In contrast, for majority members, these outcomes rely to a substantial extent on the degree to which they assume that their organization pursues a colorblind diversity ideology.

In a similar vein, Meeussen *et al.* (2014) found that the degree to which leaders of culturally diverse student groups endorsed multiculturalism was positively correlated with minority members' feelings of being accepted within the group. Conversely, the degree to which the team leaders endorsed colorblindness predicted how strongly minority members experienced conflict within the group and how much they distanced themselves from the group. In this study, majority members were not affected by the group leader's diversity ideology, suggesting that they did not feel threatened by the inclusion of cultural minority members in this specific social setting (i.e., the university).

All-Inclusive Multiculturalism

Taken together, the studies comparing the impacts of multiculturalism and colorblindness on the well being and functioning of group members in diverse groups suggest that the diversity ideology that will most benefit a member depends on the individual's status (as either a minority or majority member). In particular, the fact that multiculturalism may succeed in signaling inclusion to minority members, but may also automatically be associated with exclusion by majority members, suggests a serious dilemma for those trying to properly implement and manage diversity.

A solution here may be provided by a diversity ideology that was coined "all-inclusive multiculturalism" by Stevens and collaborators (2008). This ideology not only promotes pluralism and group differences, but also makes explicit that majority members are a valuable part of that pluralism and diversity (see also Plaut *et al.* 2011). In this vision, it is explicitly stated that the organization relies on both the minority subgroups and the majority groups in order to function well. In fact, the all-inclusive multiculturalism versus "standard" multiculturalism distinction nicely corresponds to the distinction between a focus on authenticity rather than uniqueness: in both cases, a potential negative interdependence between minority and majority inclusion is resolved or at least diminished. Authenticity means that exhibiting one's true self may mean both being and behaving mainstream as well as being and behaving different from what is prototypical. Similarly, all-inclusive multiculturalism allows equally cherishing being a minority and majority member.

Indeed, recent findings on how majority members react to an "all-inclusive" diversity ideology confirm that such an approach may be successful when trying to secure their feelings of inclusion. In two studies by Jansen *et al.* (2015), participants were confronted either with statements representing a multiculturalist ideology (i.e., focusing on the value of group-based differences and the opportunity to learn from these differences) or with statements that also explicitly mentioned the value that the majority has within the diverse group (e.g., "Our organization is happy to have employees from various cultural backgrounds; together with our Dutch employees, they can help us to secure our position jn the market"). Consistently, these studies revealed that majority members benefit from an all-inclusive multiculturalism approach. Compared with "standard" multiculturalism, this ideology elicited more positive expectations regarding being included in the organization. These inclusion perceptions, in turn, predicted the extent to which majority members supported organizational diversity efforts. Moreover, Plaut and colleagues (2011) showed that majority members' automatic association of multiculturalism with exclusion was attenuated when one's own group was explicitly included in the diversity ideology.

Cognitive Routes to Inclusion in Diverse Groups

A different approach to fostering inclusion and reducing exclusion in diverse social groups focuses on the cognitive processes that determine how group members create a link between themselves and their group. Typically, group members – at least those who identify with their groups – perceive an overlap between their mental representation of who they are individually, and how they see their group (van Veelen *et al.* 2011). This overlap can emerge in two ways: either a group member assumes that she fulfills the relevant stereotypes that are known about the group (self-stereotyping), or she assumes that characteristics that are defining for her will also apply to the group (self-anchoring). In principle, both cognitive projection processes can occur, and both predict a positive identification with the group (van Veelen *et al.* 2011). However, in diverse groups, comprising both a majority group and one or more minority groups, it matters *which* of these two cognitive processes is set in motion.

Seeking for overlap by considering how much one fulfills relevant group stereotypes should, by definition, be easier for majority than for minority members. After all, being a majority member implies being the standard rather than the exception to the rule (see also Dovidio *et al.* 2007). Taking the individual self as a starting point for defining one's fit with the group, however, should work similarly well (i.e., lead to comparable levels of identification and perceived inclusion) for minority and majority members. In principle, both minority and majority members are free to assume that traits they consider relevant for who they are as an individual person may also apply to their group as a whole. For example, a Dutch person may describe herself as a diehard soccer fan and at the same time assume that most Dutch people will feel the same. *Which* traits are perceived as self-defining, however, is not predetermined. Using the self as anchor for defining one's social group allows individuals to consider both relatively unique and broadly shared characteristics. Thus, the process of self-anchoring may facilitate perceptions of authenticity for both minority *and* majority group members; the resulting representation of the diverse groups allows its members to be different, but similarly signals that "there is nothing wrong with being normal" (Jansen *et al.* 2015).

Recent research by van Veelen and collaborators (van Veelen *et al.* 2013a, 2013b) confirmed this assumption. In the self-anchoring condition, these authors first instructed group members of a newly formed work group to describe themselves in terms of five characteristic traits and then reflect on how much these traits would also apply to the work team as a whole. In the self-stereotyping condition, participants first generated five characteristic traits of their new team and then reflected on whether these traits might also apply to themselves. Afterward, measures on group identification and appreciation of diversity were taken. The findings revealed that projecting from the individual self to the group (i.e., self-anchoring) was beneficial for minority members, but also not disadvantageous for majority members. Minority members in the self-anchoring condition assumed that the team would value diversity more and identified more with the group than in the self-stereotyping condition. For majority members, however, identification was equally high in both conditions. Moreover, majority members who did engage in self-anchoring rather than self-stereotyping perceived a higher value in team diversity. Together, these findings suggest that assigning room and relevance to the self within the group can help with reaching high levels of inclusion and positive intergroup relations for all parties within diverse social groups (for a similar argument, see also Jans *et al.* 2012). Importantly, and in line with our previous reasoning regarding the authenticity dimension, giving room for the individual self within the group allows for focusing both on unique, unshared characteristics of the self and on those that are already prototypical and shared by many others in the group.

Conclusion

Experiencing both exclusion and inclusion touches upon the fundamental human need to belong and the fact that humans are social beings. Not surprisingly, then, both types of experiences have clear implications for human functioning and human relationships. We have shown that experiencing exclusion is detrimental for the individual, but may also endanger on a broader level peaceful and harmonious interpersonal and intergroup relations by enhancing the willingness to engage in antisocial, aggressive behavior (e.g. Schaafsma & Williams 2012; Aydin *et al.* 2014). Conversely, experiencing inclusion contributes not only to higher levels of well-being but also to better group functioning (e.g., Meeussen *et al.* 2014; Vos *et al.* 2014).

Moreover, we have argued that with increasing diversity and social complexity, facilitating inclusion and preventing exclusion have become pivotal. Importantly, inclusion means more than just incorporating several subgroups within a larger social setting (i.e., structural integration; de Vroome *et al.* 2011). Rather, inclusion implies that subgroup members also feel that they are accepted and appreciated within the group, irrespective of them being different from or similar to the mainstream. In times of increasing diversity, managers and policy makers who strive for smoothly operating organizations and social systems should monitor carefully the possible differential effects that their diversity policies may have on minority and majority members. To this end, we have argued that it is relevant to focus on authenticity rather than uniqueness. While being unique is firmly associated with being different, being authentic is not. One can be true to oneself by being either different from or similar to others. It seems plausible that minority members are – on average – more concerned about whether they may safely exhibit their "true self" because they are different from the mainstream. Yet, and especially with increasing diversity, majority members' need for authenticity should not be underestimated.

The research summarized in this chapter mostly focused on the organizational context and on group members' well-being and functioning at the workplace. Severe aggression or warfare was not at stake. Rather, we focused on how positive, peaceful relations at a micro level (i.e., for individual minority and majority members in diverse groups) can be established. Yet, we think that this work can also be valuable more broadly. Our notion that inclusion implies more than the absence of exclusion resonates in the distinction between positive and negative peace (i.e., between a situation in which social justice is at least partly achieved or one in which only violence has stopped; e.g., Christie *et al.* 2001). As the recent winner of the Nobel Prize for Peace, Malala Yousafzai, put it in her speech when receiving the Freedom of Fear Award in May 2014: "peace is not only the absence of war, it is the absence of fear." In the work summarized in this chapter, the focus is on the psychological, subjectively felt experience of inclusion. This experience, a safe feeling to belong, be respected, and be allowed to be oneself, may be a pathway toward sustainable peace, that is, peace that is more than the end of fighting, but the end of fear through trusting and positive intergroup relations.

References

Andriessen, I., Nievers, E., & Dagevos, J. (2012). Op achterstand. Discriminatie van niet-westerse migranten op de arbeidsmarkt [Facing disadvantage: Discrimination of non-western immigrants on the labor market]. The Hague: Sociaal en Cultureel Planbureau.

Andriessen, I., Nievers, E., Faulk, L., & Dagevos, J. (2010). Liever Mark dan Mohammed? Onderzoek naar arbeidsmarktdiscriminatie van niet-westerse migranten via praktijktests [Better Mark than Mohammed? Research on discrimination of non-western immigrants on the labor market via practical tests]. The Hague: Sociaal en Cultureel Planbureau.

Aydin, N., Fischer, P., & Frey, D. (2010). Turning to God in the face of ostracism: Effects of social exclusion on religiousness. *Personality and Social Psychology Bulletin, 36*, 742–753.

Aydin, N., Krueger, J.I., Frey, D., Kastenmüller, A., & Fischer, P. (2014). Social exclusion and xenophobia: Intolerant attitudes toward ethnic and religious minorities. *Group Processes and Intergroup Relations, 17*, 371–387.

Barner-Barry, C. (1986). Rob: Children's tacit use of ostracism to control aggressive behavior. *Ethology and Sociobiology, 7*, 281–293.

Baumeister, R.F., & Leary, M.R. (1995). The need to belong: Desire for interpersonal attachments as a fundamental human motivation. *Psychological Bulletin, 117*, 497–528.

Bertrand, M., & Mullainathan, S. (2004). Are Emily and Greg more employable than Laskisha and Jamal? A field experiment on labor market discrimination. *American Economic Review, 94*, 991.

Bilimoria, D., Joy, S., & Liang, X. (2008). Breaking barriers and creating inclusiveness: Lessons of organizational transformation to advance women faculty in academic science and engineering. *Human Resource Management, 47*(3), 423–441. doi:10.1002/hrm.20225

Branscombe, N.R., Schmitt, M.T., & Harvey, R.D. (1999). Perceiving pervasive discrimination among African Americans: Implications for group identification and well-being. *Journal of Personality and Social Psychology, 77*, 135–149.

Brewer, M.B. (1991). The social self: On being the same and different at the same time. *Personality and Social Psychology Bulletin, 17*(5), 475–482. doi:10.1177/0146167291175001

Brewer, M.B. (1997). On the social origins of human nature. In C. McGarty & S.A. Haslam (Eds.), *The message of social psychology* (pp. 54–62). Oxford: Blackwell.

Brewer, M.B. (1999). The psychology of prejudice: Ingroup love or outgroup hate? *Journal of Social Issues, 55*, 429–444.

Brewer, M.B., & Campbell, D.T. (1976). *Ethnocentrism and intergroup attitudes: East African evidence.* Beverly Hills, CA: Sage.

Brewer, M.B., & Caporael, L.R. (2006). Social identity motives in evolutionary perspective. In R. Brown & D. Capozza (Eds.), *Social identities: Motivational, emotional and cultural influences* (pp. 135–152). Hove, UK: Psychology Press/Taylor & Francis.

Brynin, M., & Güveli, A. (2012). Understanding the ethnic pay gap in Britain. *Work Employment & Society, 26*, 574–587. doi:10.1177/0950017012445095

Cacioppo, J.T., Hawkley, L.C., Norman, G.J., & Berntson, G.G. (2011). Social isolation. *Annals of the New York Academy of Sciences, 1231*, 17–22.

Carlsson, M., & Rooth, D. (2006). *Evidence of ethnic discrimination in the Swedish labor market using experimental data* (No. 2281). Bonn: Institute for the Study of Labor.

Christie, D.J., Wagner, R.V., & Winter, D.D.N. (2001). Introduction to peace psychology. In D.J. Christie, R.V. Wagner, & D.D.N. Winter (Eds.), *Peace, conflict, and violence: Peace psychology for the 21st century* (pp. 1–13). Upper Saddle River, NJ: Prentice-Hall.

Cox, T.H. (1993). *Cultural diversity in organizations: Theory, research and practice.* San Francisco: Berrett-Koehler.

Deci, E.L., & Ryan, R.M. (2000). The "what" and "why" of goal pursuits: Human needs and the self-determination of behavior. *Psychological Inquiry, 11*(4), 227–268. doi:10.1207/S15327965PLI1104_01

De Vroome, T., Coenders, M., Van Tubergen, F.A., & Verkuyten, M.J.A.M. (2011). Economic participation and national self-identification of refugees in the Netherlands. *The International Migration Review, 45*(3), 615–638.

Dovidio, J.F., Gaertner, S.L., & Saguy, T. (2007). Another view of "we": Majority and minority group perspectives on a common ingroup identity. *European Review of Social Psychology, 18*(1), 296–330.

Dovidio, J.F., Kawakami, K., & Gaertner, S.L. (2002). Implicit and explicit prejudice and interracial interaction. *Journal of Personality and Social Psychology, 82*, 62–68.

Eisenberger, N.I., Lieberman, M.D., & Williams, K.D. (2003). Does rejection hurt? An fMRI study of social exclusion. *Science, 302*, 290–292.

Ely, R.J., & Thomas, D.A. (2001). Cultural diversity at work: The effects of diversity perspectives on work group processes and outcomes. *Administrative Science Quarterly, 46*, 229–273.

EUMC. (2006). *Muslims in the European Union. Discrimination and islamophobia.* Vienna: European Monitoring Center on Racism and Xenophobia.

Eurobarometer. (2009). *Discrimination in the EU in 2009* (Special Eurobarometer 263/Wave 65.4). Brussels: TNS Opinion & Social.

Garbarino, J. (1999). *Lost boys: Why our sons turn violent and how we can save them.* San Francisco: Jossey-Bass.

Gijsberts, M. (2004). Sociaal en Cultureel Rapport 2004: Minderheden en integratie [Social and cultural report 2004: Minorities and integration]. The Hague: Sociaal en Cultureel Planbureau.

Gil-White, F.J. (2001). Are ethnic groups biological "species" to the human brain? Essentialism in our cognition of some social categories. *Current Anthropology, 42,* 515–554.

Goodwin, S.A., Williams, K.D., & Carter-Sowell, A.R. (2010). The psychological sting of stigma: The cost of attributing ostracism to racism. *Journal of Experimental Social Psychology, 46,* 612–618.

Homan, A.C., van Knippenberg, D., van Kleef, G.A., & De Dreu, C.W. (2007). Bridging faultlines by valuing diversity: Diversity beliefs, information elaboration, and performance in diverse work groups. *Journal of Applied Psychology, 92,* 1189–1199.

Jans, L., Postmes, T., & van der Zee, K.I. (2012). Sharing differences: The inductive route to social identity formation. *Journal of Experimental Social Psychology, 48*(5), 1145–1149. doi:10.1016/j.jesp.2012.04.013

Jansen, W.S., Otten, S., van der Zee, K.I., & Jans, L. (2014). Inclusion: Conceptualization and measurement. *European Journal of Social Psychology, 44*(4), 370–385. doi:10.1002/ejsp.2011

Jansen, W.S., Otten, S., & van der Zee, K.I. (2015). Being part of diversity: The effects of an all-inclusive multicultural diversity approach on majority members' perceived inclusion and support for organizational diversity efforts. *Group Processes and Intergroup Relations, 18*(6), 817–832.

Janssens, M., & Zanoni, P. (2008, November 9). What makes an organization inclusive? Organizational practices favoring the relational inclusion of ethnic minorities in operative jobs. Paper presented at the IACM 21st Annual Conference. SSRN. Retrieved from 10.2139/ssrn.1298591

Joshi, A., & Roh, H. (2009). The role of context in work team diversity research: A meta-analytic review. *Academy of Management Journal, 52*(3), 599–627.

Leary, M.R. (2001). Toward a conceptualizaton of interpersonal rejection. In M.R. Leary (Ed.), *Interpersonal rejection* (pp. 3–20). New York: Oxford University Press.

Leary, M.R., & Cottrell, C.A. (2013). Evolutionary perspectives on interpersonal acceptance and rejection. In C.N. DeWall (Ed.), *The Oxford handbook of social exclusion* (pp. 9–19). New York: Oxford University Press.

Leary, M.R., Kowalski, R.M., Smith, L., & Phillips, S. (2001). Teasing, rejection, and violence: Case studies of the school shootings. *Aggressive Behavior, 29,* 202–214.

Lirio, P., Lee, M.D., Williams, M.L., Haugen, L.K., & Kossek, E.E. (2008). The inclusion challenge with reduced-load professionals: The role of the manager. *Human Resource Management, 47*(3), 443–461. doi:10.1002/hrm.20226

Meeussen, L., Otten, S., & Phalet, K. (2014). Managing diversity: How leaders' multiculturalism and colorblindness affect work group functioning. *Group Processes & Intergroup Relations, 17,* 629–644. doi:10.1177/1368430214525809

Mendes, W.B., Major, B., McCoy, S., & Blascovich, J. (2008). How attributional ambiguity shapes physiological and emotional responses to social rejection and acceptance. *Journal of Personality and Social Psychology, 94,* 278–291.

Mor Barak, M.E., & Cherin, D.A. (1998). A tool to expand organizational understanding of workforce diversity: Exploring a measure of inclusion-exclusion. *Administration in Social Work, 22*(1), 47–64.

Mummendey, A., & Wenzel, M. (1999). Social discrimination and tolerance in intergroup relations: Reactions to intergroup difference. *Personality and Social Psychology Review, 3,* 158–174.

Nezlek, J.B., Wesselmann, E.D., Wheeler, L., & Williams, K.D. (2012). Ostracism in everyday life. *Group Dynamics: Theory, Research, and Practice, 16,* 91–104.

Nijstad, B.A., & Paulus, P.B. (2003). Group creativity: Common themes and future directions. In P.B. Paulus & B.A. Nijstad (Eds.), *Group creativity: Innovation through collaboration* (pp. 326–339). New York: Oxford University Press.

Otten, S., & Jansen, W.S. (2014). Predictors and consequences of exclusion and inclusion at the (culturally) diverse workplace. In S. Otten, K.I. van der Zee, & M.B. Brewer (Eds.), *Towards inclusive organizations: Determinants of successful diversity management at work* (pp. 67–86). New York: Psychology Press.

Otten, S., & van der Zee, K.I. (2011). Experiencing exclusion and disadvantage at work [Uitsluiting en benadeling op de werkvloer]. In S. Otten & K.I. van der Zee (Eds.), *Werkt diversiteit? Inclusie en exclusie op de werkvloer [Does diversity work? Inclusion and exclusion at work].* Groningen, The Netherlands: Instituut ISW.

Pascoe, E., & Smart Richman, L. (2009). Perceived discrimination and health: A meta-analytic review. *Psychological Bulletin, 135,* 531–554.

Phalet, K., & Ter Wal, J. (2004). *Moslim in Nederland. Religieuze dimensies, etnische relaties en burgerschap: Turken en Marokkanen in Rotterdam* [Muslim in The Netherlands. Religious dimensions, ethnic relations and citizenship: *Turks and Moroccans in Rotterdam]* (SCP-work document No. 106c). The Hague: Sociaal en Cultureel Planbureau.

Plaut, V.C., Garnett, F.G., Buffardi, L.E., & Sanchez-Burks, J. (2011). "What about me?" Perceptions of exclusion and whites' reactions to multiculturalism. *Journal of Personality and Social Psychology, 101*(2), 337–353. doi:10.1037/a0022832

Plaut, V.C., Thomas, K.M., & Goren, M.J. (2009). Is multiculturalism or color blindness better for minorities? *Psychological Science, 20,* 444–446.

Roberson, Q.M. (2006). Disentangling the meanings of diversity and inclusion in organizations. *Group & Organization Management, 31*(2), 212–236. doi:10.1177/1059601104273064

Schaafsma, J., Nezlek, J.B., Krejtz, I., & Safron, M. (2010). Ethnocultural identification and naturally occurring interethnic social interactions: Muslim minorities in Europe. *European Journal of Social Psychology, 40,* 1010–1028.

Schaafsma, J., & Williams, K.D. (2012). Exclusion, intergroup hostility, and religious fundamentalism. *Journal of Experimental Social Psychology, 48,* 829–837.

Shelton, J.N., Richeson, J.A., & Vorauer, J.D. (2006). Threatened identities and interethnic interactions. *European Review of Social Psychology, 17,* 321–358.

Shore, L.M., Randel, A.E., Chung, B.G., Dean, M.A., Holcombe Ehrhart, K., & Singh, G. (2011). Inclusion and diversity in work groups: A review and model for future research. *Journal of Management, 37*(4), 1262–1289. doi:10.1177/0149206310385943

Stevens, F.G., Plaut, V.C., & Sanchez-Burks, J. (2008). Unlocking the benefits of diversity all-inclusive multiculturalism and positive organizational change. *Journal of Applied Behavioral Science, 44*(1), 116–133.

Tajfel, H. (1981). *Human groups and social categories.* Cambridge: Cambridge University Press.

Tajfel, H., Billig, M.G., Bundy, R.P., & Flament, C. (1971). Social categorization and intergroup behavior. *European Journal of Social Psychology, 1*, 149–177.

Turner, J.C., Hogg, M.A., Oakes, P.J., Reicher, S.D., & Wetherell, M.S. (1987). *Rediscovering the social group: A self-categorization theory.* Cambridge, MA: Basil Blackwell.

Turner, M.A., & Ross, S.L. (2003). Discrimination in metropolitan housing markets: National results from phase 3 – Native Americans. Washington, DC: Department of Housing and Urban Development.

Twenge, J.M., Baumeister, R.F., DeWall, C.N., Ciarocco, N.J., & Bartels, J.M. (2007). Social exclusion decreases prosocial behavior. *Journal of Personality and Social Psychology, 92*, 56–66.

Twenge, J.M., Baumeister, R.F., Tice, D.M., & Stucke, T.S. (2001). If you can't join them, beat them: Effects of social exclusion on aggressive behaviors. *Journal of Personality and Social Psychology, 81*, 1058–1069.

Van der Zee, K.I., & Otten, S. (2014). Organizational perspectives on diversity. In S. Otten, K.I. van der Zee, & M.B. Brewer (Eds.), *Towards inclusive organizations: Determinants of successful diversity management at work* (pp. 29–48). New York: Psychology Press.

Van der Zee, K., & Paulus, P. (2008). Social psychology and modern organizations: Balancing between innovativeness and comfort. In L. Steg, A.P. Buunk, & T. Rothengatter (Eds.), *Applied social psychology: Understanding and managing social problems* (pp. 271–290). New York: Cambridge University Press.

Van Veelen, R., Otten, S., & Hansen, N. (2011). Linking self and ingroup: Self-anchoring as distinctive cognitive route to social identification. *European Journal of Social Psychology, 41*, 628–637. doi:10.1002/ejsp.792

Van Veelen, R., Otten, S., & Hansen, N. (2013a). A personal touch to diversity: Self-anchoring increases minority members' identification in a diverse group. *Group Processes and Intergroup Relations, 16*, 671–683. doi:10.1177/13684302112473167

Van Veelen, R., Otten, S., & Hansen, N. (2013b). Enhancing majority members' pro-diversity beliefs: The facilitating effect of self-anchoring. *Experimental Psychology.* doi:10.1027/1618-3169/a000220

Verbeek, P. (2013). An ethological perspective on war and peace. In D.P. Fry (Ed.), *War, peace, and human nature: The convergence of evolutionary and cultural views* (pp. 54–77). New York: Oxford University Press.

Verkuyten, M. (2006). Multicultural recognition and ethnic minority rights: A social identity perspective. *European Review of Social Psychology, 17*, 148–184.

Verkuyten, M., & Yildiz, A.A. (2010). Religious identity consolidation and mobilization among Turkish Dutch Muslims. *European Journal of Social Psychology, 40*, 436–447.

Vos, M.W., Jansen, W.S., Otten, S., Podsiadlowski, A., & van der Zee, K.I. (2014). Colorblind or colorful? The impact of diversity approaches on inclusion and work outcomes among majority and minority employees. Manuscript under review.

Walsh, A., Beyer, J.A., & Petee, T.A. (1987). Violent delinquency: An examination of psychopathic typologies. *Journal of Genetic Psychology, 148*, 385–392.

Warburton, W.A., Williams, K.D., & Cairns, D.R. (2006). When ostracism leads to aggression: The moderating effects of control deprivation. *Journal of Experimental Social Psychology, 42,* 213–220.

Wesselmann, E.D., Cardoso, F., Slater, S., & Williams, K.D. (2012). "To be looked at as though air": Civil attention matters. *Psychological Science, 23,* 166–168.

Williams, K.D. (2001). *Ostracism: The power of silence.* New York: Guilford Press.

Williams, K.D. (2007). Ostracism: The kiss of social death. *Social and Personality Psychology Compass, 1,* 236–247.

Williams, D.R., & Chung, A.–M. (1997). Racism and health. In R. Gibson & J.S. Jackson (Eds.), *Health in black America* (pp. 71–99). Thousand Oaks, CA: Sage.

Williams, K.Y., & O'Reilly, C.A. (1998). Demography and diversity in organizations: A review of 40 years of research. *Research in Organizational Behavior, 20,* 77–140.

Wirth, J.H., Sacco, D.F., Hugenberg, K., & Williams, K.D. (2010). Eye gaze as relational evaluation: Averted eye gaze leads to feelings of ostracism and relational devaluation. *Personality and Social Psychology Bulletin, 36,* 869–882.

Wolsko, C., Park, B., & Judd, C.M. (2006). Considering the tower of Babel: Correlates of assimilation and multiculturalism among ethnic minority and majority groups in the United States. *Social Justice Research, 19,* 277–306.

4

The Peacekeeping and Peacemaking Role of Chimpanzee Bystanders

Teresa Romero

Since the first systematic study on animal postconflict reconciliation and consolation was published in 1979 (de Waal & van Roosmalen 1979), a considerable body of literature has supported the idea that these peaceful behaviors are a widespread phenomenon in social animals (Aureli & de Waal 2000; de Waal 2000; Arnold *et al.* 2010). The accumulated data have also led to the generation of hypotheses that account for the distribution and form of peacemaking and peacekeeping mechanisms within and between species (Aureli *et al.* 2002). This research has revealed surprisingly complex behavioral processes (Aureli *et al.* 2012; Thierry 2013). This chapter focuses on one such type of behavior: spontaneous peaceful interventions of individuals uninvolved in the original conflict (*bystanders*, hereafter) soon after the cessation of aggression. Although research has traditionally concentrated on peaceful reunions between the original contestants (i.e., reconciliation; de Waal & van Roosmalen 1979), a growing body of literature focuses on the effect of aggressive confrontations on bystanders and the role that these individuals play in peacemaking and peacekeeping within the group (e.g., Das 2000; Watts *et al.* 2000; Judge & Mullen 2005; Fraser *et al.* 2009; Schino & Sciarretta 2015). These interactions are of particular interest since they may require the knowledge of the social relationship among other group members and/or complex emotional abilities (Kummer 1967; de Waal & Aureli 1996). Hence, their study may provide valuable information about the evolution of socio-cognitive and emotional capacities required for conflict management and peaceful relations.

When de Waal and van Roosmalen (1979) first described the occurrence of postconflict reconciliation in a zoo colony of chimpanzees (*Pan troglodytes*) in the Netherlands, they also described the exchange of friendly behaviors, such as gentle touching, or embracing, between recipients of aggression and bystanders (de Waal & van Roosmalen 1979). The authors labeled the intervention by the bystander as "consolation," assuming that it had a calming function. In their pioneering study, though, the authors did not test the functional assumptions of the described behaviors, nor did they distinguish between interactions initiated by the bystander and those initiated by the recipient of aggression. This differentiation, as later research proved, is crucial to understanding the functional significance as well as the cognitive and emotional implications of the behavior (see Verbeek & de Waal 1997; Fraser *et al.* 2009).

Peace Ethology: Behavioral Processes and Systems of Peace, First Edition.
Edited by Peter Verbeek and Benjamin A. Peters.
© 2018 John Wiley & Sons Ltd. Published 2018 by John Wiley & Sons Ltd.

In this chapter, I review the knowledge accumulated on interactions in which an uninvolved bystander initiates friendly contact with a recent recipient of aggression (Figure 4.1). This definition excludes other forms of postconflict friendly exchanges, such as interactions between former opponents, or third-party contacts sought by the conflict participants themselves. Throughout this chapter, I will use the descriptive term *bystander affiliation*, instead of the more functionally loaded term *consolation*, because current evidence shows that this postconflict interaction is a heterogeneous phenomenon whose function depends on the social context (Fraser *et al.* 2009). In other words, consolation is only one of the many possible functions of bystander affiliation. In this regard, bystander affiliation in chimpanzees is a paradigmatic example, where different populations demonstrate the diverse use of this postconflict behavior depending on the circumstances. Chimpanzees are also the only species for which detailed functional analyses of bystander affiliation are available for populations under different social environments, allowing for intraspecies comparison. Thus, the chapter mainly focuses on studies of chimpanzees, but comparisons with findings in other apes, monkeys, and nonprimate species are made when possible. I begin by placing the conflict in the social setting, describing the negative social consequences that escalated aggression entails not only for conflict participants but also for bystanders. I then review the evidence for each of the most influential hypotheses suggested for bystander affiliation (Table 4.1) and explain their patterns at the proximate level by exploring their cognitive and emotional underpinnings. I conclude by proposing future research directions to gain a more comprehensive understanding of the functional significance and underlying mechanisms of bystander affiliation in primate and nonprimate species. Before addressing these issues, I will briefly describe the social system of the focused species: the chimpanzee.

Chimpanzee Social System

Chimpanzees live in multimale–multifemale societies, which may vary from 15 to 150 individuals (Stumpf 2007). Male chimpanzees tend to remain permanently in the community in which they are born, whereas females tend to migrate to a new group upon reaching sexual maturity (Mitani *et al.* 2002). The chimpanzee social system is characterized by a high degree of fission–fusion dynamic, which means that members of a large community are rarely all together (Aureli *et al.* 2008). Instead, chimpanzees spend most of their time in smaller subgroups with flexible membership (Nishida & Hiraiwa-Hasegawa 1987; Stumpf 2007). In both captive and wild populations, chimpanzee males create strong social bonds and long-term alliances with other males, which influence their strict, linear dominance hierarchies (e.g., de Waal 1986; Goodall 1986; Boesch & Boesch-Achermann 2000). On the other hand, chimpanzee female relationships have been described as fluid and transitory, although there is considerable diversity across populations. While at some study sites females are relatively solitary (e.g., Goodall 1986), in others they form cooperative long-term relationships with other group mates (Boesch & Boesch-Achermann 2000), which is usually the case in captive populations when female relatives are present (de Waal 1982).

(a)

(b)

Figure 4.1 An example of bystander affiliation among chimpanzees. An adult female (*center*) screams after having received a hit from a high ranking female (*right*) while a friend of the victim (*left*) closely observes the interaction (a). Soon afterwards (less than 1 min) the victim's friend approaches and hugs the victim (b). Photographs by Teresa Romero.

Table 4.1 Alternative hypotheses about the function of bystander affiliation

Consolation	Effect	Reduction of postconflict distress levels of recipients of aggression
	Prediction	Affiliation should occur between socially close individuals
	Underlying mechanisms	Some level of other-awareness or emotional perspective-taking
Mediated reconciliation	Effect	Restoration of the opponents' relationship and/or reduction of recipients' postconflict anxiety
	Prediction	Affiliation should be initiated by kin or affiliative partners of the aggressor
	Underlying mechanisms	Knowledge of third-party relationships
Self-protection	Effect	Reduction of bystanders' likelihood of becoming the target of further aggression
	Prediction	Affiliation should be directed at frequent aggressors
	Underlying mechanisms	Associative learning; responses to aversive stimuli

Framework of Conflict Management

Although aggression is not an inevitable outcome of a conflict of interest (de Waal 1996), animals may use it as a negotiation tool (Aureli *et al.* 2012). For instance, chimpanzees may threaten and/or attack group mates to gain access to a contested resource or to establish or reinforce dominance relationships. However, using aggression to settle conflicts carries unavoidable costs, such as energy expenditure and risk of injury or resource loss. Apart from these costs, for social animals, aggression may also entail other less obvious negative consequences.

The aftermath of an aggressive conflict is a risky period for both former contestants and other group members. Chimpanzees that have been a recent recipient of aggression are more likely to be involved in renewed attacks (de Waal & van Hooff 1981; Wittig & Boesch 2003; Koski *et al.* 2007a). Additionally, opponents may attack individuals uninvolved in the original conflict (de Waal & van Hooff 1981; Goodall 1986; Kutsukake & Castles 2004; Koski & Sterck 2009) as a distress alleviation mechanism or facilitator of reconciliation. For instance, in several macaque species, aggressors are more willing to exchange friendly contacts with former adversaries after the victim redirects aggression against a third party than during comparable postconflict periods without redirection (e.g., Aureli & van Schaik 1991; Kazem & Aureli 2005). Given the higher hostility that recipients of aggression face during the postconflict period, it is not surprising that they experience changes in their emotional states. Behavioral, pharmacological, and physiological evidence indicates that certain self-directed behaviors, such as self-scratching or self-grooming, provide a non-invasive, reliable index of anxiety (e.g., Schino *et al.* 1991, 1996; Barros *et al.* 2000). Following a conflict, chimpanzee victims show a significant increase in their levels of self-directed behaviors (Koski *et al.* 2007b; Fraser *et al.*

2008), suggesting an increase in their distress levels. Studies conducted in children using similar ethological methods have also found that when losing a confrontation with peers, children's self-directed behaviors and saliva cortisol levels increase compared to control periods without any conflict (Fujisawa *et al.* 2006; Butovskaya 2008).

Conflicts also influence the emotional state of uninvolved group mates. Witnessing aggression increases rates of self-directed behavior of bystanders in monkeys (Judge & Mullen 2005; Schino & Sciarretta 2015). Interestingly, bystanders' distress responses depend on the potential negative consequences of the initial conflict. Increased arousal in bystanders has been observed in primate species, with frequent redirection of aggression toward uninvolved individuals (Judge & Mullen 2005; de Marco *et al.* 2010; Schino & Marini 2014), while such increase is absent in gelada baboons (*Theropithecus gelada*), where groupwide spread of aggression is unlikely (Leone *et al.* 2010).

An additional negative consequence of fights is that opponents do not tolerate each other's proximity (e.g., Wittig & Boesch 2005). Not being able to approach and interact with a former opponent may compromise the cooperative aspects of the relationship, which in turn may affect individuals' competitive success and reproductive performance (reviewed in Cheney & Seyfarth 2012). One way to mitigate these detrimental consequences of aggression is through direct communication between the individuals in conflict. Chimpanzees that kiss and embrace their adversaries after a fight restore levels of tolerance with each other and reduce the risk of renewed aggression (Wittig & Boesch 2005; Koski *et al.* 2007b; Fraser *et al.* 2010). Human children also engage in similar reconciliations, which intriguingly resemble those of chimpanzees in both social context and morphology (e.g., touching, hugging, and kissing; Butovskaya *et al.* 2000; Verbeek *et al.* 2000; Verbeek 2008). Actually, available literature on conflict management in animals shows that reconciliation is a regular, conspicuous part of their social life (there are several comprehensive reviews on this topic for the interested reader, e.g., Aureli *et al.* 2002, 2012; Arnold *et al.* 2010). Although reconciliation is likely the most effective way to repair opponents' relationship, friendly interventions from bystanders uninvolved in the original conflict may also help to mitigate the negative consequences of aggression and provide an alternative conflict management strategy when reconciliation is either not beneficial or too risky (Aureli *et al.* 2002).

Bystander Affiliation Functioning as Consolation

In the aftermath of a conflict, chimpanzee bystanders spontaneously approach and make friendly contact with recipients of aggression. During those reunions, bystanders may groom or gently touch the victim, but it is not unusual to see them hugging and kissing the distressed partner (de Waal & van Roosmalen 1979; Fraser & Aureli 2008; Romero *et al.* 2010; Figure 4.1). Taking into account the resemblance between chimpanzee postconflict reunions and human acts of reassurance, it is not surprising that these interactions were labeled consolation the first time they were systematically studied (Table 4.1; de Waal & van Roosmalen 1979). Use of the term consolation, however, implies that the friendly contact functions to calm the recipient of aggression, and therefore should be offered to those that are in need of emotional comfort. Although emotions used to be commonplace in descriptions of animal behavior in the late nineteenth century (e.g., Darwin 1872/1998), they were carefully avoided as an explicit topic in scientific research during the next century (de Waal 2011). Apart from a lack of theoretical interest, animal emotions were considered imprecise, subjective experiences, and thus researchers faced the challenge of developing replicable, objective methods to document them. Contemporary researchers, however, regard emotions as multifaceted processes, which comprise physiological, behavioral, cognitive, and, in humans at least, subjective components (Paul *et al.* 2005). Thus defined, measures of behavioral and physiological changes become powerful tools to monitor emotional processes in animals without necessarily implying a conscious, subjective state.

Humans and chimpanzees experience similar changes in brain and peripheral skin temperature in response to emotionally charged images (Parr 2001). Similarly, changes in stress-related hormones (e.g., cortisol) and self-directed behaviors have been associated with changes in individuals' distress levels (e.g., Schino *et al.* 1991, 1996; Barros *et al.* 2000). Using these latter methods, researchers have been able to report victims' distress alleviation through the friendly contact of a bystander (Fraser *et al.* 2008), thus confirming the calming effect of consolation in chimpanzees. Moreover, bystanders approach and contact victims of serious aggression and those that vociferously scream after an attack more often than those who receive mild aggression (de Waal & Aureli 1996; Palagi *et al.* 2006), suggesting that greater distress evokes more frequent responses.

Proximate Motivation to Console Another

To be able to provide comfort to distressed parties, bystanders need to be affected by the distress of others and inclined to provide relief. Automatic responses to others' distress are among the most basic empathic reactions that can be observed in human and nonhuman animals (Preston & de Waal 2002; de Waal 2008). Humans mimic emotional states and facial expressions of others even when the expressions are presented too briefly for conscious perception (Dimberg *et al.* 2000). A reflex-like spread of fear occurs in many species (e.g., Cheney & Seyfarth 1985) when one group member is startled by, for instance, a predator. Under control experimental conditions, mice, rats, and pigeons also display distress upon perceiving distress in others (Watanabe & Ono 1986; Langford *et al.* 2006). While these responses may or may not involve any understanding of what triggered the initial reaction, emotional contagion is highly adaptive since it aids social animals in synchronizing behavioral or psychological states and facilitates collective responses that are important for their survival.

However, an automatic sharing of others' emotions would explain an increase in bystanders' arousal after witnessing a conflict, but not the subsequent actions. If chimpanzee bystanders were just matching the emotional state of their group mates, one would expect these bystanders selfishly to seek alleviation of their own negative arousal, probably turning away from the source of their distress, the victim. In contrast, the perception of a distressed individual commonly leads to a prosocial act in chimpanzees, suggesting that chimpanzees are able not only to recognize another individual's emotions, but also both to separate others' emotions from their own and to provide the appropriate response to reduce others' distress. The same behavior in human children is generally classified as an expression of empathic or sympathetic concern for the other (e.g., Zahn-Waxler *et al.* 1984; Eisenberg 2000). Given that chimpanzees are closely related to us, it is not unreasonable to assume a common motivation for both behaviors (de Waal 1999). In other words, consolation behavior in chimpanzees may also be driven by empathic processes and require some level of other-awareness or emotional perspective-taking (de Waal & Aureli 1996; Preston & de Waal 2002; de Waal 2008).

The exact level that chimpanzees' empathic abilities reach is currently under debate. Human empathy is usually described as taking the perspective of another or imagining oneself in another's position. This definition implies an understanding of others' mental states and an ability to take another's perspective, which closely relate to more general mental-state attribution skills (Preston & de Waal 2002). This cognitively demanding explanation of empathy, however, denies simpler empathic reactions that are so commonplace in our daily lives, such as automatic emotional contagion (e.g., Dimberg *et al.* 2000; de Waal 2008, 2012). Thus, another way to understand empathy is as a multilayered phenomenon that covers all ways in which one individual's emotional state affects another's, from automatic state-matching responses to empathic representations of abstract situations (Hoffman 2001; Preston & de Waal 2002). Traditionally, the literature on human developmental psychology has suggested that more complex expressions of empathy emerge in conjunction with greater cognitive capacities. Thus, the absence of certain cognitive skills in chimpanzees, or other animals (e.g., the ability to adopt another's point of view), has been interpreted as an inability to reach a certain level of empathy (e.g., empathic perspective-taking; but see de Waal 2008). However, a recent study in children has shown that while some forms of affective and cognitive empathy toward others in distress are already present before the second year of age, showing personal distress (i.e., emotional contagion) as a response to others' distress is actually rare at that age (Roth-Hanania *et al.* 2011). This evidence suggests that comfort-giving does not necessarily rest on advanced cognitive mechanisms, and thus, it is likely to be present in chimpanzees as well as in other animals. A good example of other-regarding responses is found in elephants. Elephants show a rich social organization, which includes cooperation (e.g., Plotnik *et al.* 2011), collective defensive behavior to predation threats (e.g., McComb *et al.* 2011), and assistance of group mates (e.g., Bates *et al.* 2008). Following a distress event, caused by either social or other environmental threats, Asian elephants (*Elephas maximus*) often engage in friendly exchanges with distressed partners through physical contact and vocal communication (Plotnik & de Waal 2014). Although it was difficult for researchers to differentiate between cases where individuals were reacting directly to the stimulus or to another elephant's distress, the matching of the behavior and emotional state of the first distressed individual suggested that elephants can attend and react to each other's emotional states (Plotnik & de Waal 2014).

Observational studies, such as those presented here, cannot easily differentiate between underlying mechanisms. The precise cognitive and empathetic abilities required for consolation in chimpanzees and other species thus remain hard to elucidate. Nevertheless, these studies provide crucial information on spontaneous behavioral reactions under naturally occurring aggressive and stressful episodes, allowing for the evaluation of predictions derived from the empathy-based hypothesis of consolation. If this behavior is indeed an empathic response to the distress of others, it is expected that factors that facilitate or inhibit human empathy also modulate consolation in apes. Researchers using a wide range of measurements (e.g., questionnaires, behavioral observations, and neuro-physiological data) have reported that human empathy tends to increase with age, at least up to adolescence, and that females are usually more empathic than males (e.g., Zahn-Waxler *et al.* 1992). Moreover, in humans and other animals, empathy is not equally aroused by the emotional signals of any individual, but rather is greatly facilitated by similarity, familiarity, and social closeness between individuals (Anderson & Keltner 2002; Aureli & Schaffner 2002; de Waal 2008). Findings from several independent studies on wild and captive chimpanzees have shown that consolation is disproportionally directed toward individuals that are socially close – that is, kin and affiliation partners (Kutsukake & Castles 2004; Fraser *et al.* 2008; Romero *et al.* 2010) – while the strength of the social bond between the bystander and the aggressor seems irrelevant (Romero *et al.* 2010), as predicted if empathy is the main motivation (Table 4.1). The interpretation that bystander affiliation is an empathic-driven behavior may also hold for other species in which the behavior is typically provided by close partners, such as mates in corvids (Seed *et al.* 2007; Fraser & Bugnyar 2010; Logan *et al.* 2013) or individuals with a high degree of familiarity in wolves (*Canis lupus*) (Palagi & Cordoni 2009), although the distress alleviation function of bystander affiliation remains to be demonstrated in these species.

Another point that further supports the empathy-based hypothesis of consolation in chimpanzees is the observed sex differences, which match those found in human studies. When sex differences are reported, female chimpanzees tend to offer more consolation than male chimpanzees, suggesting that females are more responsive to the distress of others (Romero *et al.* 2010). These sex differences in consolation do not reflect the typical sex differences in friendly exchanges during nonconflict situations, because in chimpanzee society males are on average more affiliative than females (Goodall 1986; Arnold & Whiten 2003; Romero *et al.* 2010). Furthermore, female chimpanzee bystanders are not always more active than male bystanders during the postconflict period. When uninvolved chimpanzees make friendly contact with former aggressors, instead of with recipients of aggression (Romero & de Waal 2011), the postconflict affiliation is provided most often by adult males and directed toward high-ranking males, whereas chimpanzee females engage less often in this behavior as both actors and recipients (Romero *et al.* 2011). These findings not only support the notion that bystander affiliation may be driven by empathic motivations when aimed to comfort distressed partners, but also highlight the flexibility of chimpanzee behavioral tactics in conflict management.

Although most research on consolation behavior has focused on chimpanzees, bonobos (*Pan paniscus*) are a species of particular interest in the debate about whether or not consolation is driven by empathic processes, as opposed to other forms of emotional responding. Chimpanzees and bonobos are our closest living relatives, and they themselves are very closely related (Prüfer *et al.* 2012). Despite this genetic similarity, their behavior and cognitive abilities differ in important ways. Bonobos show a vast repertoire of social behaviors, including play and socio-sexual contacts (i.e., mounting, genital touches, and copulation), aimed at increasing cohesiveness among group members, especially among females since bonobos live in a female-centered society (Kano 1992; de Waal & Lanting 1997). According to current empirical evidence, compared to chimpanzees, bonobos are more socially cooperative (Hare *et al.* 2007) and more tolerant (Wobber *et al.* 2010). On the other hand, chimpanzees have a better understanding of physical causality (Herrmann *et al.* 2010) and use many different types of tools to obtain food, whereas bonobos rely on tools very little (Whiten *et al.* 1999; Hohmann & Fruth 2003).

Similar to chimpanzees, bonobo bystanders spontaneously approach recent recipients of aggression and offer them friendly contacts (Palagi *et al.* 2004; Clay & de Waal 2013a), which effectively reduce victims' distress levels (Clay & de Waal 2013a; Palagi & Norscia 2013). Consistent with empathy-based predictions, bonobo bystanders are also more likely to console recipients of aggression with whom they have a close social bond (Clay & de Waal 2013a; Palagi & Norscia 2013). Clay and de Waal (2013b) further explored socio-emotional skills in this species, known to be critical in human expressions of sympathy and prosocial behaviors. Studies on developmental psychology show that children who are able to regulate the intensity and duration of their own emotions are more likely to show caring responses for others, whereas those individuals with poor managing abilities are more likely to become emotionally overwhelmed when exposed to another's distress, thus resulting in personal distress (e.g., Eisenberg *et al.* 1996; Murphy *et al.* 1999; Eisenberg & Fabes 2005; Gross & Thompson 2007; Davidov *et al.* 2013). In a forested sanctuary in RD Congo, Clay and de Waal (2013b) observed how bonobos respond to stressful events, both when the stress affected themselves (e.g., a lost fight) or others (e.g., witnessing the distress of others). They found that individuals good at handling their own distress as victims (i.e., quicker to recover) were more likely to actively respond to others' distress and provide bodily comfort to them (Clay & de Waal 2013b), suggesting that emotional competence is important for the expression of consolation in bonobos. Furthermore, in agreement with human children studies (Kagan & Snidman 2009; Zahn-Waxler *et al.* 2009), bonobos with higher social abilities were most likely to approach the victim and offer consolation rather than showing negative reactions, such as fleeing, moving away, or screaming (Clay & de Waal 2013b).

The demanding cognitive requirements of consolation (i.e., the ability to experience another individual's emotions while separating them from one's own) have been suggested as the reason for the observed distribution of the behavior among primates. Despite ample attention to postconflict interactions in more than 30 primate species, both in captivity and in the field (Aureli & de Waal 2000; Arnold *et al.* 2010), friendly contacts from uninvolved bystanders to recent recipients of aggression are very rarely reported in monkeys (reviewed in Fraser *et al.* 2009). Spontaneous bystander affiliation in monkeys seems to be absent, even in situations in which subjects should be highly motivated to interact with the recipient of aggression. For instance, a study on a captive group of Japanese macaque (*Macaca fuscata*) shows that mothers do not initiate friendly contacts toward their offspring after they receive aggression, nor do they show signs of increased distress, which led the authors of the study to suggest that Japanese macaque mothers may be unable to understand their offspring's need for social comfort (Schino *et al.* 2004). This apparent dichotomy in the occurrence of consolation has been interpreted as the ability (in the case of humans and great apes) or the inability (in the case of monkeys) to express sympathetic concern for others (de Waal & Aureli 1996). This idea, however, has been recently challenged by a study on a different macaque species, the tonkean macaque (*Macaca tonkeana*; Palagi *et al.* 2014). Macaques live in large multi-male–multifemale groups that vary on a gradient ranging from more intolerant to more tolerant social systems (Thierry *et al.* 2008; Balasubramaniam *et al.* 2012). While Japanese macaques are among the most intolerant macaque species, with strict dominance hierarchies and high degrees of nepotism, tonkean macaques belong to the more tolerant grade, with less strict dominance relationships, frequent retaliation during conflicts, and high frequencies of friendly exchanges. In the aftermath of a conflict, uninvolved tonkean macaques spontaneously approach and exchange friendly behaviors with recent recipients of aggression, which reduces victims' distress levels (Palagi *et al.* 2014). Moreover, bystander macaques often direct friendly contacts toward their close associates and especially toward those that experience higher levels of distress, suggesting that tonkean macaques, like chimpanzees, are capable of empathically reacting to others' distress and providing consolatory behaviors (Palagi *et al.* 2014). Since Japanese and tonkean macaques are very closely related, it is unlikely that they show different cognitive and empathic abilities. It is more likely that expressions of consolation during the postconflict period are constrained by the degree of tolerance in their social systems (Social Constraint Hypothesis; de Waal & Aureli 1996). In species, or groups where groupwide spread of aggression is the norm, bystanders may restrain themselves from contacting recipients of aggression for fear of being the target of renewed aggression. This explanation is strengthened by qualitative observations reporting approaches and contacts by infant rhesus macaques (*Macaca mulatta*) toward peers showing intense distress (e.g., screaming after punishment by an adult or after a fall; de Waal & Aureli 1996), while the adult rhesus macaques, fully integrated in the dynamics of their strict societies, do not provide consolation during postconflict periods.

Bystander Affiliation Functioning as Mediated Reconciliation

Not every study investigating the determinants and function of bystander affiliation in chimpanzees has found support for the empathy-based hypothesis. In a community of wild chimpanzees living in Tai National Park (Cote d'Ivoire), recipients of aggression may receive friendly contacts from their friends, but they are more likely to be approached by aggressors' close associates (Wittig & Boesch 2010). Furthermore, not only recipients of aggression, but also aggressors, receive friendly contacts from their opponents' close associates. Since an increase in distress levels has not been observed in chimpanzee aggressors and close bonding partners are better at reducing distress levels (Aureli & Fraser 2012), it is unlikely that such contacts serve to reassure former opponents. Instead, these interactions have been interpreted as part of the relationship-mending tactics observed in chimpanzees (de Waal 1982). After a conflict, recipients of aggression might be deterred from contacting former aggressors when the likelihood of renewed attacks is high (Wittig & Boesch 2003), but they may allow the proximity of uninvolved bystanders since they pose a lower risk. In such cases, bystanders may make friendly contact with one of the opponents in an attempt to mediate the resolution of the conflict between the former adversaries (Table 4.1). Actually, in the chimpanzee community living in Tai Forest, friendly exchanges between a close associate with one of the opponents and the other opponent produce effects similar to those of direct reconciliation, namely the restoration of tolerance and cooperation levels between former opponents (Wittig & Boesch 2010). Furthermore, aggressors were more likely to approach and reconcile with the recipient of aggression after a good friend of the victim had affiliated with them (Wittig & Boesch 2010). This interconnection between reconciliation and bystander affiliation has also been found in a colony of captive bonobos, where bystander affiliation generally precedes the occurrence of reconciliation (Palagi *et al.* 2004), suggesting a similar facilitation function.

Cognitive Underpinnings of Mediated Reconciliation

That bystander affiliation can substitute for direct reconciliation implies that the recipient of the postconflict behavior recognizes the bond between the bystanders and his or her former opponent and connects the current friendly contact with the recent aggression, therefore treating the bystander's action as a "proxy" for reconciliation. Both experimental and observational studies have revealed that primates are able to form abstract concepts of kinship (e.g., Dasser 1988) and recognize the kin and rank relationships of third individuals (e.g., Bachmann & Kummer 1980; Cheney & Seyfarth 1999; Parr & de Waal 1999). Chimpanzees' social bonds, however, are not necessarily based on kinship, and in some communities the majority of bonding partners are not close kin (e.g., Langergraber *et al.* 2009). Additionally, such bonds may be transient and unstable (de Waal 1982), although in some cases they last for years (Mitani 2009). Furthermore, in contrast with rank relationships, which can be inferred from a single or few observations such as the display of ritualized submissive signals, friendly bonds are less conspicuous since they are based on how often two individuals exchange friendly behaviors and cooperate with each other, rather than if they do it or not. Thus, to assess others' social bonds, chimpanzees need to observe repeated interactions between each dyad and monitor changes over time. There is little information about the mechanisms underlying the assessment of third-party relationships, but it is unlikely that it is only based on the observation and memorization of multiple interactions of different dyads and the computation of their relative frequencies. It seems more likely that some form of transitive inference and/or emotional mediation is involved, as it has been suggested in order to assess relationships involving oneself (Aureli & Schaffner 2002) or complex social patterns in other animals (e.g., reciprocity of support in coatis [*Nasua nasua*]; Romero & Aureli 2008).

Since the ability to recognize triadic relationships is also found in monkeys and nonprimate species, there is no reason why bystander affiliation could not function as triadic reconciliation in these species. Although contacts between opponents and bystanders have been documented in a number of monkey species (reviewed in Das 2000; Fraser *et al.* 2009), most studies did not distinguish whether interactions were initiated by the opponent or by the bystander, or whether they involved the recipient of aggression or the aggressor, limiting our understanding of the behavior. Evidence that bystander affiliation can function as mediation for reconciliation in monkeys is found in a playback experiment with free-ranging female chacma baboons (*Papio hamadryas ursinus*). In this species, aggressors use friendly grunt vocalizations to reconcile with their opponents (Silk *et al.* 1996). Female recipients of aggression are more likely to tolerate their aggressor's presence after hearing the playback of a friendly vocalization from one of their aggressor's relatives than after hearing the vocalization from other group members (Wittig *et al.* 2007), suggesting that kin may function as mediators, reconciling with the victim on behalf of their relatives.

Bystander Affiliation Functioning as Self-Protection

A question that arises from the previous two functional interpretations of bystander affiliation is "What benefits do bystanders derive from the interactions?" By reducing the stress of a recent victim of aggression, or by restoring the relationship of former opponents, bystander affiliation confers benefits to others while carrying the risk for the performer of getting drawn into the original conflict. As such, bystander affiliation fits the definition of altruistic behavior (cf. Hamilton 1964) and may become part of an exchange system among group mates (de Waal 2008). Actually, chimpanzees do not simply provide consolation to distressed others with whom they frequently exchange friendly behaviors, but selectively offer this benefit to those partners that console them in return, suggesting that consolation is an integrated part of mutually beneficial relationships (Romero *et al.* 2010).

Alternatively, some researchers have suggested that bystander affiliation may provide immediate benefits for the third party (Table 4.1). For instance, bystanders may gain indirect benefits by calming the recipient of aggression and thus reducing tension in the group as a whole (Kutsukake & Castles 2004; Palagi *et al.* 2006). More directly, friendly behaviors directed toward a recent conflict participant might provide direct protection to bystanders, reducing their probabilities of becoming the target of redirected aggression (Koski & Sterck 2009). However, in contrast with monkey species in which aggression frequently spread far beyond the two original opponents (Das 2000; Kazem & Aureli 2005; Romero *et al.* 2009), the associated risk of further aggression among chimpanzees is rather low (de Waal & van Hooff 1981; Arnold & Whiten 2001; Romero & de Waal 2010). Nevertheless, individuals that frequently receive attacks from recipients of aggression are susceptible to become the targets of redirected aggression, and thus it would be beneficial for those individuals to reduce the aggressive tendencies of potential aggressors.

Simple Proximate Mechanisms for a Self-Oriented Behavior

Of the three main functional hypotheses suggested for bystander affiliation, self-protection is the only function that entails direct benefits for the performer. Furthermore, the proximate mechanisms involved do not need to be particularly complex. Individual recognition, associative learning, and responses to aversive stimuli, which are present in most social species, suffice to prompt this reaction. Despite this, studies investigating the function of bystander affiliation in great apes have failed to find support for the self-protection hypothesis (Romero & de Waal 2010; Wittig & Boesch 2010; Clay & de Waal 2013a; Palagi & Norscia 2013), with one exception (Koski & Sterck 2009). In a group of chimpanzees housed at the Burgers' Zoo in Arnhem, the Netherlands, recipients of aggression often receive postconflict friendly contacts from their typical targets of redirected aggression, and the contact effectively decreases bystanders' chances of receiving further aggression (Koski & Sterck 2009). It is worth noting that chimpanzees living at this zoo experience considerably higher risk of receiving redirected aggression during the postconflict period (i.e., 10.8% of postconflict periods; Koski & Sterck 2009) than chimpanzees living in groups where the use of bystander affiliation as self-protection seems to be absent (e.g., < 0.5% of postconflict periods;

Romero & de Waal 2010). The relative frequency of redirected aggression thus seems to explain the prevalence of the self-protection function of bystander affiliation in chimpanzees.

Among monkeys, friendly acts directed to potential aggressors also serve as a self-protection strategy. In mandrills (*Mandrillus sphinx*), recipients of aggression frequently redirect aggression toward uninvolved bystanders. Following aggression, victims receive friendly contacts primarily from those bystanders that are likely to be attacked by them, and the friendly exchange is associated with a reduction of redirection (Schino & Marini 2012). It is surprising, though, that if friendly contacts from bystanders to recent victims of aggression can be used as a self-protection strategy, this phenomenon is so rarely reported in monkeys and nonprimate species in which redirected aggression is common. One possibility is that when the probability of receiving aggression, or when the intensity of the attack, is too high, avoiding a former conflict participant is a better (i.e., safer) strategy. For instance, meerkats (*Suricata suricatta*) tend to avoid dominant aggressors instead of offering affiliation or showing submission, since the two latter options do not reduce their risk of receiving further aggression (Kutsukake & Clutton-Brock 2008). Unfortunately, there is insufficient comparative data to ascertain when the benefits of using bystander affiliation as a self-protection strategy would outweigh its costs.

Conclusions and Unanswered Questions

More than three decades ago, a pioneering work described how chimpanzees uninvolved in the original conflict offer consolatory embraces to recent victims of aggression (de Waal & van Roosmalen 1979). There is now increasing evidence that bystanders are not indifferent to others' conflicts and that they may play an important role in peacemaking and peacekeeping through their friendly, aggressive, or policing interventions. Detailed functional analysis of one such intervention (i.e., bystander affiliation in chimpanzees) has also revealed that behaviors that superficially appear to be rather similar serve very different functions and may require different cognitive and emotional capacities. The scarce relevant data on monkeys and nonprimate species, though, do not allow us to know whether these different functions and proximate mechanisms are shared across species. Another aspect that should receive more attention is the patterning of behaviors used for bystander affiliation. For instance, explicit behaviors, such as kissing or embracing, might be used only for particular functions (e.g., consolation) or among particular types of dyads (e.g., dyads with a less secure relationship), and thus the use and effectiveness of behaviors typically considered as equivalent (i.e., friendly behaviors) may not be uniform. At the same time, evaluation of the cognitive processes involved in bystander affiliation should be more widely undertaken, integrating data from naturalistic observations and experiments. In this regard, more effort should be put toward examining individuals' emotional responses and how variation in emotional signaling affects others' responses. Better empirical evidence for subtle signals and vocalizations would be valuable since they are known to convey a rich array of emotional information. We know very little about the developmental trajectories of conflict management behaviors in general and of bystander affiliation in particular. Similarly, the study of individual behavioral flexibility and its link to the differential use of bystander affiliation has been largely neglected. I encourage further research on these topics in the years to come, since their study could help us to gain a more complete understanding of the relative importance of the emotional and cognitive components in the expression of bystander affiliation.

References

Anderson, C., & Keltner, D. (2002). The role of empathy in the formation and maintenance of social bonds. *Behavioral and Brain Sciences, 25*, 21–22.

Arnold, K., Fraser, O.N., & Aureli, F. (2010). Postconflict reconciliation. In C.J. Campbell, A. Fuentes, K.C. MacKinnon, S.K. Bearder, & R.M. Stumpf (Eds.), *Primates in perspective* (pp. 608–625). Oxford: Oxford University Press.

Arnold, K., & Whiten, A. (2001). Post-conflict behaviour of wild chimpanzees (*Pan troglodytes Schweinfurthii*) in the Budongo Forest, Uganda. *Behaviour, 138*, 648–690.

Arnold, K., & Whiten, A. (2003). Grooming interactions among the chimpanzees of the Budongo Forest, Uganda: Tests of five explanatory models. *Behaviour, 140*, 519–552.

Aureli, F., Cords, M., & van Schaik, C.P. (2002). Conflict resolution following aggression in gregarious animals: A predictive framework. *Animal Behaviour, 64*, 325–343.

Aureli, F., & de Waal, F.B.M. (Eds.). (2000). *Natural conflict resolution*, Berkeley, CA: University of California Press.

Aureli, F., & Fraser, O. (2012). Distress alleviation in monkeys and apes: A window into the primate mind? In F.B.M. de Waal & P.F. Ferrari (Eds.), *The primate mind: Built to connect with other minds* (pp. 246–265). Cambridge, MA: Harvard University Press.

Aureli, F., Fraser, O.N., Schaffner, C.M., & Schino, G. (2012). The regulation of social relationships. In J.C. Mitani, J. Call, P.M. Kappeler, R.A. Palombit, & J.B. Silk (Eds.), *The evolution of primate societies* (pp. 531–551). Chicago: University of Chicago Press.

Aureli, F., & Schaffner, C.M. (2002). Relationship assessment through emotional mediation. *Behaviour, 139*, 393–420.

Aureli, F., Schaffner, C.M., Boesch, C., Bearder, S.K., Call, J., Chapman, C.A., Connor, R.C., Di Fiore, A., Dunbar, R.I.M., & Henzi, S.P. (2008). Fission-fusion dynamics. *Current Anthropology, 49*, 627–654.

Aureli, F., & van Schaik, C.P. (1991). Post-conflict behaviour in long-tailed macaques (*Macaca fascicularis*) II. Coping with the uncertainty. *Ethology, 89*, 101–114.

Bachmann, C., & Kummer, H. (1980). Male assessment of female choice in hamadryas baboons. *Behavioral Ecology and Sociobiology, 6*, 315–321.

Balasubramaniam, K.N., Dittmar, K., Berman, C.M., Butovskaya, M., Cooper, M.A., Majolo, B., Ogawa, H., Schino, G., Thierry, B., & De Waal, F.B.M. (2012). Hierarchical steepness, counter-aggression, and Macaque Social Style Scale. *American Journal of Primatology, 74*, 915–925.

Barros, M., Boere, V., Huston, J.P., & Tomaz, C. (2000). Measuring fear and anxiety in the marmoset (*Callithrix penicillata*) with a novel predator confrontation model: Effects of diazepam. *Behavioural Brain Research, 108*, 205–211.

Bates, L.A., Lee, P.C., Njiraini, N., Poole, J.H., Sayialel, K., Sayialel, S., Moss, C.J., & Byrne, R.W. (2008). Do elephants show empathy? *Journal of Consciousness Studies, 15*, 204–225.

Boesch, C., & Boesch-Achermann, H. (2000). *The chimpanzees of the Taï Forest: Behavioural ecology and evolution*. Oxford: Oxford University Press.

Butovskaya, M.L. (2008). Reconciliation, dominance and cortisol levels in children and adolescents (7–15-year-old boys). *Behaviour, 145,* 1557–1576.

Butovskaya, M., Verbeek, P., Ljungberg, T., & Lunardini, A. (2000). The multicultural view of peacemaking among young children. In F. Aureli & F.B.M. de Waal (Eds.), *Natural conflict resolution* (pp. 243–258). Berkeley, CA: University of California Press.

Cheney, D.L., & Seyfarth, R.M. (1985). Vervet monkey alarm calls: Manipulation through shared information? *Behaviour, 94,* 150–166.

Cheney, D.L., & Seyfarth, R.M. (1999). Recognition of other individuals' social relationships by female baboons. *Animal Behaviour, 58,* 67–75.

Cheney, D.L., & Seyfarth, R.M. (2012). The evolutionary origins of friendship. *Annual Review of Psychology, 63,* 153–177.

Clay, Z., & de Waal, F.B.M. (2013a). Bonobos respond to distress in others: Consolation across the age spectrum. *PLoS ONE, 8,* e55206. doi:10.1371/journal.pone.0055206

Clay, Z., & de Waal, F.B.M. (2013b). Development of socio-emotional competence in bonobos. *Proceedings of the National Academy of Sciences USA, 110,* 18121–18126.

Darwin, C. (1998). *The expression of the emotions in man and animals.* Oxford: Oxford University Press. (Original work published 1872)

Das, M. (2000). Conflict management via third parties: Post-conflict affiliation of the aggressor. In F. Aureli & F.B.M. de Waal (Eds.), *Natural conflict resolution* (pp. 263–280). Berkeley, CA: University of California Press.

Dasser, V. (1988). Mapping social concepts in monkeys. In R.W. Byrne & A. Whiten (Eds.), *Machiavellian intelligence: Social expertise and the evolution of intellect in monkeys, apes and humans* (pp. 85–93). Oxford: Clarendon Press.

Davidov, M., Zahn-Waxler, C., Roth-Hanania, R., & Knafo, A. (2013). Concern for others in the first year of life: Theory, evidence, and avenues for research. *Child Development Perspectives, 7,* 126–131.

de Marco, A., Cozzolino, R., Dessi-Fulgheri, F., & Thierry, B. (2010). Conflicts induce affiliative interactions among bystanders in a tolerant species of macaque (*Macaca tonkeana*). *Animal Behaviour, 80,* 197–203.

de Waal, F.B.M. (1982). *Chimpanzee politics.* New York: Harper.

de Waal, F.B.M. (1986). The integration of dominance and social bonding in primates. *The Quarterly Review of Biology, 61,* 459–479.

de Waal, F.B.M. (1996). Conflict as negotiation. In W.C. McGrew, L.F. Marchant, & T. Nishida (Eds.), *Great ape societies* (pp. 159–172). New York: Cambridge University Press.

de Waal, F.B.M. (1999). Anthropomorphism and anthropodenial: Consistency in our thinking about humans and other animals. *Philosophical Topics, 27,* 255–280.

de Waal, F.B.M. (2000). Primates: A natural heritage of conflict resolution. *Science, 289,* 586–590.

de Waal, F.B.M. (2008). Putting the altruism back into altruism: The evolution of empathy. *Annual Review of Psychology, 59,* 279–300.

de Waal, F.B.M. (2011). What is an animal emotion? *Annals of the New York Academy of Sciences, 1224,* 191–206.

de Waal, F.B.M. (2012). The antiquity of empathy. *Science, 336,* 874–876.

de Waal, F.B.M., & Aureli, F. (1996). Consolation, reconciliation, and a possible cognitive difference between macaques and chimpanzees. In A.E. Russon, K.A. Bard, & S.T. Parker

(Eds.), *Reaching into thought: The minds of the great apes* (pp. 80–110). Cambridge: Cambridge University Press.

de Waal, F.B.M., & Lanting, F. (1997). *Bonobo: The forgotten ape.* Berkeley, CA: University of California Press.

de Waal, F.B.M., & van Hooff, J.A.R.A.M. (1981). Side-directed communication and agonistic interactions in chimpanzees. *Behaviour, 77,* 164–198.

de Waal, F.B.M., & van Roosmalen, A. (1979). Reconciliation and consolation among chimpanzees. *Behavioral Ecology and Sociobiology, 5,* 55–66.

Dimberg, U., Thunberg, M., & Elmehed, K. (2000). Unconscious facial reactions to emotional facial expressions. *Psychological Science, 11,* 86–89.

Eisenberg, N. (2000). Empathy and sympathy. In M. Lewis & J.M. Haviland-Jones (Eds.), *Handbook of emotion* (pp. 677–691). New York: Guilford Press.

Eisenberg, N., & Fabes, R.A. (2005). Emotion regulation and children's socio-emotional competence. In L. Balter & C.S. Tamis-LeMonda (Eds.), *Child psychology: A handbook of contemporary issues* (pp. 357–384). New York: Psychology Press.

Eisenberg, N., Fabes, R.A., Murphy, B., Karbon, M., Smith, M., & Maszk, P. (1996). The relations of children's dispositional empathy-related responding to their emotionality, regulation, and social functioning. *Developmental Psychology, 32,* 195–209.

Fraser, O.N., & Aureli, F. (2008). Reconciliation, consolation and postconflict behavioral specificity in chimpanzees. *American Journal of Primatology, 70,* 1114–1123.

Fraser, O.N., & Bugnyar, T. (2010). Do ravens show consolation? Responses to distressed others. *PLoS ONE, 5,* e10605. doi:10.1371/journal.pone.0010605

Fraser, O.N., Koski, S. E., Wittig, R.M., & Aureli, F. (2009). Why are bystanders friendly to recipients of aggression? *Journal of Communicative and Integrative Biology, 2,* 1–7.

Fraser, O.N., Stahl, D., & Aureli, F. (2008). Stress reduction through consolation in chimpanzees. *Proceedings of the National Academy of Sciences USA, 105,* 8557–8562.

Fraser, O.N., Stahl, D., & Aureli, F. (2010). The function and determinants of reconciliation in *Pan troglodytes. International Journal of Primatology, 31,* 39–57.

Fujisawa, K.K., Kutsukake, N., & Hasegawa, T. (2006). Peacemaking and consolation in Japanese preschoolers witnessing peer aggression. *Journal of Comparative Psychology, 120,* 48–57.

Goodall, J. (1986). *The chimpanzees of Gombe: Patterns of behavior.* Cambridge, MA: Harvard University Press.

Gross, J.J., & Thompson, R.A. (2007). Emotion regulation: Conceptual foundations. In J.J. Gross (Ed.), *Handbook of emotion regulation* (pp. 3–24). New York: Guilford Press.

Hamilton, W.D. (1964). The genetical evolution of social behavior. I & II. *Journal of Theoretical Biology, 7,* 1–52.

Hare, B., Melis, A.P., Woods, V., Hastings, S., & Wrangham, R. (2007). Tolerance allows bonobos to outperform chimpanzees on a cooperative task. *Current Biology, 17,* 619–623.

Herrmann, E., Hare, B., Call, J., & Tomasello, M. (2010). Differences in the cognitive skills of bonobos and chimpanzees. *PLoS ONE, 5,* e12438. doi:10.1371/journal.pone.0012438

Hoffman, M.L. (2001). *Empathy and moral development: Implications for caring and justice.* Cambridge: Cambridge University Press.

Hohmann, G., & Fruth, B. (2003). Culture in bonobos? Between-species and within-species variation in behavior. *Current Anthropology, 44,* 563–571.

Judge, P.G., & Mullen, S.H. (2005). Quadratic postconflict affiliation among bystanders in a hamadryas baboon group. *Animal Behaviour, 69*, 1345–1355.

Kagan, J., & Snidman, N. (2009). *The long shadow of temperament*. Cambridge, MA: Harvard University Press.

Kano, T. (1992). *The last ape: Pygmy chimpanzee behavior and ecology*. Palo Alto, CA: Stanford University Press.

Kazem, A.J.N., & Aureli, F. (2005). Redirection of aggression: Multiparty signalling within a network? In P.K. McGregor (Ed.), *Animal communication networks* (pp. 191–218). Cambridge: Cambridge University Press.

Koski, S.E., de Vries, H., van den Tweel, S.W., & Sterck, E.H.M. (2007a). What to do after a fight? The determinants and inter-dependency of post-conflict interactions in chimpanzees. *Behaviour, 144*, 529–555.

Koski, S.E., Koops, K., & Sterck, E.H.M. (2007b). Reconciliation, relationship quality, and postconflict anxiety: Testing the integrated hypothesis in captive chimpanzees. *American Journal of Primatology, 69*, 158–172.

Koski, S.E., & Sterck, E.H.M. (2009). Post-conflict third-party affiliation in chimpanzees: What's in it for the third party? *American Journal of Primatology, 71*, 409–418.

Kummer, H. (1967). Tripartite relations in hamadryas baboons. In S.A. Altmann (Ed.), *Social communication among primates* (pp. 63–71). Chicago: University of Chicago Press.

Kutsukake, N., & Castles, D.L. (2004). Reconciliation and post-conflict third-party affiliation among wild chimpanzees in the Mahale Mountains, Tanzania. *Primates, 45*, 157–165.

Kutsukake, N., & Clutton-Brock, T.H. (2008). Do meerkats engage in conflict management following aggression? Reconciliation, submission and avoidance. *Animal Behaviour, 75*, 1441–1453.

Langergraber, K., Mitani, J., & Vigilant, L. (2009). Kinship and social bonds in female chimpanzees (*Pan troglodytes*). *American Journal of Primatology, 71*, 840–851.

Langford, D.J., Crager, S.E., Shehzad, Z., Smith, S.B., Sotocinal, S.G., Levenstadt, J.S., Chanda, M.L., Levitin, D.J., & Mogil, J.S. (2006). Social modulation of pain as evidence for empathy in mice. *Science, 312*, 1967–1970.

Leone, A., Mignini, M., Mancini, G., & Palagi, E. (2010). Aggression does not increase friendly contacts among bystanders in geladas (*Theropithecus gelada*). *Primates, 51*, 299–305.

Logan, C.J., Emery, N.J., & Clayton, N.S. (2013). Alternative behavioral measures of postconflict affiliation. *Behavioral Ecology, 24*, 98–112.

McComb, K., Shannon, G., Durant, S.M., Sayialel, K., Slotow, R., Poole, J., & Moss, C. (2011). Leadership in elephants: The adaptive value of age. *Proceedings of the Royal Society B: Biological Sciences, 278*, 3270–3276.

Mitani, J.C. (2009). Male chimpanzees form enduring and equitable social bonds. *Animal Behaviour, 77*, 633–640.

Mitani, J.C., Watts, D.P., & Lwanga, J.S. (2002). Ecological and social correlates of chimpanzee party size and composition. In C. Boesch, G. Hohmann, & L.F. Marchant (Eds.), *Behavioural diversity in chimpanzees and bonobos* (pp. 102–111). New York: Cambridge University Press.

Murphy, B.C., Shepard, S.A., Eisenberg, N., Fabes, R.A., & Guthrie, I.K. (1999). Contemporaneous and longitudinal relations of dispositional sympathy to emotionality, regulation, and social functioning. *The Journal of Early Adolescence, 19*, 66–97.

Nishida, T., & Hiraiwa-Hasegawa, M. (1987). Chimpanzees and bonobos: Cooperative relationships among males. In B.B. Smuts, D.L. Cheney, R.M. Seyfarth, R.W. Wrangham, & T.T. Struhsaker (Eds.), *Primate societies* (pp. 165–177). Chicago: University of Chicago Press.

Palagi, E., & Cordoni, G. (2009). Postconflict third-party affiliation in *Canis lupus*: Do wolves share similarities with the great apes? *Animal Behaviour, 78*, 979–986.

Palagi, E., Cordoni, G., & Tarli, S.B. (2006). Possible roles of consolation in captive chimpanzees (*Pan troglodytes*). *American Journal of Physical Anthropology, 129*, 105–111.

Palagi, E., Dall'Olio, S., Demuru, D., & Stanyon, R. (2014). Exploring the evolutionary foundations of empathy: Consolation in monkeys. *Evolution and Human Behavior, 35*, 341–349.

Palagi, E., & Norscia, I. (2013). Bonobos protect and console friends and kin. *PLoS ONE, 8*, e79290. doi:10.1371/journal.pone.0079290

Palagi, E., Paoli, T., & Tarli, S.B. (2004). Reconciliation and consolation in captive bonobos (*Pan paniscus*). *American Journal of Primatology, 62*, 15–30.

Parr, L.A. (2001). Cognitive and physiological markers of emotional awareness in chimpanzees (*Pan troglodytes*). *Animal Cognition, 4*, 223–229.

Parr, L.A., & de Waal, F.B.M. (1999). Visual kin recognition in chimpanzees. *Nature, 399*, 647–648.

Paul, E.S., Harding, E.J., & Mendl, M. (2005). Measuring emotional processes in animals: The utility of a cognitive approach. *Neuroscience & Biobehavioral Reviews, 29*, 469–491.

Plotnik, J., & de Waal, F.B.M. (2014). Asian elephants (*Elephas maximus*) reassure others in distress. *PeerJ, 2*, e278. doi.org/10.7717/peerj.278

Plotnik, J.M., Lair, R., Suphachoksahakun, W., & de Waal, F.B.M. (2011). Elephants know when they need a helping trunk in a cooperative task. *Proceedings of the National Academy of Sciences USA, 108*, 5116–5121.

Preston, S.D., & de Waal, F.B.M. (2002). Empathy: Its ultimate and proximate bases. *Behavioral and Brain Sciences, 25*, 7–71.

Prüfer, K., Munch, K., Hellmann, I., Akagi, K., Miller, J.R., Walenz, B., Koren, S., Sutton, G., Kodira, C., & Winer, R. (2012). The bonobo genome compared with the chimpanzee and human genomes. *Nature, 486*, 527–531.

Romero, T., & Aureli, F. (2008). Reciprocity of support in coatis (*Nasua nasua*). *Journal of Comparative Psychology, 122*, 19–25.

Romero, T., Castellanos, M.A., & de Waal, F.B.M. (2010). Consolation as possible expression of sympathetic concern among chimpanzees. *Proceedings of the National Academy of Sciences USA, 107*, 12110–12115.

Romero, T., Castellanos, M.A., & De Waal, F.B.M. (2011). Post-conflict affiliation by chimpanzees with aggressors: Other-oriented versus selfish political strategy. *PLoS ONE, 6*, e22173. doi:10.1371/journal.pone.0022173

Romero, T., Colmenares, F., & Aureli, F. (2009). Testing the function of reconciliation and third-party affiliation for aggressors in hamadryas baboons (*Papio hamadryas hamadryas*). *American Journal of Primatology, 71*, 60–69.

Romero, T., & de Waal, F.B.M. (2010). Chimpanzee (*Pan troglodytes*) consolation: Third-party identity as a window on possible function. *Journal of Comparative Psychology, 124*, 278–286.

Romero, T., & de Waal, F.B.M. (2011). Third-party postconflict affiliation of aggressors in chimpanzees. *American Journal of Primatology, 73*, 397–404.

Roth-Hanania, R., Davidov, M., & Zahn-Waxler, C. (2011). Empathy development from 8 to 16 months: Early signs of concern for others. *Infant Behavior and Development, 34*, 447–458.

Schino, G., Geminiani, S., Rosati, L., & Aureli, F. (2004). Behavioral and emotional response of Japanese macaque (*Macaca fuscata*) mothers after their offspring receive an aggression. *Journal of Comparative Psychology, 118*, 340–346.

Schino, G., & Marini, C. (2012). Self-protective function of post-conflict bystander affiliation in mandrills. *PLoS ONE, 7*, e38936. doi:10.1371/journal.pone.0038936

Schino, G., & Marini, C. (2014). Redirected aggression in mandrills: Is it punishment? *Behaviour, 151*, 841–859.

Schino, G., Perretta, G., Taglioni, A.M., Monaco, V., & Troisi, A. (1996). Primate displacement activities as an ethopharmacological model of anxiety. *Anxiety, 2*, 186–191.

Schino, G., & Sciarretta, M. (2015). Effects of aggression on interactions between uninvolved bystanders in mandrills. *Animal Behaviour, 100*, 16–21.

Schino, G., Troisi, A., Perretta, G., & Monaco, V. (1991). Measuring anxiety in nonhuman primates: Effect of lorazepam on macaque scratching. *Pharmacology Biochemistry and Behavior, 38*, 889–891.

Seed, A.M., Clayton, N.S., & Emery, N.J. (2007). Postconflict third-party affiliation in rooks, *Corvus frugilegus. Current Biology, 17*, 152–158.

Silk, J.B., Cheney, D.L., & Seyfarth, R.M. (1996). The form and function of post-conflict interactions between female baboons. *Animal Behaviour, 52*, 259–268.

Stumpf, R.M. (2007). Chimpanzees and bonobos: Diversity within and between species. In C.J. Campbell, A. Fuentes, K.C. MacKinnon, M. Panger, & S.K. Bearder (Eds.), *Primates in perspective* (pp. 321–344). New York: Oxford University Press.

Thierry, B. (2013). Identifying constraints in the evolution of primate societies. *Philosophical Transactions of the Royal Society B: Biological Sciences, 368*, 20120342.

Thierry, B., Aureli, F., Nunn, C.L., Petit, O., Abegg, C., & de Waal, F.B.M. (2008). A comparative study of conflict resolution in macaques: Insights into the nature of trait covariation. *Animal Behaviour, 75*, 847–860.

Verbeek, P. (2008). Peace ethology. *Behaviour, 145*, 1497–1524.

Verbeek, P., & de Waal, F.B.M. (1997). Post-conflict behavior of captive brown capuchins in the presence and absence of attractive food. *International Journal of Primatology, 18*, 703–725.

Verbeek, P., Hartup, W.W., & Collins, W.A. (2000). Conflict management in children and adolescents. In F. Aureli & F.B.M. de Waal (Eds.), *Natural conflict resolution* (pp. 34–53). Berkeley, CA: University of California Press.

Watanabe, S., & Ono, K. (1986). An experimental analysis of "empathic" response: Effects of pain reactions of pigeon upon other pigeon's operant behavior. *Behavioural Processes, 13*, 269–277.

Watts, D.P., Colmenares, F., & Arnold, K. (2000). Redirection, consolation and male policing: how targets of aggression interact with bystanders. In F. Aureli & F.B.M. de Waal (Eds.), *Natural conflict resolution* (pp. 281–301). Berkeley, CA: University of California Press.

Whiten, A., Goodall, J., McGrew, W.C., Nishida, T., Reynolds, V., Sugiyama, Y., Tutin, C.E.G., Wrangham, R.W., & Boesch, C. (1999). Cultures in chimpanzees. *Nature, 399*, 682–685.

Wittig, R.M., & Boesch, C. (2003). The choice of post-conflict interactions in wild chimpanzees (*Pan troglodytes*). *Behaviour, 140*, 1527–1559.

Wittig, R.M., & Boesch, C. (2005). How to repair relationships: Reconciliation in wild chimpanzees (*Pan troglodytes*). *Ethology, 111*, 736–763.

Wittig, R.M., & Boesch, C. (2010). Receiving post-conflict affiliation from the enemy's friend reconciles former opponents. *PLoS ONE, 5*, e13995. doi:10.1371/journal.pone.0013995

Wittig, R.M., Crockford, C., Wikberg, E., Seyfarth, R.M., & Cheney, D.L. (2007). Kin-mediated reconciliation substitutes for direct reconciliation in female baboons. *Proceedings of the Royal Society B: Biological Sciences, 274*, 1109–1115.

Wobber, V., Wrangham, R.W., & Hare, B. (2010). Bonobos exhibit delayed development of social behavior and cognition relative to chimpanzees. *Current Biology, 20*, 226–230.

Zahn-Waxler, C., Cole, P.M., Welsh, J.D., & Fox, N.A. (2009). Psychophysiological correlates of empathy and prosocial behaviors in preschool children with behavior problems. *Development and Psychopathology, 7*, 27–48.

Zahn-Waxler, C., Hollenbeck, B., & Radke-Yarrow, M. (1984). The origins of empathy and altruism. In M.W. Fox & L.D. Mickley (Eds.), *Advances in animal welfare science* (pp. 21–41). Springer Netherlands.

Zahn-Waxler, C., Radke-Yarrow, M., Wagner, E., & Chapman, M. (1992). Development of concern for others. *Developmental Psychology, 28*, 126–136.

5

The Experiential Peacebuilding Cycle: Grassroots Diplomacy, Environmental Education, and Environmental Norms

Saleem H. Ali and Todd Walters

Achieving and sustaining peace is a complex process. Behavioral processes and systems that catalyze peacebuilding can take antagonists past the threshold of impasse. In his book *Islamic Peace Paradigms*, Bangura (2005, p. 5) outlines various means to pursue peace at the international level, including "peace through power and coercion, peace through world order, peace through nonviolence, peace through conflict resolution, and peace through personal and community transformation." He proposes that at the individual level, personal transformation and conflict resolution skills are essential to begin the process of community transformation and achieving peace through nonviolent change. Building on this trajectory of thought, we demonstrate how environmental issues can provide a higher purpose for cooperation and peacebuilding despite their frequent framing as drivers of conflict (Ali 2003, 2007). The causal mechanism by which ecology may be linked to integrative peace rather than to distributive conflict has much to do with the process of instilling environmental norms through experiential educational systems.

There is an emerging literature on experiential environmental education and peacebuilding (e.g., Bruch *et al.* 2016; Deudney & Matthew 1999; Fall 2005). In reviewing this literature, we found it difficult to find clear linear linkages between environmental activities and peacebuilding. Here, we propose a dialectical and cyclical view of environmental peacebuilding education, where both experience and learning contribute to peaceful outcomes. We call our model the *Experiential Peacebuilding Cycle* (EPC). We believe that our iterative process model of environmental education offers an alternative to the high politics of war and security that currently prevail. Of course, our approach necessitates parallel efforts at conflict resolution through conventional means of issue-based diplomacy, including at the grassroots level, as we demonstrate in this chapter. To illustrate our model, we discuss examples of experiential environmental peacebuilding and education programs that we have been involved with in Iraq, in the Balkans, and among Muslim educators in Indonesia and the United States. Before we introduce our model and discuss the case studies, we review some of the theoretical foundations for our model.

Peace Ethology: Behavioral Processes and Systems of Peace, First Edition.
Edited by Peter Verbeek and Benjamin A. Peters.
© 2018 John Wiley & Sons Ltd. Published 2018 by John Wiley & Sons Ltd.

Theoretical Foundations

Following Lederach's framework for analyzing peacebuilding (e.g., Lederach 1997), conflict transformation requires empowerment of agents of change, and this is a core component of the experiential peacebuilding process, especially in terms of helping participants understand the shared meaning of their experience. Such empowerment is often the goal of small-scale peacebuilding initiatives that target people with influence in divided communities, so that they will be able to apply what they have learned in their communities. This is often done through joint cooperative initiatives or projects that allow a transformed group of people from each side of the divide to have a platform to reach out and extend their transformation to others in their communities. In our understanding, this approach to peacebuilding rests on four criteria: (1) the responsibility and judgment of being empowered; (2) the conscious use of symbolism and ritual to create meaningful images, metaphors, and experiences; (3) the transference of individual transformation back into the community; and (4) the reframing of community relationships so as to counter prevailing stereotypes and inspire a positive notion of the "other." We explore the theoretical background for these key conditions of experiential peacebuilding in more detail in this chapter.

Focusing first on *empowerment*, Freire (2002) noted that two of its central components are responsibility and judgment, and he underscores the learning process and its utility in encouraging students to become agents of change:

> Participants empower themselves by taking responsibility for their own learning (actively engaging as teachers as well as students), by increasing their understanding of the communities in which they live, and by understanding how they as individuals are affected by current, and potential, policies and structures. Equipped with this greater understanding and with new confidence in themselves, participants can develop policies and structures that better meet their needs, and strategies for bringing those policies into being. (pp. 80–81)

It is easy to see why empowerment is such an essential component of the transformative peacebuilding process. Empowerment motivates agents of change to pursue the changes that they feel will have a positive impact on their own lives and their communities, and not to be forced into commitments that they do not believe in and do not want to be a part of.

Freire (2002) emphasizes that empowerment can come through the process of experiential education, that is, not just through learning and understanding, but also through teaching and developing new policies, structures, and strategies in order to effect positive change in one's life and the lives of fellow community members.

An effective way to create an impactful experience for potential "agents of change" is through the transformative power of *ritual*. Schirch (2005) identifies the qualities that make a ritualistic experience so powerful:

> Ritual has three specific characteristics. First, it occurs in a unique social space, set apart from everyday life. Second, communication operates through symbols and emotions rather than relying primarily on words or rational thought. In ritual, individuals learn by doing and utilize nonverbal communication. Third, ritual confirms and transforms people's worldviews, identities, and relationships with others. (p. 2)

Utilizing the ritual qualities of Outdoor Experiential Education (OEE) creates a powerful transformational tool for peacebuilding: cooperating in a unique setting, communicating through symbols and emotions, learning by doing, and transforming worldviews, identities, and relationships. Schirch (2002) notes, "Creating and performing rituals help people in conflict to relate to one another and engage oppressive social structures that need to be changed. By offering tools that stimulate the mind, body, and senses, ritual enables parties to get beyond hatred and violence" (p. 2).

Schirch (2002) cautions that ritual is only meant to complement other peacebuilding tools and processes, such as dialogue and mediation. Based on Schirch's work and that of others, we believe that the use of ritual can be an effective tool for peacebuilding, as it can help enhance the transference of lessons learned in the OEE environment back into home communities.

In terms of *metaphor*, Bacon's (1993) reflections on the methodology of Outward Bound inform our work. Bacon notes,

> Instructors must carefully choose and actively interpret course activities as metaphors to further educational and therapeutic goals that lie outside the present experience. Archetypal endeavors and relationships are to be programmed as patterns of behavior to help participants to change their present behaviors so that they more closely resemble those encapsulated in the metaphor. This is accomplished subconsciously through the design of the course activities and consciously by verbally comparing and contrasting participant patterns of behavior with the ideal archetypal pattern through processing the experience. (1993, p. 23)

Consciously using metaphor in peacebuilding provides an image and promotes mutually beneficial reciprocal relationships. Take, for example, the whitewater rafting metaphor of paddling in the same direction, communicating, and trusting each other in order to run the rapids successfully. Applied to people in conflict, this "We are all in the same boat" metaphor highlights the importance of working for the common goal of peace by communicating with all the stakeholders and trusting that individuals and groups will uphold their ends of the agreement.

Another useful metaphor is "When it rains, everyone gets wet," meaning that some adversities are indiscriminate and impact all people, regardless of age, gender, culture, or religion. In the same way, conflict directly or indirectly impacts all members of society; there is no escaping it. Using metaphor, groups in conflict can conceive of shared goals, critically contrast their stereotypes against the ideal potential relationship, and cooperate for successful transformation.

In a personal interview with one of the authors, Bauman (2004) gave a great example of the conscious use of metaphor in the evaluation of his program on the conflict between Arabs and Israelis:

> Language is a sign of structural inequality. And they didn't realize that, but on course (expedition) what happened is, that the Arab students had to speak in English, and that is their third language. And so when an Arab-Israeli would get upset they would speak in Arabic, but the Israelis, the Jewish students would get mad, because they couldn't understand, and they would get paranoid, and that would invoke fear. And you have this on a very small scale, and in a very controlled

environment, it is the perfect opportunity to talk about language and structural inequalities, and power dynamics and all the things that we talk about when we talk about conflict resolution. It's just that those things are really hard to grasp unless you have an example. And so another good term you can use instead of "teachable moment" is the "conscious use of metaphor."

Taking symbolic acts and the conscious use of metaphor to a greater level of complexity brings us back to ritual, ceremonies, and pledges. These are composed of specific actions and can be either consciously designed or spontaneous, but they always have a specific goal or purpose. They can create an energy that allows participants to accomplish things together that would be impossible individually. John MacDonald, founder of the Institute for Multitrack Diplomacy, identified this in his work with youth from the former Yugoslavia and described an elaborate chanting ritual that he participated in with the youth before they each broke an arrow to symbolize destroying weapons that cause violence so they could no longer be used against each other. This ritual was used at the start of an expedition to create a "clean slate" upon which to build a new relationship between participants from different ethnic groups, including Bosniaks, Serbs, Roma, and Albanians who were perpetrators and victims in the Balkans conflict of the 1990s.

We also believe that *transference* is critical for peacebuilding, and we see two points of impact. First, transference can be internal as individuals seek to apply knowledge learned in one environment to a different environment. Second, transference can be communal, as transformed individuals return to their communities and seek to transfer their knowledge, feelings, and personal transformation to the other members of the community. The application of transference in the field of peacebuilding is not yet fully understood, as the following quote taken from a meta-analysis of peer mediation in educational settings illustrates:

> The suggestion that students who are trained in conflict resolution strategies apply those skills to settings that are external to the school environment in which they are learned implies that training students can have longer–lasting impact and affect wide audiences (siblings, families and the community at large). Unfortunately, the lack of data on this point does not permit inclusion of this feature as part of the meta-analysis. The potential transference of skills points to a target area for future research studies on the impact of mediation in schools. (Burrell *et al.* 2003, p. 9)

This quote highlights both aspects of transference and suggests the need for innovative peacebuilding projects that allow for proper evaluation of the role of transference. OEE is geared toward transference, particularly in the hard technical wilderness skills necessary for students to become teachers in their own right, and also in the soft skills of communication, leadership/followership, and trust building that can create great depth in interpersonal relationships in a relatively short period of time. OEE students carry these skills back to their relationships with loved ones, family members, friends, and their communities and utilize their new skills in new social settings to solve problems, lead groups, resolve conflicts, and improve communication.

The Experiential Peacebuilding Cycle

Building on the theoretical and practical foundations of *empowerment, ritual, metaphor and symbolism*, and *transference*, we propose the following EPC and illustrate it with examples of our work in the field.

We define *experiential peacebuilding* as an interactive process that harnesses the unique qualities of outdoor activities and combines them with proven experiential learning techniques to foster positive intergroup experiences within a neutral environment. This framework serves as a platform upon which to learn and practice peacebuilding skills. Outdoor experiential activities can be customized into various types of peacebuilding experiences based upon the specific objectives. The framework follows from the Action Learning Cycle (Kolb & Fry 1975), which has participants do something, reflect upon what they did, use their analysis to come up with a new plan to do better the next time, and test it by doing it again. Facilitated debriefings help to focus the reflection process and draw out the peacebuilding lessons.

Examples include daylong excursions, extended wilderness expeditions, and ropes/ challenge course initiatives. Different activities promote different aspects of peacebuilding and lend themselves to different skills trainings. For example, a day-long whitewater rafting excursion requires participants to cooperate, communicate, and trust each other to navigate the rapids safely, and participants can explore the metaphor "We are all in the same boat" in the debriefing. An extended wilderness expedition into a remote backcountry environment provides time to develop deep relationships and dialogue, to build trust, and to hone skills through practical training. A ropes/challenge course presents participants with opportunities for communication, active listening, collaborative problem solving, and a leadership/followership dynamic. All programs are designed to increase in intensity over time and reinforce lessons learned though facilitated debriefings.

Experiential learning consists of interactive activities that stimulate the mind, body, emotions, and attitudes of the participants. The unique methodology combines intellectual, physiological, emotional, and psychological aspects of learning. It creates an experience that taps into multiple layers of consciousness and memory and stimulates positive transformation. Dunning (2004) refers to the process as "kinesthetic encoding" – "when a concept is connected to something that we physically experienced, the potential for internalizing the given concept is much greater" (p. 23). After the experience comes the task of reflecting upon it and discussing the lessons it taught the group. This is followed by reflection on how best to transfer and apply the new knowledge back to the everyday environment.

Inevitably, when the lessons learned are brought back into a familiar environment, the participants are again confronted by the habits, prejudices, and ways of thinking that they left behind. This "re-entry shock" must be overcome to ensure successful transference. One way to address this is through a post-program, participant-developed, community service project (cf. Wessells & Kostelny, this volume, Chapter 9) that helps participants maintain their newly developed relationships and transfer to other members of their community the lessons learned, attitudes changed, and stereotypes broken down. In this role, they serve as living examples of people with the courage and skills to bridge the divide between conflicted communities.

Examples from the Field

Iraq

In 2010, Todd Walters, coauthor of this chapter, served as the Program Director of an Iraqi Youth Summer Leadership Program for a nongovernmental organization (NGO) called Agricultural Cooperative Development International and Volunteers in Overseas Cooperative Assistance (ACDI/VOCA), funded by the US Agency for International Development (USAID). The program is a model of grassroots-level diplomacy between different groups within the same state, Iraq, and was based in Kirkuk, Erbil, and Rawandouz in the northern parts of the country. The experiences of those 75 days, working with a diverse group of Iraqi facilitators, translators, staff, and youth participants, provided a unique opportunity to test hypotheses regarding the efficacy of the techniques included in the EPC. In our analyses of the program, we focused on the following two specific questions:

1) How can individuals and communities sustain a vision and practice of peace and resilience in the midst of crisis and disaster?
2) How are communities finding new forms of expression? And how is that reducing the space between the communities and the government?

The 2010 Iraqi Youth Leadership and Local Governance Summer Camp set out to create a space to practice teamwork and collaborative problem solving; to share and appreciate the cultural, ethnic, and religious diversity of peers; to understand what it means to be an active and engaged citizen in a new democracy; and to envision a shared future where the participants could become leaders for peaceful change. During the first week, both Iraqi men and women participated in the program and trained as experiential peacebuilding facilitators (hereafter, *facilitators*). During the second week, these 27 facilitators worked with an enrollment of 150 young Iraqi men. During the third week, the facilitators worked with 100 young Iraqi women. All participants were between 16 and 25 years old and were chosen by their communities to participate in the program. They came from Salah ad Din, Ninawa, Kirkuk, and Diyala provinces and from ethnic and religious backgrounds that included Turkmen, Kurds, Sunni and Shia Arabs, Christians, Muslims, and secular youth.

The facilitators were assigned to five groups representing artificial provinces given the names "A" through "E." The groups mirrored each other in composition and were composed of male (in week 2) and female (in week 3) Arabic and Kurdish speakers from various religious backgrounds and geographic regions. The age of the group members varied. The diversity of backgrounds of the facilitators was a key strength and point of uniqueness of the program. Elements of the natural world where the learning took place and the lessons were being taught provided a consistent ecosystem-based context, while other elements of the sense of place were in flux.

The young Iraqi people engaged in a series of experiential learning activities about newly developed Iraqi local government structures and election procedures. They also learned how to participate in local community development and mobilize their peers for action on issues of importance to them. In sum, they learned what it means to be an active and engaged citizen in a democracy and how to participate more effectively in community decision-making structures. The ultimate aim of the program was for the participants to harness their own experiential learning and apply what it taught them toward the peaceful transformation of their home communities.

Experiential peacebuilding activities progressively build skills of communication and teamwork, honing nuances based on the knowledge developed during each activity. The facilitated debriefing of each activity stimulated learning as the facilitators helped the participants to analyze what made an activity successful or why they struggled. Sharing experiences about what they went through together as a group helped to cement the bonds between the participants and helped them to prepare for and be more successful in their subsequent challenges as a group.

The learning objectives of the camp focused on building leadership skills, effective communication, and cross-group collaboration skills, and on enhancing the confidence levels of all participants in order for them to be able to address shared problems or challenges in their home communities. The camp curriculum focused on developing these skills while teaching about the new democratic structure in Iraq and how to function effectively within that system as an engaged citizen. It focused on elections and voting procedures, parliamentary procedure, and how to write resolutions, as well as public speaking skills, campaigning, and coalition building.

Each of the five groups representing the artificial provinces (A–E) included participants from the actual Iraqi provinces Salah ad Din, Ninawa, Kirkuk, and Diyala. During the first day, the facilitators described their artificial province, giving it strengths and weaknesses, and creating province posters depicting these dynamics. Some even gave a name to their province, like "Province of Hope" or "Province of Economic Prosperity." After some initial icebreaker activities, the participants developed newly shared identities tied to their artificial province and began to root for it and support "their" province – which was in fact composed of representatives from all four real Iraqi provinces mentioned earlier.

The facilitators worked together in their artificial provinces to discuss and resolve issues of leadership and practiced democratic processes such as electing provincial representatives and a governor for their province. They also organized a camp-wide campaign to try to have their provincial governor elected as the overall camp governor. Their work together in their artificial provinces was critical toward creating a new "temporary identity" that allowed the facilitators the opportunity to break down stereotypes and develop teamwork across groups that previously had little or no interaction with each other. The significance of the moment when they started to become a part of a new group and no longer primarily identified themselves simply by their ethnic group or actual geographic province was profound indeed.

The facilitators modeled how to come together as a diverse group, and they were effective in communicating the ideas of forming new friendships and working together with new friends. During a dancing event, there was an appreciation of cultural heritage and cultural nuances through the playing of different ethnic songs so different groups could dance according to their customs and different ethnic groups could mix and learn each other's dances to each other's music.

The election activities occurred over three days. The facilitators had a chart of positions common at the Qada, Nahiya, and provincial levels: Provincial Council Chair, Director General for Agriculture, Director General for Education, Director General for Health, and Province Governor. Each artificial province first nominated provincial representatives and province governor candidates and held campaigns and elections for each position. Next, they supported their elected province governor in the race for Camp Governor. For each of the two election cycles, the candidates practiced lobbying, creating coalitions and banners, presenting speeches, and registering and casting their ballots. They watched as the ballots were publicly counted and learned about the importance of the privacy of their vote, and when the winner was announced, an acceptance speech as well as gracious concession speeches were given in Arabic, Kurdish, and Turkmen.

Representatives from the Provincial Reconstruction Teams (PRTs) and from USAID Erbil witnessed the Final Day Election Process for the Camp Governor – from voter registration to campaign speeches and slogans, to actual vote casting and public counting of the ballots, and the announcement of the winner and her acceptance speech. They were asked to serve as independent international election monitors and took their responsibility seriously.

Sports activities were woven into the curriculum using a local gymnasium for basketball (girls) and the local 6v6 football pitch (boys). Volleyball nets were also set up on the flat grass outside the office villa. The sports teams were composed of young Arabs, Kurds, and Turks, all playing together, communicating, and practicing teamwork in an arena that they were all familiar with. The competition with the other teams spurred collaboration within the teams, and the artificial provinces grew more tightly knit. This also served as an excellent icebreaker between the participants and the staff because the sixth team entered in all the tournaments was composed of facilitators and staff members. Being on the field competing and having fun together, as well as demonstrating sportsmanship, allowed participants from all groups, facilitators, and staff to appreciate the power of sport to bridge divides.

Field Day Competitions were organized in the form of tug-o-war, relay races, three-legged races, sack races, egg balancing, and a water balloon toss. These new games enabled teams from the five artificial provinces to compete for medals and to learn how to work together both physically and mentally. Celebrating success and cheering each other up after losses were important parts of these events. Witnessing how someone else was feeling when winning or losing served as a powerful empathic tool and helped participants to relate to each other across perceived divisions. It developed bonds that were deep enough to help overcome previous stereotypes.

The Arts and Culture Committee provided an opportunity for campers with special talents to showcase them and to receive training and feedback from facilitators who had expertise in these areas. Campers practiced photography, calligraphy, and pencil artwork, along with song, dance, and poetry. The best examples were shown during the cultural exhibition at the Final Awards Ceremony. The songs, dances, and poetry readings brought real energy and focus to the start of the Final Awards Ceremony, which helped to create a special atmosphere of Iraqi cultural sharing and cultural pride along with the display and recognition of special talents. One of the facilitators wrote a camp anthem for the camp that was sung at the boys' camp as well as a song of love and hope that was sung at the girls' camp, and singing these songs brought out pride and unity.

At the boys' camp, a morning was spent on a challenging hike through a local sand-stone canyon. Seventy-five of the 150 boys chose to participate, and ten of the facilitators came along to help supervise them. With the help of local guides from the Mayor of Rawandouz's security team, they wound their way along a narrow path on a steep hillside along the side of a canyon with a flowing river at the bottom. They passed traditional shepherd summer camps and two springs bubbling out of the side of the canyon, and they learned about numerous local plants, nuts, berries, and medicinal herbs. The terrain was challenging, and the young men often had to help each other up and down difficult sections of the trail. This reinforced the notion of caring for others and making sure that everyone could achieve the shared goal of reaching the end of the hike. Another positive aspect of the hike was the small-group conversations that are stimulated by hiking together along a single-track path. The young men discussed their first few days at the camp and shared their surprises and what they had learned about being an active and engaged citizen in a new democracy.

There was time set aside for individual reflection and time for group processing and sharing. The elected Camp Governors of the 2009 program were invited for a visit to the boys' and girls' camps to serve as role models and provide inspiration for the 2010 participants. They floated among the artificial provinces, shared what they had learned, provided pointers and feedback to the campers, and participated in activities. They also delivered inspirational speeches on what they did when they returned to their home communities after being elected Governor.

Both the boys' and the girls' camps celebrated their hard work during the first three days of the camp with an afternoon visit to Bekhal Falls. The visit gave them a chance to relax and have a few hours of fun wading in the cool water, climbing on the rocks stacked in different parts of the waterfall, and taking many photographs with their new friends and facilitators. In this safe space, the young men and women bought ice cream, super-soaker water toys, and memorabilia. Laughter and ear-to-ear smiles were the order of the day. The group stood out with their white T-shirts with the Youth Summer Camp (YSC) logo on the front, and they generated interest from tourists from around Iraq who were visiting the falls. The participants served as genuine ambassadors for the program as they explained why they were there and what they were learning as members of the YSC. By this point, the artificial provinces had had a chance to form, friendships had been built, and the T-shirts served as a visual reminder of the unity and equality of all the campers, regardless of their artificial province or actual ethnic group. The campers wore their T-shirts with pride for what it stood for and what they were a part of.

The 2010 camp culminated in an energy-filled, action-packed cultural presentation and awards ceremony, where each participant received a certificate highlighting his or her successful participation in the camp, and all recognized their facilitators and camp officials with loud ovations and applause. The night was filled with an outdoor celebration banquet and traditional dances to the tunes of a local DJ.

One of the concrete outcomes of the two camps came from the attendance of two members of the Iraqi government who flew in from Baghdad to attend. Two Director Generals and their Directors from the Ministry of Youth and Sports observed the camp, including the campaigns for the Camp Governor, and they requested joint collaboration between the camps and the Iraqi Youth Parliament to promote and develop additional summer camps on democracy in other regions of the country and to expand program activities between ACDI-VOCA and the Ministry of Youth and Sports through community youth centers in all the major cities. They considered the educational and peacebuilding experience to be transformative and wanted to expand access to it to a broader cross section of Iraqi youth. Likewise, US government officials working for USAID confirmed their perceived value of the camps by committing funding for future years in a private meeting with the author during the celebration at the culmination of the camp. Finally, the Iraqi Director General from the province of Salah ad Din noted that they had two open seats for the Iraqi National Youth Parliament and that he would extend invitations to the girls who were from Salah ad Din and participated well in the Camp Governor election process – as he felt the leadership roles they achieved at the camp would provide good experience to become members of the Iraqi National Youth Parliament.

In the case of present-day Iraq, how can individuals and communities sustain a vision and practice of peace and resilience in the midst of crisis and disaster? And how can communities and the government relate to each other and work together toward a democratic and peaceful country? We believe that the more people know about the new system of governance, how it is structured, how it works, and who is responsible for what, the greater the chance is that the system will work as intended. Elected officials need to learn how to represent the voice of the people in their community and realize that the community can elect someone else. The more people understand how the system works and what their role in it is, the greater the trust in the system will become, effectively reducing the space between the communities and the government.

The purpose of the final experiential learning activity was to review what the young participants had learned during the week and to begin thinking about their return home as leaders for peaceful change in their respective communities. To be successful leaders, they would have to be good at communicating with many different people. We utilized a "re-entry skit" as an opportunity to practice such transference conversations. With the facilitators acting as parents, friends, or a difficult member of the community, the campers who participated in the skit were asked a series of prepared questions representing the stereotypes and attitudes they could encounter when they returned home. Once a participating camper had completed the skit, the facilitators debriefed with all the campers. They asked, answered, and discussed questions such as "Was what he or she said in response to the questions effective?" "What would you have said differently?" and "How can you be a leader for change in your local community, with your family, with your friends, in the entire community?" In this way, the participants had a chance to reflect on how best to convey to the members of their home communities what they had learned from the week of experiential peacebuilding with residents of four different provinces from diverse ethnic and religious backgrounds, and how they had been transformed by the experience.

As they boarded the buses the next morning, the Director of the Camp and coauthor of this chapter, Todd Walters, addressed each bus before they left – challenging each of the young people to expand on the impact of the camp and take the lessons they had learned home to their families, friends, and communities. "Can you do this?" he asked. "YES!" they responded loudly in unison. The future of Iraq would soon be in their hands – or would it?

As the current civil war and the rise of the Islamic State of Iraq and Syria/the Levant (ISIS/ISIL) suggest, there are clearly limits to the overall impact of such peacebuilding efforts when external agents overwhelm any impact that experiential peacebuilding may have had. In this regard, the lessons from Iraq also suggest that when the conflict itself is deeply rooted in doctrinal differences, it is important to engage more deliberately within the particular ideological narrative. Thus, in Iraq, the sectarian aspects of the conflict between Shias and Sunnis may suggest the need to integrate some common theological elements for future experiential environmental peacebuilding efforts. Such shared theological elements can potentially provide a superordinate goal that does not avoid the religious aspects of the conflict. While this was not the focus of the 2010 program in northern Iraq, such an approach has been tried in the context of environmental education within the Islamic education community in the United States, as we describe in the next subsection.

United States

Within the Islamic community in the United States, the coauthor of this chapter, Saleem Ali, initiated an environmental service learning program at Berkeley's Zaytuna College. The inaugural session was held in the first week of January 2013. The program exemplifies the idea that environmental service learning can be a peace process. Zaytuna College was chosen as the venue since it is the first Muslim liberal arts college in the United States. Imam Dawood Yasin, the former Muslim chaplain of Dartmouth College, served as program director for the multiday event, which was cosponsored by Zaytuna College, the Institute for Environmental Diplomacy and Security at the University of Vermont, and the Center for Islamic Studies at Berkeley's Graduate Theological Union. A group of educators from Islamic schools across the nation were invited to take part in a series of experiential education activities with the aim that they would return back to their schools and replicate the lessons in their own schools. Elaine Pasquini documented the program for the *Washington Report on Middle East Affairs* (Pasquini 2013), and the quotations that follow are from her article.

The scriptural drawings of the program attempted to transcend sectarian divide between Muslim communities, and, as Anna Gade from the University of Wisconsin noted, focused on "ways people of faith are drawn and compelled to address issues of environmental change and the ways they are turning to faith-based approaches." Quoting from the Qur'an, the Islamic scholar and cofounder of Zaytuna College Imam Zaid Shakir noted, "Ecological consciousness is something integral to our religion. Historically, we didn't need a concerted effort to focus people's attention on environmental issues. This was something that our religion did and we didn't have to talk about it. It was something we lived. The Qur'an calls our attention to nature and to be in harmony with nature." Noting that environmental issues are affecting the heart of the Muslim world, Rhamis Kent of the Permaculture Research Institute explained, "Islamic tradition for protecting the environment is explicitly pointed out in the teachings of the Qur'an."

Imam Afroz Ali, managing director of Seekers Guidance, an online learning platform for Islamic education, and founder and president of the Al-Ghazzali Centre for Islamic Sciences and Human Development based in Sydney, Australia, noted how the Australian Religious Response to Climate Change had brought together all faith communities and helped to heal race relations in Australia. Despite past disagreements over issues such as the Arab–Israeli conflict or burka attire, the Muslim and non-Muslim communities had a common concern around global environmental change processes, which provided a common conversational element. Ibrahim Abdul-Matin, author of *Green Deen*, stated, "What Islam teaches about protecting the planet, is not to beat people over the head about environmental issues, but rather to approach the matter in a way people can relate to in terms of their daily life experiences of interacting with the environment." Mosques opening their vacant land for organic farming, riverside clean-up drives, and energy efficiency campaigns are examples of this.

In addition to motivational learning from speakers, the participants were taken to urban gardens started by Muslims and non-Muslims alike and shown permaculture techniques that were helping to connect children to the origin of their food as an act of worship. Permaculture is a concept that draws from the fields of ecological design, ecological engineering, environmental design, construction, and integrated water resource management. Permaculture seeks to develop sustainable architecture, regenerative and self-maintained habitat, and agricultural systems modeled after natural ecosystems (Mollison 1991). Islamic learning is highly metaphorical and allegorical. Many surahs of the Quran use elaborate natural metaphors, which are often employed to convey multiple meanings. The program attempted to instill the importance of conveying such allegories of meaning about the natural world as an act of community connectivity.

It is also important to recognize that there can be entrenched prejudice within particular theological practices that still needs to be addressed in order for lasting experiential peace education to take root. For example, in the context of Islamic schools, we were only able to undertake experiential programs with relatively moderate teaching staff who was amenable to allowing women teachers to interact with men in a professional context. Although appropriate decorum was followed, in more absolutist and extreme versions of Islamic practice the experience itself would not have been allowed. Thus, for experiential environmental education to be effective, there must be concomitant efforts at theological moderation, particularly in the Islamic tradition, which is particularly uncompromising in many contexts. It may be argued that this inertia itself is a result of constantly feeling threatened by cultural dilution, which has led to a form of cultural paranoia that itself prevents experiential peacebuilding from taking root. The process of building trust to allow for environmental experiential education to flourish will thus be an incremental process.

Consciously designed experiential environmental peacebuilding can also take a long form, such as transboundary expeditions, accredited by university systems for undergraduate and graduate credit, that explore the environmental peacebuilding activities in postconflict regions through a combination of experiential learning, cross-cultural exchange, and established academic practices. In the next section, we discuss an example of this in the Balkans.

The Balkans

Coauthor J. Todd Walters, the Founder and Executive Director of International Peace Park Expeditions, uses proven experiential learning methodologies, combined with traditional academic methodologies and cutting-edge technology, to teach the theory and the practice of environmental peacebuilding.

One example of his work is a field course entitled "Environmental Peacebuilding and Sustainability: A Peace Park in the Balkans?" accredited by International Peace Park Expeditions. The course takes place in Washington, DC; Kosovo, Albania; and Montenegro each summer and provides students with a unique, experiential learning opportunity in one of the last biodiverse areas of Europe – a place where the legacy of the conflict of the 1990s is being overcome with entrepreneurship, ecotourism, and environmental peacebuilding.

The course starts with three days in Washington, DC, where the students gain a foundational understanding of how environmental security thinking led to the development of the field of environmental peacebuilding – identifying ways in which the environment can present opportunities for cross-group collaboration for mutual benefit among people, communities, and countries impacted by conflict. In Washington, the students meet with and hear presentations from some of America's foremost scholars and policy analysts on environmental peacebuilding, sustainability, and transboundary protected areas management, including officials at the National Park Service's Office of International Affairs, USAID's Conflict Management and Mitigation Division, Conservation International's Policy Center for the Environment and Peace, and the Environmental Change and Security Program at the Woodrow Wilson International Center for Scholars. Students also visit an embassy from one of the countries they will soon visit for a meeting with diplomats to discuss regional environmental cooperation, the value of international educational exchange, and the country's goals for cooperation with their neighboring countries and the United States.

Upon arrival in the Balkans, the course focuses on community-driven sustainable development, cross-border environmental conservation, ecotourism, and postconflict natural resource management. Community-driven sustainable development focuses on the local people and their desires and priorities for their own development in a way that will be sustainable in the long term both environmentally and with livelihood opportunities. Cross-border environmental conservation refers to the fact that ecosystems do not stop at political borders, and the goal of conservation of ecosystems, particularly biodiversity hotspots, requires cross-border collaboration between communities, civil organizations, and governments.

Ecotourism refers to travel to a place to experience all that location can offer environmentally, culturally, and educationally, while consciously ensuring that visitors do not damage that environment and encouraging its protection. Postconflict natural resource management refers to the need to establish effective policies and governance mechanisms to manage resources like water, forests, gems and minerals, and oil and natural gas in ways that help to stimulate economic growth while not destroying the environment in which those natural resources are found. Students trek sections of the award-winning Peaks of the Balkans Trail, which links Muslim, Catholic, and Orthodox enclaves as well as Slavic and numerous Albanian tribes in three adjoining national parks, each showcasing the border region's stunning natural beauty. The cross-border trails were previously used by Roman soldiers and Kosovar refugees and were recently named by the World Tourism and Travel Council as the World's Top Sustainable Destination, highlighting the layers of history that the landscape holds. The students enjoy the hospitality of the local people in all three countries through home stays, and they experience the unique "highlander mountain culture" and its traditional organic farming and rustic cuisine, traditional clothing, music, and centuries-old architecture and rugged way of life.

The students learn to set wildlife camera traps as part of a transboundary endangered species conservation initiative between Kosovo, Albania, Macedonia, and Montenegro called the Balkan Lynx Project. GPS data are gathered on transboundary treks and shared with mountaineering organizations from all three countries. Lodging is in ecotourism guesthouses with local families, allowing students to learn directly from the people who practice environmental peacebuilding every day.

The students experience firsthand how the environment is harnessed as a tool to stimulate cross-border cooperation between individuals, communities, and countries and how entrepreneurship and ecotourism are providing an alternative model for sustainable development and livelihood creation. International Peace Park Expeditions not only teaches about environmental peacebuilding, but also actively contributes to it by supporting local entrepreneurs and the local economy involved in transboundary cooperation, by linking organizations and people across the borders in personal and professional relationships, and by encouraging cross-cultural experiential learning to build peace, understanding, and appreciation of differences between all people. The more bonds that can be built between communities, civil society organizations, and governments, the greater the resilience will be to help the region to prevent a return to conflict – because a critical mass of people will oppose it, as they will now have personal friendships, collaborative business opportunities, and cooperative scientific and educational ventures that a return to conflict would irreparably damage.

The environment serves as common ground upon which Muslims and Christians, Orthodox and secular, Western and non-Western, and young and old from different ethnicities, cultures, and nations can come together to learn. Through that experiential process, they can undergo intrapersonal transformation of stereotypes and beliefs, which then fosters interpersonal friendships and trust across artificial boundaries, a sense of common humanity, and a notion of belonging to a common environment, the protection of which serves as a superordinate goal uniting people despite their differences.

People of all backgrounds are coming together across political boundaries to protect and manage their critical water resources – where the source may exist in one country but the river flows across the border into another country. Rangers and National Park Directors are cooperating to address the issue of illegal logging and promote the development of long-term livelihoods built upon collaborative sustainable forest management. Cross-border ecotourism initiatives are being financed, developed, and created between international organizations, local entrepreneurs, and national governments to stimulate economic opportunity in these remote border regions and to generate income that encourages the cooperation across political boundaries and the preservation of critical ecosystems. Scientists and local environmental organizations are studying species whose habitats extend across borders and finding ways to protect those habitats and conserve the species. Participants can bring their changed attitudes and perspectives back into their communities, organizations, and governments and serve as leaders for positive peaceful change.

These OEE environmental opportunities possess a symbolism that can be celebrated and promoted at the national and international levels through the press and social media to amplify the message and to shift the narrative away from maintaining entrenched attitudes and divisions and toward trustbuilding and peacebuilding. Examples include national newspaper interviews featuring International Peace Park Expeditions staff and students and the local partner organizations that help make our transboundary expeditions possible. The Balkan Lynx Project has been on national television in all four countries showcasing a cross-border scientific conservation effort. And, as mentioned in this subsection, a diverse group from Kosovo, Albania, and Montenegro collaboratively applied for and won the World Tourism and Travel Council's award for Top Sustainable Destination of 2014, and has subsequently received write-ups highlighting the effort from Lonely Planet, on the front page of the *New York Times* Travel section, and on BBC.com. All of this media recognition reverberates through social media, as students from the expedition and participants and organizations involved in these environmental peacebuilding efforts share the links of the news with their friends and supporters, spreading the peace process that the program fosters.

Conclusion

The emergent field of experiential peacebuilding is now gaining greater currency as a tool for developing programs and activities in conflict-affected communities. Through our experience working with various experiential peacebuilding efforts and co-teaching programs in the field, we feel more confident that there is a process by which peace develops in such contexts. Environmental issues provide a perfect context for experiential peacebuilding because the conservation of natural systems can provide a neutral and necessary medium of inquiry. The Experiential Peacebuilding Cycle, as introduced in this chapter and shown in Figure 5.1, emerged from the work that we have undertaken with our colleagues in the field.

The cycle is a fusion of the "group performance sequence" that builds on the classic work of Tuckman (1965) and the "experiential learning cycle" developed by Kolb and Fry (1975). The former describes how groups go through dynamic stages of development from *forming* (getting to know one another) to *storming* (ironing out disagreements and increasing understanding), then *norming* (developing trust and establishing processes to communicate and share information effectively), and finally *performing* (where they can effectively collaborate to solve problems). The latter highlights the process by which groups and individuals take action, reflect upon that action, develop a new plan of action to perform better, and transfer their new strategy into a second attempt at improving their ability to complete the original action.

It is important to note that the cycle starts with a problem–solution proposition, often focused on a common environmental concern. The *immediate causation* imperative is in synch with this problem-solving proximate goal that starts the cycle. Information sharing is the next necessary step. A mutually agreed-upon repertoire of scientific knowledge and a peer review process for vetting and contesting knowledge can help to provide a "norming" mechanism to advance through the cycle. Trust usually emerges through the epistemic exchange that occurs in this process. There is likely to be greater connectivity between interest groups when a community of collective learning forms, particularly around ecological factors. However, as we learned from examples in the Islamic schools, this common knowledge and learning often need to incorporate ideological metaphors from scripture that hold salience for the communities in question. For example, in the village of Bantul in a remote part of central Java, a small religious school, a *pesantren* (the local word for a madrassa, or religious school), is graduating environmentalists whose commitment to the earth is not based on Western conservation texts, but rather predicated in values derived from Islam. The head of the school, Nasruddin Anshari, frequently uses the refrain "One earth for all" just as much as he does the usual Islamic invocation of "Allahu Akbar" ("God is great"). Indonesia's *pesantren* have come under scrutiny in recent years due to their perceived connections to terrorist incidents such as the Bali bombings in 2005. Even then-US presidential hopeful Barack Obama felt obliged to distance himself from his childhood days in Indonesia because of a rumor that he too had attended a *pesantren*, since both his father and stepfather were Muslims. Yet the transformation that is taking place at Pesantren Lingkungan Giri Ilmu would certainly please most constituencies in the West, as the children here are learning about the importance of preserving their ecosystem as a mark of worshiping God.

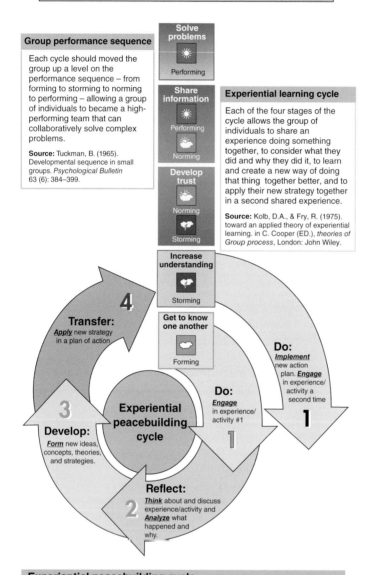

Figure 5.1 Experiential peacebuilding cycle. *Source*: Walters *et al.* (2014). Reproduced with permission of International Peace Park Expeditions.

Facilitators who do not espouse particular beliefs or only focus on the empirical scientific aspects of environmental learning are sometimes uncomfortable dealing with metaphysical narratives that may emerge in the peacebuilding process. In such contexts, it is important to stress "understanding" rather than "belief" to any detractors of particular ideological viewpoints.

The human connection of getting to know each other regardless of different values and beliefs comes through the experiential process, particularly when participants experience basic human needs for food and water as well as the constraints of exhaustion and fatigue in the field. At this stage, the activities constructed for the experiential peacebuilding cycle, such as those that we described for the program in Iraq, become most efficacious. After the activity itself, the time to reflect on the process and consider its broader ramifications for our own survival is essential. Whether such a reflection is done tacitly or explicitly depends on the cultural context in which the process is being undertaken. It is important for the facilitators to allow time for such reflection.

Innovation must be an important aspect of the peacebuilding experience. The reflection should aim for new solutions to the particular problem that initiated the experiential process. If the problem was cultivation in a water-scarce environment, the collective learning process should clearly document some particular permaculture technique that the participants adapted and applied. There could also be social innovations whereby a new paradigm for interaction or engagement emerges. In conflict zones, social innovations may seem mundane, but they provide meaning in the uncertainty of the moment.

As described in this chapter, *transference* is an important aspect of the experiential peacebuilding cycle, and it is of particular importance to sustain the process beyond the first phase. As Figure 5.1 suggests, the cycle is like a spiral that must lead to refinement in each loop toward a more robust peace. External calamities or trust-eroding events such as violent conflicts or an attack in the community may interrupt or terminate the spiral. Yet, through each opportunity to physically engage and collectively experience the process, participants may become better able to withstand such shocks and build more resilient relationships. Sustainable peace is more likely when participants appreciate their mutual dependence and the benefits of cooperation. Further research on applying such processes in various operational arenas such as international diplomatic contexts is needed. As the theoretical insights and tentative findings from experiential processes in educational programs suggest, cooperating to solve shared environmental problems may be a neglected process for developing and harnessing peace.

References

Ali, S.H. (2003). Environmental planning and cooperative behaviour. *Journal of Planning Education and Research*, 23(2), 165–176.

Ali, S.H. (2007). *Peace parks: Conservation and conflict resolution*. Cambridge, MA: The MIT Press.

Bacon, S.B. (1993). *The conscious use of metaphor in Outward Bound*. Denver: Colorado Outward Bound School.

Bangura, A.K. (2005). *Peace paradigms*. Dubuque, IA: Kendall Hunt Publishing.

Bauman, P.A. (2004). *Outward bound peace missions* (proposal requesting support from Outward Bounds' Board of Directors). Albany, NY: Outward Bound.

Bruch, C., Muffett, C., & Nichols, S.S. (Eds.). (2016). *Governance, natural resources, and post-conflict peacebuilding.* London: Earthscan.

Burrell, N., Zirbel, C., & Allen, M. (2003). Evaluating peer mediation outcomes in educational settings: A meta-analytic review. *Conflict Resolution Quarterly, 21*(1), 7–26.

Deudney, D., & Matthews, R. (1999). Bring Nature back in: Geopolitical theory from the Greeks to the global era. In D. Deudney & R. Matthews (Eds.), *Contested grounds: Security and conflict in the new environmental politics.* Albany, NY: SUNY Press.

Dunning, S. (2004). *A call for adventure based conflict resolution.* Master's thesis, George Mason University International Conflict Analysis and Resolution Program, Washington, DC.

Freire, P. (2002). *Pedagogy of the oppressed.* New York: Continuum.

Kolb, D.A., & Fry, R. (1975). Toward an applied theory of experiential learning. In C. Cooper (Ed.), *Theories of group process.* Chichester: John Wiley.

Lederach, J.P. (1997). Reconciliation: The building of relationship. In *Building Peace: sustainable reconciliation in divided societies.* Washington, DC: US Institute of Peace Press.

Mollison, B. (1991). *Introduction to permaculture.* Tasmania, Australia: Tagari.

Pasquini, E. (2013, April). Zaytuna College hosts historic workshop on environmental education in Islamic schools. *Washington Report on Middle East Affairs*, 48–49.

Schirch, L. (2005). *Ritual and symbol in peacebuilding.* Bloomfield, CT: Kumarian Press.

Tuckman, B.W. (1965). Developmental sequence in small groups. *Psychological Bulletin, 63*(6), 384–399.

Walters, T., Smither, M., & Dowlatabadi, T. (2014). Experiential learning cycle. In *Experiential peacebuilding activities manual* (No. 6). Washington, DC: International Peace Park Expeditions.

Additional Resources

Brynen, R., & Milante, G. (2013). Peacebuilding with games and simulations. *Simulation & Gaming, 44*(1), 27–35.

Carius, A. (2007). Environmental peacebuilding: Conditions for success. *ECSP Report, 12.*

Carney, S. (2005). Team building activities for teens: Group and classroom games to promote communication skills. Retrieved from http://youth-activities.suite101.com/article.cfm/team_building_activities_for_teens

Crocker, C.A., Hampson, F.O., & Aall, P. (2005). *Grasping the nettle: Analyzing cases of intractable conflict.* Washington, DC: US Institute of Peace Press.

Fisher, R.J. (1993). Developing the field of interactive conflict resolution: Issues in training, funding and institutionalization. *Political Psychology, 14*(1), 123–138.

Interaction Associates. (1997). *Facilitative leadership.* Boston, MA: Interaction Associates.

Jeong, H. (2006). Peacebuilding in post-conflict societies: Strategy and process. *Journal of Peace Research, 43*(4), 495–496.

Kreisberg, L. (2001). Mediation and the transformation of the Israeli Palestinian conflict. *Journal of Peace Research, 38*(3), 373–392.

Orlick, T. (2006). *Cooperative games and sports.* Champaign, IL: Human Kinetics,

Pedler, M. (1997). *Action learning in practice.* Aldershot, UK: Gower Publishing.

Rohnke, K., Wall, J., Tait, C., & Rogers, D. (2012). *The complete ropes course manual* (4th ed.). Dubuque, IA: Kendall Hunt.

Shapiro, I. (2005). Theories of change. In G. Burgess & H. Burgess (Eds. & Co-dirs.), *The Beyond Intractability Knowledge Base Project*. Boulder: Conflict Research Consortium, University of Colorado. Retrieved from http://www.beyondintractability.org/essay/theories_of_change/

Team Craft. (N.d.). Retrieved from http://www.teamcraft.com

Teampedia. (N.d.). Tools for Teams. Retrieved from http://www.teampedia.net

Teamwork & Teamplay. (N.d.). Activities. Retrieved from http://www.teamworkandteamplay.com/resources.html

Vince, R., & Martin, L. (1993). Inside action learning: An exploration of the psychology and politics of the action learning model. *Management Education & Development, 24*(3), 205–215.

Walters, J.T. (2006). *Outdoor experiential education: A new tool for peacebuilding.* Master's thesis, American University International Peace & Conflict Resolution Program, Washington, DC.

Wilderdom. (N.d.). Index to group activities, games, exercises & initiatives. Retrieved from http://wilderdom.com/games/

Wolpe, H., & McDonald, S. (2013). Democracy and peacebuilding: Re-thinking the conventional wisdom. *The Round Table, 97*(394), 137–145.

Part Two

Development

6

The Developmental Niche for Peace

Darcia Narvaez

A Baseline for a Cooperative and Peaceful Society: Small-Band Hunter-Gatherers

Prior to the spread of agriculture, small-band hunter-gatherer societies (SBHG) were universal. After the spread of agriculture, SBHG continued to exist side by side with settled agricultural communities, indicating a stable social structure (Ingold 1999) and "a persistent and well-adapted way of life" (Lee 1998, p. 61). SBHG conversion to agriculture was generally by force (Gowdy 1998). Lee and Daly (1999, p. 3) defined the hunter-gatherer lifestyle as "subsistence based on hunting of wild animals, gathering of wild plant foods, and fishing, with no domestication of plants, and no domesticated animals except the dog." *Small-band hunter-gatherers* refers to "immediate-return" societies characterized by worldviews and subsistence systems that have few possessions and emphasize quick returns on individual effort (e.g., eating meat from a successfully hunted animal within days instead of drying and saving it for later) (Woodburn 1998). These contrast with delayed-return societies that emphasize waiting for the return of one's invested energy (e.g., harvesting crops several months after planting). Anthropologists have summarized the characteristics of SBHG societies and, unless otherwise noted, their reviews are used here (Ingold 1999; Lee & Daly 1999; Fry 2006).

There seems to be something about SBHG society that makes for a peaceful character and social life. Adulthood and adult-generated society are considered endpoints of development, so it can be worthwhile to examine adult behavior and culture and then work backward for how a society constructs them. Three caveats: (1) It should be noted that in many cases, prior to examination, studied SBHG were displaced from their historic landscapes by European settlements and governments. In the last few centuries, most traditional societies, originally "comparatively free from material pressures" (Marshall, 1961, p. 243, cited by Sahlins, 1998, p. 11), "were selectively stripped by Europeans before reliable report could be made of indigenous production" (Sahlins 1998, p. 11). These events shifted longstanding stable social systems into stressed, disequilibrated systems, which we now know can make people more self-centered. Despite these disequilibria, many SBHG still exhibit the characteristics noted by explorers upon early

Peace Ethology: Behavioral Processes and Systems of Peace, First Edition.
Edited by Peter Verbeek and Benjamin A. Peters.
© 2018 John Wiley & Sons Ltd. Published 2018 by John Wiley & Sons Ltd.

contact. (2) Some illustrations are from hunter-gatherers who are less nomadic or more complex in social structure but who nevertheless exhibit some of the SBHG characteristics. (3) The present tense is used to describe SBHG, even though some of these societies today may be altered by colonization and dominant cultural incursion.

Several social elements of SBHG are highlighted here: social intensity, group fluidity and mobility, personal autonomy, egalitarianism, sharing, concern for public opinion, placefulness, trust, and sustainability (for more detail, see Narvaez 2013, 2014).

Social Intensity

Anthropologists and resident scholars have noted the intense social fabric of SBHG and the prevalence of positive social interactions (e.g., Everett 2009). They enjoy social company and social activities, doing virtually nothing alone. Although anyone can leave at will to go off alone, in most cases band members enjoy being with others virtually all the time (anthropologists complain that they were even followed into the jungle when they relieved themselves; e.g., Dentan 1968). "To them the worst fate that could befall anyone was being left alone and solitary.... Not even monkeys walk alone" (Bowen 1964, pp. 173–174). The Ju/wasi "are extremely dependent emotionally on the sense of belonging and companionship," notably sitting shoulder to shoulder and ankle across ankle in the vast Kalahari (Marshall 1976, p. 287). Touching a friend or loved one releases oxytocin which is related to greater relaxation and prosociality (Zak 2012).

Work is light and made enjoyable through socializing, singing, and joking. Generally only adults do much work (typically only a few hours per day), and even then it is voluntary. Gowdy (1998, p. xv) points out that among the !Kung, people worked between ages 20 and 40 and "spent their abundant leisure time eating, drinking, playing, and socializing – in short, doing the very things we associate with affluence." Such a lifestyle fosters relaxation, and studies show that relaxation (e.g., the relaxation response) is fundamental to good health (Kabat-Zinn 1990). Among SBHG are social practices that we now know calm stress responses and foster the production of bonding hormones, allowing for prosocial emotions to be activated, for example, through practices such as laughter, joking, singing, and dancing, which are commonplace. These social activities benefit immune systems, well-being, and health (e.g., Valentine & Evans 2001; Hayashi *et al.* 2007).

Group Fluidity and Mobility

Mobility is cherished, with constant movement of people in and out of a group; whether at one site or shifting sites, people move, together or apart (Woodburn 1998). In many cases group membership shifts daily, and researchers find that their confidants can disappear without warning, gone for days or weeks at a time (Bowen 1964). When conflicts are not resolved through consensus or compromise, movement away from the conflict often takes place. Disagreeing parties may split the group or take their family members to another group. "Relocating was always preferable to conflict," and sometimes "the stress" of conflict "is too great and the accusations too terrible" (Thomas 2006). Mobility allows for segregation from conflicting parties without economic or social penalty (Woodburn 1998). This is a demonstration of the high autonomy assumed by and given to individuals.

Personal Autonomy

To complement intense sociality, every individual has personal autonomy (which also relates to mobility). No one tells others what to do. Individuals choose themselves what to do (e.g., to go hunting or gathering or not). In fact, some community members may never help with hunting – for which there is no sanction (Sahlins 1972). Psychological research refers to autonomy as a basic human need. For example, when children in modern societiesare are not given meaningful choices in school, they are more likely to misbehave (Deci & Ryan 1985). Among SBHG, parent–child relations are not coercive. For example, when inoculations were being promoted by a health organization, Wolff (2001) reported that a Senoi mother asked her child if he wanted the vaccine, and when he said no, she did not press; that was that. Autonomy is so valued that young children are allowed to play with dangerous objects at will (e.g., sharp spears and axes), and even babies are encouraged to start learning to use them (Hewlett & Lamb 2005).

Some bands place greater emphasis on self-reliance than others. In these cases, individuals are responsible for their own food gathering (as among the Ju/wasi of the Kalahari Desert in present-day Namibia: berries, roots, and slow game): "You find it, you own it. And you carry it" (quoted in Thomas 2006, p. 193). The Piraha of the Amazon did not expect others to help them with their responsibilities, but they also did not change their activities to help someone sick or dying. However, they would come to the aid of others in immediate need (Everett 2009).

Egalitarianism

Though humanity's ancestral primate lineage included hierarchical dominance patterns, the most recent human ancestor was egalitarian, emerging at least 100,000 years ago (Boehm 1999). From SBHG descriptions, all members experience purposeful social egalitarianism (Gowdy 1998; Fry 2006). There are no culturally imposed social hierarchies but instead a collective dominance over inegalitarian behavior that maintains the autonomy of every member. Egalitarianism and personal autonomy are fiercely enforced to the degree that a person who bossed someone else would quickly hear about it. Coercion could break a relationship.

In studies of egalitarianism across species, negative reactions to perceived unequal treatment are common (e.g., Brosnan & de Waal 2014). Wilkinson and Pickett (2009) compiled data across and within nations that show that the more unequal the society is, the worse individuals fare on multiple measures (see Hyslop & Morgan, this volume, Chapter 13).

Sharing

Sharing is a way of life for SBHG, so much so that it is considered good manners and does not require a thank you (van der Post 1961). "Selfishness and favoritism were death" (Marshall quoted in Thomas 2006, p. 243). As Lee (1988, p. 252) noted, when a cabin of hungry Iroquois encounters hungry others, they share what they have "without waiting to be asked, although they expose themselves thereby to the same danger of perishing as those whom they help at their own expense so humanely and with such greatness of soul" (from Lafitau 1724/1974). Sharing is so fundamental that when sharing is not forthcoming, individuals may loudly insist on their claim to a portion of the food that others have acquired (Barnard & Woodburn 1997). Generally, band members cannot build up savings of anything because people will ask for the excess, and one must give it away or experience gossip and complaints. Other animals act this way too, "inviting themselves to the picnic," so to speak, as they scavenge, coming by to get their share of a carcass or attractive foodstuff (Dirkmaat 2012).

Sharing indicates mutual dependence, a form of communalism, and is a means for maintaining equality. Lee (1988) calls the economics of hunter-gatherers *primitive communism*: "A deep-rooted egalitarianism, a deep-rooted commitment to the norm of reciprocity, a deep-rooted desire for ... *communitas* – the sense of community" (p. 268). Lee describes it concretely this way:

> A useful way of looking at primitive communism is to visualize a ceiling of accumulation of goods above which nobody can rise, with the corollary that there is also a floor below which one cannot sink. The ceiling and the floor are dialectically connected; you cannot have one without the other. If there is any food in the camp, everybody in the camp is going to get some of it. (p. 267)

Lee points out, "All theories of justice revolve around these principles, and our sense of outrage at the violation of these norms indicates the depth of its gut-level appeal" (Lee 1988, p. 268), a reaction that our monkey and ape cousins also demonstrate (Brosnan & de Waal 2014).

Concern for Public Opinion

When Darwin (1871/1981) described the development of humanity's moral sense through the tree of life, he identified a concern for public opinion as a fundamental characteristic. Among SBHG, "anything other than peace and harmony in human relations makes the Ju/wasi uneasy" (Marshall quoted in Thomas 2006, p. 193). "They are also extremely dependent emotionally on the sense of belonging and on companionship. Separation and loneliness are unendurable to them" (Marshall 1976, p. 287). In such tightknit societies, concerns about one's reputation and place in the community are effective means of social control of personal behavior. For example, when a person is considered to be in danger of ego inflation or self-aggrandizement, like a hunter credited with a good hunt, ridicule and joking are used to keep his ego in check. In SBHG societies around the world, a similar type of joking is used – a rough humor that includes playful putdowns and teasing, a method of enforcing humility (Lee 1988). "Leveling" comes to the fore when there is a good hunting outcome. The members of the hunting party "insult the meat" by complaining to the hunter credited for bringing it in, denigrating the size or quality of the meat (e.g., "look how small it is – maybe we should find something bigger, like a rabbit"). In fact, the bigger and better the prize, the more insults given. Such leveling is thought to assist in maintaining the values of humility but also sharing. "Public opinion was police, judge and jury, and its expression in ridicule or ostracism made other punishment unnecessary" (Bowen 1964, p. 209). When someone is too aggressive (e.g., killing another) or a repeat offender, he can be expelled.

Placefulness, Trust, and Sustainability

Hunter-gatherers feel like they belong to where they are. They have a "sense of place" on the earth. This includes a sense of fitting in with their environment. They have deep personal knowledge and collective knowledge about local environs, and often laugh at visitors' lack of perception (e.g., "Don't you recognize this bush?").

As immediate-return societies, SBHG live as if there is no need to save foodstuff for a rainy tomorrow as they feast on the food they hunt or gather, eating it up. Though some call it confidence, from a psychological viewpoint I would call this trust – trust in Nature, the natural world, to provide what is needed but also trust in one's capacity to deal with what comes. Many of these bands see the world as a giving place (Bird-David 1990). The Mbuti of the Ituri Forest in Republic of the Congo consider the forest like a mother, telling unborn and young children, "The forest is good, the forest is kind; mother forest, father forest" (Turnbull 1983, p. 34).

Typically, and at least due in part to these beliefs, SBHG live in ways that do not destroy their habitat. Of course, the bands are small. But, like other migrating animals who return to feeding grounds cyclically, they move on before irreversible destruction occurs. They live off renewable flows of resources instead of exhaustible stocks (Gowdy 1998). Native Americans, representing different types of societies shifting over time, describe the "honorable harvest" in which nonhumans are respected by asking permission of plants and animals for their lives, employing principles such as taking no more than half of a plant resource (Kimmerer 2013). These attitudes contrast with those of moderns who have cultural narratives of humans against nature and have anxiety about survival in the future, and who often lack a sense of place (Shepard 1982).

Living sustainably also means playing the "sacred game" of moving between being predator and prey and living in wily ways with predators. There is no immediate inclination to kill predators for human safety as sometimes occurs in modern communities (Thomas 2006). This does not mean people are not wary. Like hunter-gatherers generally do, the Ju/wasi sleep lightly at night with intermittent sleeping during daylight hours, making vigilance easy to carry out.

Sahlins (1998) suggested that hunter-gatherers have limited wants because scarcity is not apparent. Barnard and Woodburn (1997) speak of limited production targets instead of limited wants. That is, the least amount of effort is put out to meet the basic amount of need (they think agriculturalists have to work too hard). Members of agricultural societies consider hunter-gatherers to "not be using" or be "doing nothing" with the land on which they forage and hunt (Thomas 2006).

The characteristics of hunter-gatherer societies are integrated into a whole. Gowdy (1998, p. xiii) refers to "the indissoluble connection between social and environmental harmony." Simple and complex hunter-gatherer societies "were environmentally sustainable because they were egalitarian, and they were egalitarian because they were environmentally sustainable" (p. xiii). Mobility (along with the few possessions that make it possible) makes sustainability more attainable.

Adults' Personalities in Small-Band Hunter-Gatherer Societies

The social practices described thus far in this chapter are rarely found in modern civilizations, which may matter for developing personalities and dispositions. Anthropologists have described personalities among adult members of SBHG societies from around the world as self-controlled and patient, kind and generous, open and flexible, and nonviolent. (Note that we do not really have psychological research data to examine personality differences, so these are generalizations based on behavioral observations reported by explorers and ethnographers.)

Self-Control and Patience

SBHG are commonly perceived to maintain equanimity (unless under the influence of alcohol provided by outsiders). Among the Ju/wasi, self-control of negative feelings was expected, such as not showing hunger, envy, or anger. Sometimes, self-control occurred to a remarkable degree. For example, Thomas (2006) reports on a Ju/wasi girl who stepped on a hyena trap set by a visiting Western biologist, impaling her foot. She stood for hours in pain, alone, in an area with predators, waiting for assistance; when her wound was being dressed, she acted as if it were nothing. Thomas (2006, p. 232) also notes that although the ancient lifestyle started to break down when Western civilization encroached with all its degenerative cultural border aspects (alcohol, disease, poverty, and hunger), they were able to curb their violence: "No matter how provoked, they rarely acted out their discomfort, nor did they vent it on one another.... Not for nothing did they call themselves the Harmless People, the Pure People, when the harm and impurities were jealousy, violence, and anger."

Self-control, or self-regulation, is fundamental for life, for all animals. It represents a biosocial adaptation critical for both physiological and social functioning. Self-regulation facilitates getting along with others in part because it ensures that the focus in social relations is not solely on the self (Eisenberg 2000; Narvaez & Gleason 2013). Built on physiological mechanisms, self-regulation in social terms refers to the capacity to control negative emotions or self-protective mechanisms like fear (e.g., Porges & Carter 2010), and these capacities appear to be especially strong in SBHG.

Self-control extends to eating. SBHG do not necessarily eat every day. In fact, they can go several days without eating (to the consternation of Western visitors). But they take advantage when foodstuffs are available. For example, when Piraha visit the city, they stuff themselves on the first meal and perhaps the second, but by the third meal they complain, saying, "Are we eating again?" (Everett 2009).

Kindness and Generosity

Although culturally expected with community members, generosity extends to outsiders and is usually shown without reluctance. As noted in this chapter, even those with very little share what they have with others. Possessions are shared and not hoarded. The accumulation of objects is not associated with status (Marshall 1961). Among indigenous American groups, generosity is explicitly mentored in the young who are encouraged to practice it, such as by giving away one's favorite possession (Jacobs 1998). The Piraha, though oriented toward self-reliance, helped linguist Dan Everett who was struggling to carry his share of wood through the jungle (to repair Everett's home). Each man originally carried about 50 pounds, so the man who helped Everett carried 100 pounds, without complaint (Everett 2009).

Openness and Flexibility

Groups exhibit flexible lifestyles. For example, sleeping takes place anytime of the day and is more intermittent both day and night (Sahlins 1998). Among the Piraha, naps take up to two hours each at day or night, and there is loud talking all night. Activities can take place at any time (e.g., fishing), so that if a night-fishing expedition is successful everyone gets up to share the meal (Everett 2009).

Hunter-gatherers are known for their non-exclusive, open social systems (Bird-David 1988). Lee (1979) noted the !Kung's conscious attempt to maintain a boundary-less universe. This openness makes them flexible in relations with outsiders, adopting their codes when interacting but not changing their social practices in any fundamental way (Bird-David 1998).

Nonviolence

Typically, there is little aggression among hunter-gatherer adults, and any act of aggression by a young child is quickly interrupted (Thomas 2006). Among the Semai, which are settled hunter-gatherers of the Malay Peninsula, to make someone unhappy by frustrating her desires is to put her at risk of being hurt from having an accident (Dentan 1968). When asked about hitting a child, the Semai say, "How would you feel if it died?" and when asked about hitting another adult, the response is "Suppose he hits you back?" (Dentan 1968, p. 58). Although there is occasional physical violence due to sexual jealousy, the usual response to conflict is discussion to consensus; if that does not work, the group splits up to join or form a different band (Fry 2006).

I have reviewed some of the common characteristics within SBHG, many of which run counter to prevailing views of our "savage" cousins or even views of human nature. What are the mechanisms for building such highly social, egalitarian societies with adults displaying these kinds of peaceful personalities? It begins with childrearing.

A Peacebuilding Nest: The Evolved Developmental Niche

Westerners have noted the pleasant nature of children in SBHG societies. For example, Thomas (2006, p. 200) states, "No culture can ever have raised better, more intelligent, more likable, more confident children." Attributing this to childrearing, she noted that Ju/wasi childhoods were "serene," "pleasant," and "free of hurt and punishment." Indeed, one of the distinctive features of SBHG society is the nest provided for the young.

Intensive parenting initially emerged among social mammals, having evolved to fixation between 30 and 40 million years ago (Konner 2005). Parenting intensified further through human evolution as humans became bipedal and babies had to be born earlier to fit the shrunken birth canal (Trevathan 2011). Humans are born extremely immature (with 25% of their adult brain), with many systems under final development in the first years postnatally and the longest maturational period (over 20 years) of any animal. Thus, intense parenting evolved to match up with the maturational schedule of human young. Based on new insights from neurobiological research, I suspect that early childhood experiences might be a key source of the differences in adult personality between moderns and SBHG (for extensive detail, see Narvaez 2014). Because the child cannot self-regulate at birth, caregivers act as external regulators while physiological systems develop their parameters and thresholds (Montagu 1957; Schore 1996). Attentive care in early life contributes to a species-typical trajectory. "Departure from [these practices] since the end of the hunting-gathering era constitute a discordance and may have psychological and biological consequences that merit further study" (Konner 2005, p. 63). Indeed, among human societies until recent times, there has been great consistency in caregiving, and these practices are associated with social attachment capacities and well-being (Hrdy 2009).

Hunter-gatherer childhoods, as summarized across studied groups, include natural childbirth, 2–5 years of breastfeeding, responsiveness to child needs, nearly constant positive touch and no punishment, multiple adult caregivers, positive social support, and extensive free play (Hewlett & Lamb 2005). Each of these characteristics has been documented to influence child development and adult well-being (Narvaez, Panksepp *et al.* 2013). Recent research in neurobiology shows how influential early caregiving environments are on lasting neurobiological functions, both physiologically and psychologically. Research into each component (including experimental studies with mammals and observational or post-hoc comparison studies with neglected humans) shows that the evolved developmental niche (EDN) fosters physiological, psychological, and social health. For example, epigenetic effects (up or down gene regulation) occur for hundreds of genes in rat pups as a consequence of mother affection in the first few days of life – high nurturance sets up genes for controlling anxiety, whereas low-nurturing care does not, with similar indications in humans (Meaney 2010). Vagal tone (the function of the tenth cranial nerve, the vagus nerve), which maintains the self-regulation of multiple body systems, is trained by responsive parenting – for example, calming baby distress (Porges 2011). Indeed, Kochanska (2002) has done extensive research demonstrating that mutually responsive caregiving leads to children with not only greater self-regulation but also greater empathy, conscience, and prosocial capacities.

The EDN fosters a healthy, well-functioning neurobiology and self-regulation from the first hours of life (see reviews in Narvaez, Panksepp *et al.* 2013; Narvaez *et al.* 2014). My colleagues and I investigate the EDN and its effects on well-being and morality. To put it in a nutshell, even after controlling for demographic variables and maternal responsiveness, maternal EDN practices for 3–5-year-old children are related to more of the good stuff: empathy, conscience, self-control, cooperation, and intelligence, and less of the bad stuff: depression and aggression (for details, see Narvaez, Gleason *et al.* 2013; Narvaez, Wang *et al.* 2013). Furthermore, adults who report EDN-consistent childhoods are more likely to be mentally healthy and have prosocial moral capacities and orientations (Narvaez, Wang & Cheng 2016). It appears that the EDN fosters basic universals in the type of implicit procedural social knowledge that guides thought, emotion, and action. In other words, after experiencing the EDN in childhood, the uniqueness of a particular culture is the frosting on the cake of a common human nature.

Human Peacefulness

Until recently in human evolutionary history, the EDN was universally experienced as there were no alternatives (e.g., no infant formula, and no daycare centers or schools where children are sent away from loved ones). The contention here is that the EDN for young children occurs during a time when thresholds and physiological and psychological habits for the social life are established. Mammalian social-emotional needs for touch and responsive care are vital not only in the first years of life for structuring body systems that are developed from calming touch (facilitating calming hormones like oxytocin), but throughout life. Indeed, from observational reports of SBHG, affectionate touch among all ages is pervasive, a practice that would facilitate the flow of calming hormones like oxytocin (Narvaez 2014). The peaceful, prosocial adult personalities of SBHG appear to be biosocial outcomes of the EDN in early life and beyond. The EDN evolved to foster not only optimal physiological development but also the psychosocial underpinnings of human sociality, a key to human adaptation. Early life is when the implicit social procedural knowledge is established that guides future behavior and forms the basis for lifelong sociality. In a way, EDN-consistent early care provides a "cultural commons" for the development of peaceable character.

The sociomoral developmental core that develops in SBHG environments includes social effectivity (effectiveness plus self-efficacy) and a communal autonomy. Psychologically, these are dispositions that contribute to collaborative morality, which represents human nature's higher capacities. First, because the individual is grounded in intensely supportive community experience from the beginning of life, fundamental skills for basic sociality are developed, such as emotional presence and interpersonal responsiveness (intersubjectivity). Consistent, regular practice builds a sense of social efficacy as well as pleasurable sensations for these experiences, resulting in social effectiveness. All these experiences contribute to procedural emotional intelligence – automatic social knowhow or tacit knowledge. These roots are established implicitly by the way that the infant is treated in early life when brain networks and systems are established. Second, the individual develops deep personal knowledge about how her actions affect others (autonomy space). When the surges for autonomy occur (e.g., the "terrible twos" in North America but not terrible in most cultures), the individual seeks to test his or her capacities. Within the close community of the SBHG, these surges are shaped by prosocial guidance from adults and older children in the community. So, if a child takes a stick and aims it at someone, that person makes a game of it, laughing it off and redirecting the aggression. When children are raised with these experiences, they turn into adults who create a cooperative society that continues raising children in this manner.

We can see two moral inheritances that emerge from empathic effectivity and a communal autonomy space. These underlie inherited moral mindsets from which one takes action. One is what I call the *engagement ethic*, built on the types of experiences that SBHG adults provide their children. Loving care provides extensive experiences of intersubjectivity and related types of experiences (as noted in this chapter). These are critical for establishing conscious and unconscious capacities for relational communication and connection. An active engagement ethic entails being emotionally present with the Other, co-constructing the moment together in a playful dance. From all accounts, SBHG spend most of their social time in this orientation. A second ethical mindset, communal imagination, is the capacity to deliberate and imagine alternate possibilities within an inclusive framework. Imagination capacities generally rely on later-evolved brain areas such as executive functions in the prefrontal cortex that allow for stepping outside the present moment to consider alternative possibilities, a capacity that may be one of humanity's greatest gifts. Good care during the early sensitive period facilitates strong linkages between executive functions in the prefrontal cortex and control of self-preservational systems in evolutionarily older parts of the brain (Schore 2003a, 2003b). Then, when imagination is deployed, it automatically fits the parameters of a communal autonomy space. These capacities contribute to a peaceable nature and thus reflect a species-typical outcome. However, deficiencies in early caregiving from a community of adults can lead to species-atypical outcomes.

What Happens When the EDN Is Missing or Degraded?

In the last 1% of human evolutionary history (the last 10,000 years or so), humans have forgotten or rejected their evolutionary history and dismantled the EDN for young children (and especially in the last 200 years or so with medicalized birth, formula feeding, stranger daycare, etc.). Childrearing practices that do not match the evolved baseline can be called *undercare*. (The term *undercare* is used to distinguish it from *neglect* as understood in the legal and medical communities; neglect does not take into account all evolved caregiving practices.) *Undercare* specifically refers to the absence of one or more of the caregiving practices characteristic of the EDN identified thus far. Undercare is to care as malnourishment is to nourishment. Malnourishment in the extreme is starvation. Undercare in the extreme is trauma and abuse. But there are many degrees in between nourishment and malnourishment, and between full care and undercare. It does not take that much violation of the evolved early caregiving environment to have long-term effects because undercare causes stress (e.g., lack of touch increases stress hormones), which in early life can be toxic to brain development (Champagne 2014). Although it might not be as visible, undercare leads to various suboptimal outcomes such as autoimmune and emotional disorders. The stress response, whose parameters and thresholds are established early in life, can be set to be hyperreactive (Lupien *et al.* 2009). Undercare promotes a "stressed brain" with poor social brain networks that are otherwise developed with good care, underpinning moral capacities.

Without the EDN, human nature appears to misdevelop in multiple ways (toward aggression and self-centeredness) depending on the type, duration, intensity, and timing of undercare. Extensive research with animals and humans demonstrates that lifelong deficits can occur when early experience is suboptimal (e.g., Narvaez, Panksepp *et al.* 2013). If a child does not receive intensive social support during sensitive periods when brain and body systems are established, the foundations for health and well-being are undermined. Instead of developing prosocial networks and self-regulation from companionship care, the child develops a stress-reactive physiology sensitized to personal distress with relative inflexibility. Stress reactivity undermines not only higher order thinking and capacities for intimacy but also empathic response and communal imagination (Narvaez 2014). When things go wrong in early life, brain and body systems develop suboptimally, and the submoral components previously described are misaligned. Systems governing sociality are misformed. For example, the right hemisphere does not receive the experience it expects during the period when it is developing rapidly, undermining the development of neural networks critical for physiological self-regulation (e.g., the vagus nerve, which influences all systems of the body; Porges 2011) and for sociality (Schore 2001a, 2001b). These misdevelopments undermine empathy and imagination, propelling the individual to rely on rigid scripts and ideologies for self-calming instead of a set of flexible social skills developed in a species-typical early life.

Within cultures where the EDN is frayed or absent, during autonomy surges a child may not be surrounded by a close, prosocial community to shape autonomy. Worse, the child may be (mis)guided by aggressive adults. In these cases, through punishment, disrespect, or encouragement, the child will be caught in maintaining aggressive selfishness, growing an egocentric autonomy space. Instead of curtailing energies to account for the needs of others, self-aggrandizing energies will be encouraged without sensitivity to their destructive power toward relationships, peoples, or other species.

Without appropriate care in early life, mammals can grow up with erratic biological systems that are easily thrown into disarray when unpredictable things happen. When the EDN is inadequate, social capacities, which depend on early experience, are underdeveloped and innate survival systems are sensitized to protect the individual from harm. And when there are impaired capacities to self-regulate, the stress response will persist. The personality may grow to rely on survival systems that become easily activated under perceived threat. Self-protection can become the default system for social life. Self-protectionist ethics rely on the survival systems present at birth: mammalian emotions systems of fear, rage, and panic. The stress response is related to the functioning of these systems, so when the stress response becomes habitual, these primitive systems will dominate personality (also as a result of the right hemisphere and prefrontal controls being underdeveloped). The primitive systems are rigid, so the individual will demonstrate social inflexibility, lack of social creativity, and a reliance on routines and precedent.

Besides a general safety orientation to the social life that seeks to stay safely in control, there are two subtypes of this basic safety ethic that can operate "in the moment." One is anger-based and aggressive (combative), where one feels enough strength and power to take action against the threat (one-up). In fact, with a dispositional combative safety mindset, one feels less than adequate unless one is dominant; hence, the bulldoggedness of some personalities in the face of challenge. This externalizing, or pushing away of oth-ers with hostility or aggression, can become habitual in social situations as a learned form of self-regulation. The other safety subtype is fear-based appeasement based in a dissoci-ated state (emotional detachment from the immediate situation); the individual is cut off from external and internal stimuli. In this case, one feels paralyzed or too weak to take action and so withdraws physically and/or emotionally. Energy is internalized toward anxiety and depression. This, too, can become habitual in social situations as a way to cope in a perceived hostile environment. How much one resorts to using these innate instincts for self-protection and rigidity in moral decisions and actions is initiated during the preverbal years of life (although one can revamp one's orientations to a degree later; Gilbert 2010; Narvaez 2014). The early years have lasting effects, affecting one's imagina-tion and cognitive sensitivity to others. In our research with adults, my colleagues and I have found that EDN-consistent childhood experiences are related to more secure attach-ment (biosocially comfortable and skilled sociality) and better mental health, perspective taking, and moral engagement. In contrast, EDN-inconsistent experiences are related to insecure attachment (biosocial discomfort and awkwardness with sociality), worse men-tal health, personal distress, and self-protective morality (see Narvaez *et al.* 2016).

Imagination can take different forms, and the forms that become habitual can reflect early experience as well (Schore 2001b). As noted in this chapter, communal imagination is our heritage, formed in supportive environments. But in species-atypical environments, imagination can partner with self-protective concerns. When early needs are thwarted, imagination may develop in self-protectionist directions. When the stress response is activated from a perceived threat, which becomes habitual in the ill-formed imagination, imagination can be hijacked for self-aggrandizement or revenge. Angry and aggressive emotions underlie combative morality that, when enhanced with imagination capacities, fuels vicious imagination and planned social aggression. Anxiety is directed toward grasping or controlling the Other in an ongoing, ranking, "us-versus-them" orientation. Because of heightened fear and panic (overt or covert), the individual will demonstrate intolerance of perceived outgroup members and fear of contamination; such individuals are unable to live in the present moment with the Other, whom they quickly categorize. Ideologies that tout the superiority of one's group are attractive ("safety nests") and reinforce insensitive beliefs and behavior, characteristics of ethnocentric monoculturalism (Sue *et al.* 1999).

The other form of safety-based imagination is emotionally detached and dissociated. Emotional distancing and fearful orientation underlie a withdrawing morality that, when enhanced with imagination capacities, becomes a *detached imagination*, morally disengaged (Bandura 1999) and distanced from affiliative emotions and consequences of actions. Detached imagination is encouraged in emotionally desiccated environments (e.g., babies left to cry and sleep alone with little touch) that perhaps offer some support but not enough to promote empathy and social intimacy, supporting instead narcissism and moral egoism (the sense of disconnection to others, human and nonhuman, and an emphasis on self-reliance and defiance), which have gained traction in recent centuries among civilized peoples. Many Western cultural narratives fuel detached imagination. Moral disengagement, something expected in Western science, economics, and business, can lead to great harm from taking actions without concern for long-term consequences, lacking deep sociality (empathic effectivity roots) and a broad cooperative autonomy with all of Life. Instead, one dismisses responsibility or concern for those deemed inferior – whether other human groups, animals, plants, or whole ecosystems – and uses them for one's own ends.

Inegalitarianism is at the heart of safety-based imagination, usually valuing one set of humans over another. Safety-based imagination is governed by past conditioning of distrust of one's softer emotions and distrust of the goodwill of others, making receptivity to and emotional presence with others difficult. Prior experience is imposed on the present. It is detached from present reality and uses categorizations and stereotyping as a replacement for flexible adaptation, for example by employing rigid scripts for social interactions instead of flexible adaptedness and the co-creation of the relationship in the moment. Categorizing others based on a worldview or ideology engenders miscalculations, illusions that increase hostility and the chance of conflict (Narvaez 2014). Detached and vicious imaginations can lead to war and environmental degradation but also to impositional altruism, imposing one's will on others "for their own good." Clearly, these are characteristics not tolerated or cultivated in SBHG.

In modern cultures, the only SBHG social characteristic that is supported is personal autonomy but only to a limited degree as individuals have chains of obligations put on them from which they cannot escape into the bush or jungle. The rest of the characteristics have been eliminated with the population surges needed for agriculture (decreasing social intensity, group fluidity, and mobility), the hierarchies that developed from unequal distribution of goods (undermining egalitarianism and sharing), and agriculture itself, which alienates humanity from Nature (sense of place, trust in nature, and sustainable living).

What Can We Do Today?

Whereas evolution has set us up for a "moral sense" (Darwin 1871/1981), early experience, in very deep neurobiological ways, influences what *type* of moral sense develops. Even our imaginations are rooted in biology and are shaped by social experience (Emde *et al.* 1991). Although humans evolved to be prepared for communal morality, we are realizing now that its roots must be cultivated carefully during sensitive periods (such as the first few years of life) and maintained with social support. Caregivers who follow the EDN promote pleasurable social experience. Extensive joyful interaction promotes brain development on all levels (neurochemical, circuitry, and integration; Schore 2001a). Caregiving environments that match up with human evolved needs shape dispositions for humanity's fullest moral capacities, what I call the *ethics of engagement* and *communal imagination*. But these prepared inheritances appear to be epigenetic and plastically dependent on caregiving practices that evolved to match the maturational schedule of the baby.

Nevertheless, humans may be distinctive from other animals for their capacity to self-author or self-transform. Individuals can remake themselves, and so can communities and cultures (e.g., Norway, once violent, became a peaceful nation; Fry 2006). One of humanity's gifts is the ability to transcend the present moment and imagine possibility. With the ability to make narratives of meaning that govern action choices, humans can purposefully shift consciousness, moving to a greater awareness of the Whole.

Gowdy (1998) suggested that, from an economic perspective, the modern world could learn from the social and ecological harmony of hunter-gatherer societies (whom he called, after Sahlins [1972], "uneconomic man"). Specifically, modern societies could organize around "social security; living off renewable flows rather than exhaustible stocks; sexual equality; cultural and ecological diversity based on bioregionalism; and social rather than private capital" (Gowdy 1998).

From a developmental psychological perspective, here are three things that adults can do. First, redirect societal energies to restore the EDN for the raising of the young. Denying babies what they evolved to need risks building self-centered adults and societies that are destructive to life on the planet. Parents often treat their children the way they were treated. (In animal studies, a low-nurturing mother breeds a daughter who is even worse as a mother, and so parenting can spiral downward over generations; Meaney 2010.) Of course, what one believes to be appropriate childrearing depends on what one believes a child's nature to be and what entails a good adult. And what we think about human moral nature and its possibilities is highly influenced by our cultural inheritances, the culture's stories about who we are, as well as beliefs and attitudes about the etiology and development of morality. All these influence how adults treat children during sensitive periods when the biopsychosocial self, including moral personality, is being formed. Because these behaviors are guided by belief about human beings in general, it is important to understand humanity's heritages: that we are naturally cooperative under species-typical environments and that our intelligence can be much broader and include receptive intelligence toward nonhumans.

Second, and concretely, we can help families and communities "get the neurobiology right" in their child raising. This requires educating all societal members not only about the evolved developmental niche but also about child development; understanding the brain and its functions, including the stress response, mindfulness, and positive moods;

and understanding how environments influence worldview, thoughts, and behavior (Narvaez 2014). This requires that hospitals become baby-friendly, an initiative of the World Health Organization (WHO & UNICEF 2009) (e.g., no separation of baby from mother, and no feeding of formula or sugar water), and provide soothing introductions to the world (no pain-inducing procedures that are unnecessary or can wait). It also requires workplaces and communities to be reorganized to emphasize optimal caregiving in early life, including such things as paid parental leave for a year or more as most countries provide, and arranging workplaces so that babies can be brought to work (when well cared for, they do not cry) and young children can be cared for in quality workplace daycare.

Third, alter social contexts not only for babies but for everyone else. Immersing oneself and one another in socially supportive environments primes good moods and prosocial values. Psychological research provides guidance on how to change the schemas and filters we use to understand life events (Narvaez 2005). Contexts prime our emotions, actions, and imaginations (Bargh 1989; Narvaez & Lapsley 2005). Intense repeated experience builds schemas (generalized knowledge structures) that guide perception, interpretation, and behavior (Bargh 1989; Narvaez & Lapsley 2005). Making stories about and practicing cooperation (instead of competition or aggression) routinely will ensure that they become automatic filters and frames for everyday experience (Hogarth 2001; see also Ali & Walters, this volume, Chapter 5). This means altering social discourse that emphasizes competition as natural over cooperation, toughness over empathy, and so on. Self-aggrandizing vicious imagination should be nipped in the bud. Detached imagination should be used for a limited set of situations. Always consult the wisdom of those who see the big picture over multiple generations. Attention should be routinely drawn to communal values, and the identity narratives we tell one another should be those that foster an inclusive communal imagination (e.g., no us-against-them rhetoric but inclusive we-members-of-Earth-community discourse).

We need to enfold attitudes and skills regarding conflict resolution and compassionate behavior into everyday discourse and action at home, school, and work. We need to foster skills for self-calming and help one another learn to control the stress response and practice relational attunement (Narvaez 2014). We can avoid putting children and ourselves into desensitizing experiences, such as violent media. We can adopt a developmental approach to people's misbehavior under age 30, using functionally appropriate forms of gossip and leveling to suppress a safety ethic in each other. We can make empathy and openness part of our expectations for leaders and parents.

In short, instead of accepting slipping baselines for childrearing, human potential, and normative social life, we can challenge them with evidence, and practice alternative ways to be and live.

Conclusion

We can see how the EDN can be analyzed at multiple levels beyond developmental issues, the focus of the chapter: In terms of immediate causation, emotion systems must be well developed to function optimally. Social life requires a flexibly responsive set of emotion systems and their regulation. With poor development during early life, stress reactivity can become built in, undermining executive controls and impairing relational attunement and receptive intelligence; one will be attracted to ideologies of aggression or withdrawal, domination or submission (Tomkins 1965). In terms of function, when humans fulfill their human essence, which is a peaceful nature built epigenetically and maintained through social support, they can live mindfully and sustainably like other animals in natural conditions. In terms of evolution, barring the need to employ survival systems to survive (eat, reproduce, and react to threat cues), peaceful processes are predominant in nature. Species coexist and mutually help one another (Margulis 1998).

Recent neurobiological research has confirmed what ancestral societies knew intuitively from generations of practice and attention – that meeting a baby's needs with EDN-consistent practices fosters a well-functioning brain, body, and person. We understand now that this occurs through epigenetic and plasticity effects during the time period when thresholds and system parameters are being established. The body/brain has to work well for the person to develop optimal morality. One can imagine that a child who was too anxious, stress reactive, or aggressive would have made a difficult member of the community and would not have survived in SBHG communities. Cooperation was of utmost importance in a physically challenging environment. When the EDN is not provided (undercare), empathy can be undermined and autonomy uncontrolled by empathy. Nevertheless, cultures and individuals can revamp their capacities for peaceable existence with immersion in environments and activities that foster intuitions for engagement and communal imagination.

References

Bandura, A. (1999). Moral disengagement in the perpetration of inhumanities. *Personality and Social Psychology Review*, *3*(3), 269–275.

Bargh, J.A. (1989). Conditional automaticity: Varieties of automatic influence in social perception and cognition. In J.S. Uleman & J.A. Bargh (Eds.), *Unintended thought* (pp. 3–51). New York: Guilford.

Barnard, A., & Woodburn, J. (1997). Property, power and ideology in hunter-gathering societies: An introduction. In T. Ingold, D. Riches, & J. Woodburn (Eds.), *Hunters and gatherers, Vol. 2: Property, power and ideology* (2nd ed., pp. 4–31). Oxford: Berg.

Bird-David, N. (1990). The giving environment: Another perspective on the economic system of hunter-gatherers. *Current Anthropology*, *31*, 189–196.

Bird-David, N. (1998). Beyond "The original affluent society": A culturalist reformulation. In J. Gowdy (Ed.), *Limited wants, unlimited means: A reader on hunter-gatherer economics and the environment* (pp. 115–138). Washington, DC: Island Press.

Boehm, C. (1999). *Hierarchy in the forest: The evolution of egalitarian behavior.* Cambridge, MA: Harvard University Press.

Bowen, E.S. (1964). *Return to laughter.* New York: Doubleday Anchor.

Brosnan, S.F., & de Waal, F.B.M. (2014). Evolution of responses to (un)fairness. *Science*, *346*(6207). doi:10.1126/science.1251776

Champagne, F. (2014). The epigenetics of mammalian parenting. In D. Narvaez, K. Valentino, A. Fuentes, J. McKenna, & P. Gray (Eds.), *Ancestral landscapes in human evolution: Culture, childrearing and social wellbeing* (pp. 18–37). New York: Oxford University Press.

Darwin, C. (1981). *The descent of man.* Princeton, NJ: Princeton University Press. (Original work published 1871)

Deci, E.L., & Ryan, R.M. (1985). *Intrinsic motivation and self-determination in human behavior.* New York: Plenum Publishing.

Dentan, R.K. (1968). *The Semai: A nonviolent people of Malaya.* New York: Harcourt Brace College Publishers.

Dirkmaat, D. (Ed.). (2012). *A companion to forensic anthropology.* New York: Wiley-Blackwell.

Eisenberg, N. (2000). Emotion, regulation, and moral development. *Annual Review of Psychology, 51,* 665–697.

Emde, R.N., Biringen, Z., Clyman, R., & Oppenheim, D. (1991). The moral self of infancy: Affective core and procedural knowledge. *Developmental Review, 11,* 251–270.

Everett, D. (2009). *Don't sleep, there are snakes: Life and language in the Amazonian jungle.* New York: Vintage.

Fry, D.P. (2006). *The human potential for peace: An anthropological challenge to assumptions about war and violence.* New York: Oxford University Press.

Gilbert, P. (2010). *Compassion-focused therapy.* London: Routledge.

Gowdy, J. (Ed.). (1998). *Limited wants, unlimited means: A reader on hunter-gatherer economics and the environment.* Washington, DC: Island Press.

Hayashi, T., Tsujii, S., Iburi, T., Tamanaha, T., Yamagami, K., Ishibashi, R., Hori, M., Sakamoto, S., Ishii, H., & Murakami, K. (2007). Laughter up-regulates the genes related to NK cell activity in diabetes. *Biomedical Research, 28*(6), 281–285.

Hewlett, B.S., & Lamb, M.E. (2005). *Hunter-gatherer childhoods: Evolutionary, developmental and cultural perspectives.* New Brunswick, NJ: Aldine.

Hogarth, R.M. (2001). *Educating intuition.* Chicago: University of Chicago Press.

Hrdy, S. (2009). *Mothers and others: The evolutionary origins of mutual understanding.* Cambridge, MA: Belknap Press.

Ingold, T. (1999). On the social relations of the hunter-gatherer band. In R.B. Lee & R. Daly (Eds.), *The Cambridge encyclopedia of hunters and gatherers* (pp. 399–410). New York: Cambridge University Press.

Jacobs, D.T. (1998). *Primal awareness.* Rochester, VT: Inner Traditions.

Kabat-Zinn, J. (1990). *Full catastrophe living: Using the wisdom of your body and mind to face stress, pain, and illness.* New York: Delta.

Kimmerer, R.W. (2013). *Braiding sweetgrass: Indigenous wisdom, scientific knowledge and the teachings of plants.* Minneapolis, MN: Milkweed Editions.

Kochanska, G. (2002). Mutually responsive orientation between mothers and their young children: A context for the early development of conscience. *Current Directions in Psychological Science, 11*(6), 191–195.

Konner, M. (2005). Hunter-gatherer infancy and childhood: The !Kung and others. In B. Hewlett & M. Lamb (Eds.), *Hunter-gatherer childhoods: Evolutionary, developmental and cultural perspectives* (pp. 19–64). New Brunswick, NJ: Transaction.

Lafitau, J.-F. (1974). *Custom of the American Indian* (Vol. 1, Trans. and eds. W. Fenton & E. Moore). Toronto: The Champlain Society. (Original work published 1724)

Lee, R.B. (1979). *The !Kung San: Men, women, and work in a foraging society.* Cambridge: Cambridge University Press.

Lee, R. (1988). Reflections on primitive communism. In T. Ingold, D. Riches, & J. Woodburn (Eds.), *Hunter and gatherers, Vol. 1: History, evolution and social change* (pp. 252–268). Oxford: Berg.

Lee, R. (1998). What hunters do for a living, or, how to make out on scarce resources. In J. Gowdy (Ed.), *Limited wants, unlimited means: A reader on hunter-gatherer economics and the environment* (pp. 43–64). Washington, DC: Island Press.

Lee, R.B., & Daly, R. (Eds.). (1999). *The Cambridge encyclopedia of hunters and gatherers.* New York: Cambridge University Press.

Lupien, S.J., McEwen, B.S., Gunnar, M.R., & Heim, C. (2009). Effects of stress throughout the lifespan on the brain, behaviour and cognition, *Nature Reviews Neuroscience, 10*(6), 434–445.

Margulis, L. (1998). *Symbiotic planet: A new look at evolution.* Amherst, MA: Sciencewriters.

Marshall, L. (1961). Sharing, taking, and giving: Relief of social tensions among !Kung Bushmen. *Africa, 31,* 231–249.

Marshall, L. (1976). *The !Kung of Nyae Nyae.* Cambridge, MA: Harvard University Press.

Meaney, M.J. (2010). Epigenetics and the biological definition of gene X environment interactions. *Child Development, 81*(1), 41–79.

Montagu, A. (1957). *Anthropology and human nature.* New York: MacMillan.

Narvaez, D. (2005). The Neo-Kohlbergian tradition and beyond: Schemas, expertise and character. In G. Carlo & C. Pope-Edwards (Eds.), *Nebraska Symposium on Motivation, Vol. 51: Moral motivation through the lifespan* (pp. 119–163). Lincoln: University of Nebraska Press.

Narvaez, D. (2013). Development and socialization within an evolutionary context: Growing up to become "a good and useful human being." In D. Fry (Ed.), *War, peace and human nature: The convergence of evolutionary and cultural views* (pp. 341–358). New York: Oxford University Press.

Narvaez, D. (2014). *Neurobiology and the development of human morality: Evolution, culture and wisdom.* New York: W.W. Norton.

Narvaez, D., & Gleason, T. (2013). Developmental optimization. In D. Narvaez, J. Panksepp, A. Schore, & T. Gleason (Eds.), *Evolution, early experience and human development: From research to practice and policy* (pp. 307–325). New York: Oxford University Press.

Narvaez, D., Gleason, T., Wang, L., Brooks, J., Lefever, J., Cheng, A., & Centers for the Prevention of Child Neglect. (2013a). The evolved development niche: Longitudinal effects of caregiving practices on early childhood psychosocial development. *Early Childhood Research Quarterly, 28*(4), 759–773.

Narvaez, D., & Lapsley, D.K. (2005). The psychological foundations of everyday morality and moral expertise. In D.K. Lapsley & C. Power (Eds.), *Character psychology and character education* (pp. 140–165). Notre Dame, IN: University of Notre Dame Press.

Narvaez, D., Panksepp, J., Schore, A., & Gleason, T. (Eds.). (2013b). *Evolution, early experience and human development: From research to practice and policy.* New York: Oxford University Press.

Narvaez, D., Valentino, K., Fuentes, A., McKenna, J., & Gray, P. (2014). *Ancestral landscapes in human evolution: Culture, childrearing and social wellbeing.* New York: Oxford University Press.

Narvaez, D., Wang, L, & Cheng, A. (2016). Evolved Developmental Niche History: Relation to adult psychopathology and morality. *Applied Developmental Science, 4,* 294–309. http://dx.doi.org/10.1080/10888691.2015.1128835

Narvaez, D., Wang, L., Gleason, T., Cheng, A., Lefever, J., & Deng, L. (2013c). The evolved developmental niche and sociomoral outcomes in Chinese three-year-olds. *European Journal of Developmental Psychology, 10*(2), 106–127.

Porges, S.W. (2011). *The polyvagal theory: Neurophysiological foundations of emotions, attachment, communication, self-regulation.* New York: Norton.

Porges, S.W., & Carter, C. (2010). Neurobiological bases of social behavior across the life span. In M.E. Lamb, A.M. Freund, & R.M. Lerner (Eds.), *The handbook of life-span development. Vol. 2: Social and emotional development* (pp. 9–50). Hoboken, NJ: John Wiley & Sons.

Sahlins, M. (1972). *Stone age economics.* New York: Aldine de Gruyter.

Sahlins, M. (1998). The original affluent society. In J. Gowdy (Ed.), *Limited wants, unlimited means: A reader on hunter-gatherer economics and the environment* (pp. 5–42). Washington, DC: Island Press.

Schore, A.N. (1996). The experience-dependent maturation of a regulatory system in the orbital prefrontal cortex and the origin of developmental psychopathology. *Developmental Psychopathology, 8,* 59–87.

Schore, A.N. (2001a). Effects of a secure attachment relationship on right brain development, affect regulation, and infant mental health. *Infant Mental Health Journal, 22,* 7–66.

Schore, A.N. (2001b). The effects of early relational trauma on right brain development, affect regulation, and infant mental health. *Infant Mental Health Journal, 22,* 201–269.

Schore, A.N. (2003a). *Affect dysregulation & disorders of the self.* New York, NY: Norton.

Schore, A.N. (2003b). *Affect regulation and the repair of the self.* New York, NY: Norton.

Shepard, P. (1982). *Nature and madness.* Athens: University of Georgia Press.

Sue, D.W., Bingham, R.P., Porche-Burke, L., & Vasquez, M. (1999). The diversification of psychology: A multicultural revolution. *American Psychologist, 54,* 1061–1069.

Thomas, E.M. (2006). *The old way: A story of the first people.* New York: Picador.

Tomkins, S. (1965). Affect and the psychology of knowledge. In S.S. Tomkins & C.E. Izard (Eds.), *Affect, cognition, and personality.* New York: Springer.

Trevathan, W.R. (2011). *Human birth: An evolutionary perspective* (2nd ed.). New York: Aldine de Gruyter.

Turnbull, C.M. (1983). *The human cycle.* New York: Simon and Schuster.

Valentine, E., & Evans, C. (2001), The effects of solo singing, choral singing and swimming on mood and physiological indices. *British Journal of Medical Psychology, 74,* 115–120.

Van der Post, L. (1961). *The heart of the hunter: Customs and myths of the African Bushman.* San Diego, CA: Harvest/Harcourt Brace & Co.

Wilkinson, R., & Pickett, K. (2009). *The spirit level: Why greater equality makes societies stronger.* New York: Bloomsbury.

Wolff, R. (2001). *Original wisdom.* Rochester, VT: Inner Traditions.

Woodburn, J. (1998). Egalitarian societies. In J. Gowdy (Ed.), *Limited wants, unlimited means: A reader on hunter-gatherer economics and the environment* (pp. 87–110). Washington, DC: Island Press.

World Health Organization (WHO) & UNICEF. (2009). *Baby-Friendly Hospital Initiative: Revised, updated and expanded for integrated care.* Geneva: World Health Organization. Available from http://www.who.int/nutrition/publications/infantfeeding/bfhi_trainingcourse/en/

Zak, P. (2012). *The moral molecule: The source of love and prosperity.* New York: Dutton Adult.

7

Children's Peacekeeping and Peacemaking
Cary J. Roseth

Children, like all social species, must balance their need for personal and social resources (Roseth *et al.* 2011). After all, the benefit of controlling a toy likely decreases when depriving peers results in there being nobody left to play with (Vaughn & Santos 2007). Social competence therefore requires the development of conflict management skills that maintain peaceful relations (*peacekeeping*) as well as repair peer relationships after conflict occurs (*peacemaking*).

This chapter provides a brief overview of contemporary understanding of the group- and individual-level processes by which children keep and maintain peaceful relations. I bring an ethological perspective to the study of children's peacekeeping and peacemaking, which means that I connect developmental questions to proximal considerations of what makes these processes occur (causation) and distal considerations of how they influence survival and reproductive success (ultimate function). Because this perspective differs in important ways from developmental psychology's traditional approach, I begin the chapter by briefly contrasting the two perspectives. I then review the empirical literature on children's peacekeeping and peacemaking, detailing what is known and unknown about the development of these processes and their capacity for change. Topics include the strategies that children use to control resources, social dominance, reconciliation, and third-party intervention (e.g., peers and teachers). I focus on children's peacekeeping and peacemaking with peers and refer interested readers to previous reviews about conflict management with siblings and parents (e.g., Verbeek *et al.* 2000; Ross *et al.* 2006).

Theoretical Roots

Historically, developmental psychologists have classified children's behavior in terms of *social desirability*, or the extent to which a given behavior enables children to become integrated into society (Pellegrini 2008; Hawley 2014; Roseth 2016). As such, developmental psychologists have emphasized intent, form, and function to identify behavioral subtypes. For example, the term *prosocial*, defined as "voluntary behavior intended to benefit another" (Eisenberg *et al.* 2006, p. 646), is used to describe behaviors

Peace Ethology: Behavioral Processes and Systems of Peace, First Edition.
Edited by Peter Verbeek and Benjamin A. Peters.
© 2018 John Wiley & Sons Ltd. Published 2018 by John Wiley & Sons Ltd.

such as cooperation, helping, and sharing. The assumption is that prosocial behavior enhances social relations and benefits society and therefore represents an adaptive response to optimal ecological conditions (Eisenberg *et al.* 2006). Likewise, *aggression* is labeled an "antisocial" behavior and defined as "behavior aimed at harming or injuring another person or persons" (Dodge *et al.* 2006, p. 722; see also Parke & Slaby 1983). The assumption is that aggression damages social relations and imposes costs on society and therefore represents a maladaptive response to suboptimal conditions (Dodge *et al.* 2006).

An ethological perspective begins with different assumptions (Bateson 2015). For starters, an ethological perspective does *not* make any assumptions about what counts as socially desirable behavior, or what counts as an adaptive or maladaptive response to different ecological conditions. Instead, guided by evolutionary theory, ethology begins with the assumption that social conflict is inevitable where resources are limited and individuals differ in need, and consequently natural selection should have favored processes that limit the costs of conflict or repair damage to social relationships afterward (Aureli & de Waal 2000). What motivates behavioral processes (e.g., intent) is therefore secondary to their ultimate function.

An ethological perspective also makes no assumptions about how form relates to proximal function. Because different ecologies make different demands on different species, ethologists assume that species have evolved their own ways of meeting particular demands within particular ecologies. A behavior's form and function therefore cannot be understood in isolation from its natural social ecology. This volume's definition of aggression as a "behavior through which individuals, families, groups, communities, or nations pursue active control of resources and the social environment at the expense of others" reflects this perspective. What defines aggression is not its form, its intent, or the suboptimal conditions that triggered it, but its function (to control resources) and social cost (expense of others). What differentiates typical from atypical is the extent to which a behavior is exhibited by a species within its ecology. Aggression commonly shown by members of a species is context-dependent and species-typical, while uncommon aggressive behavior (e.g., extreme violence) is considered species-atypical. What differentiates is species typicality, not the form of behavior or social judgments about what does and does not benefit society.

An ethological perspective does not imply that age-related change depends solely on biology, nor that biology predetermines developmental outcomes. It is also *not* the case that an evolutionary perspective assumes that biology and the environment can be cleanly separated (Bateson 2015). While this dichotomous, nature–nurture view was popular in early ethology (e.g., Tinbergen 1951), we now understand that ontogeny is shaped dynamically by continuous *trans*actions between biology and experience (Shonkoff & Phillips 2000). This is relevant to the question of when peace processes first emerge in children's behavioral repertoire. Human infants may be biologically prepared for peace processes, but the form these take will depend on experience so that they fit within the social ecology in which they emerge. From an ethological perspective, an important issue is the extent to which peace processes can change in response to different environmental conditions. While natural selection should have favored ontogenetic plasticity, the extent to which processes can change should also be constrained by their evolutionary function.

In summary, developmental psychology emphasizes the social desirability of children's behavior, assuming, for example, that "aggressive behavior during conflict represents a mismanaged conflict, sometimes a destructive one" (Shantz & Hartup 1992, p. 4), and, as such, is inimical to successful social development. Developmental psychology also tends to emphasize behaviors' intent, form, and proximal function, the assumption being that behavioral subtypes reflect different developmental precursors, pathways, and outcomes. An ethological perspective challenges these assumptions by proposing that behavior cannot be understood in isolation of the particular ecology in which a particular species is embedded. For ethologists, ontogeny and proximal function are also linked to ultimate function. As we shall see in this chapter's review of literature about children's peacekeeping and peacemaking, this has important implications for the way we think about when and how these processes emerge and the extent to which they can change in response to different ecological conditions.

Children's Peacekeeping

One of the processes by which children maintain peaceful relations is by reducing the frequency and potential costs of social conflict. Such *peacekeeping* processes manage conflict but do not necessarily result in harmonious relations. Thus, children's peacekeeping processes tend to focus on the avoidance of obstacles to peace (i.e., *negative peace*) rather than on active processes serving to make peace (i.e., *positive peace*) (Verbeek 2008). This focus is in line with developmental psychology's emphasis on the antecedents and immediate consequences of children's conflict.

Developmental Trends

One of the challenges of reviewing research on children's peacekeeping is that different researchers employ different definitions of conflict. Studies of aggressive conflict often focus on an individual child's behavior (e.g., Sluckin & Smith 1977), assuming that a conflict of interest gives rise to aggression, or that aggression inevitably sets up a conflict situation. Such a focus underestimates the frequency of conflicts involving nonaggressive behavior. Other researchers define conflict as involving "mutual resistance" between two or more individuals (Shantz & Shantz 1985, p. 4; see also Hay 1984). This approach ensures that conflict refers to more than a single behavior, but it risks underestimating conflicts involving delayed or less overt forms of opposition, such as when a child yields a toy to a peer but later, in retaliation, damages the peer's social standing in the group (e.g., Galen & Underwood 1997). This particular dyadic definition also fails to clarify whether initial opposition is sufficient to define conflict. After all, one child might respond to another's initial opposition by giving up on the issue (Maynard 1985). In these cases, there is no "opposition-to-opposition" and, arguably, no dyadic conflict.

My point is not to argue in favor of one definition over another but, instead, to highlight the different ways that researchers conceptualize conflict and the limitations associated with them. "What counts" as conflict is not always clear (Hay 1984; Shantz 1987), and therefore any inferences about behavioral processes that reduce the frequency of conflict are limited by a researcher's focus, the choice of observational method, and various ecological factors (e.g., group size, and open versus closed field).

Longitudinal studies from Canada, New Zealand, and the United States suggest that the frequency of aggression decreases over the lifespan, peaking for boys and girls at about four years of age and decreasing thereafter (Dodge *et al.* 2006). Developmentally, this suggests that the majority of aggressive children do not become aggressive adults. In fact, research suggests that only 20% of aggressive two-year-old children tend to become aggressive adolescents, and, of these, only 10% go on to have major problems as adults (Broidy *et al.* 2003).

The frequency of children's conflict varies as a function of peer group dynamics. In our own work, we found that the frequency of observed aggression is highest at the start of a new school year, then decreases throughout the year for both preschoolers (Pellegrini, Long, *et al.* 2007; Pellegrini, Roseth, *et al.* 2007; Roseth *et al.* 2007, 2011) and adolescents (Pellegrini & Long 2002). Evolutionary-oriented researchers typically explain this trend in terms of children's *social dominance relationships*, the idea being that the frequency of aggression declines as individuals recognize the costs and benefits of using aggression in conflicts of interest (Bernstein 1981; Hawley 1999; Hinde 1980). Developmentally, this implies the ability to understand relationships between self and others as well as between others, and preliminary longitudinal evidence has linked preschoolers' social dominance status to theory of mind (Pellegrini *et al.* 2011).

From an ethological perspective, it is interesting to consider whether these social-cognitive abilities are antecedent to aggression's decline or develop as a consequence of aggressive conflict. The latter is consistent with the idea that the "function" of children's peacekeeping can have proximal *and* distal dimensions. That is, rather than assuming that deficits in children's verbal ability, perspective taking, and social skills underlie young children's aggression, an evolutionary account suggests that young children may exhibit aggression because it has proximal and distal utility. Proximately, young children's aggression may be used for personal goal attainment and negotiation (e.g., de Waal 2000; Pellegrini 2008; Hawley 2014). Insofar as the development of peacekeeping depends on experience, it may also stimulate the experiences that young children require to identify and avoid obstacles of peace, and in so doing shape subsequent peacekeeping development.

Another reason why children may be more likely to keep peace with age is that the issues giving rise to conflict change. While object disputes (e.g., possession of a toy) are common among very young children (e.g., Dawe 1934; Eisenberg & Garvey 1981; Hay 1984; Hartup *et al.* 1988; Killen & Turiel 1991; Chen *et al.* 2001), conflicts involving social concerns become more common during the preschool years (e.g., Dawe 1934; Strayer & Strayer 1976; Bakeman & Brownlee 1982; Chen *et al.* 2001). For example, Ljungberg *et al.* (2005) found that the most frequent source of conflict among preschool boys was disagreements over games (31%), followed by object disputes (25%) and physical (18%) and psychological harm (15%). These age-related trends are attributed to changes in children's social and cognitive abilities, including their increased ability to evaluate social situations (Astington 1993), growing mastery of expressive speech (Dunn & Slomkowski 1992), and increased ability to understand intentions (Astington 1993). The premise is that the issues giving rise to conflict shift as children become less egocentric (Piaget 1966) and their individual motives begin to interact with social motives.

For older children, a promising approach to studying the issues that trigger conflict is examining children's *goals* in conflict situations. For example, research on seven- to

12-year-olds using hypothetical situations (e.g., Chung & Asher 1995, 1996; Erdley & Asher 1996; Delveaux & Daniels 2000) and children's self-reports of actual conflicts (e.g., Murphy & Eisenberg 1996, 2002) indicates that relational goals are correlated with peaceful strategies, while instrumental goals are correlated with aggressive strategies.

Given the different methodologies, it is difficult to determine whether these findings reflect developmental changes. Observational and self-report data do not always align, and individuals can pursue multiple goals within a single conflict situation. For example, two peers might seek to control an object but differ in their relational goals (e.g., one seeks to maintain harmony while the other seeks retaliation). Different combinations of goals might also lead to different strategies (e.g., peaceful or coercive) and, perhaps, different conflict outcomes. This example highlights that any one goal (e.g., resource control) must be understood in the context of other goals (e.g., social goals). It also suggests that research examining children's conflict goals must allow for the possibility of multiple goals and goal interactions.

Finally, it would be a mistake to assume that pursuing individual goals always undermines social goals. When combined with peaceful processes, pursuing personal concerns can also clarify mutual expectations and build trust, and in so doing result in a more positive relationship than if conflict had never occurred (Deutsch 1973; de Waal 1996). Conflict may also involve *mixed* motives, such as when one child gives up a toy to a dominant peer to avoid injury or gain favor (e.g., Strayer & Santos 1996), or a dominant child takes another's toy when duplicates are available (e.g., Hay & Ross 1982). As detailed in the next section, such social status concerns not only give rise to conflict but also are associated with different conflict strategies and outcomes.

Conflict Strategies

Across ages, a consistent finding in observational research is that aggression is absent in the majority of children's conflicts. Among toddlers, studies report aggression in only 25% of conflicts (e.g., Hay & Ross 1982) to 36% of conflicts (e.g., Chen *et al.* 2001). Among preschoolers, Laursen and colleagues found that aggression was used in 33% of conflicts (Hartup *et al.* 1988; Laursen & Hartup 1989), while Chen *et al.* (2001) found that "high insistence" behaviors (e.g., using physical force, inflicting physical and/or psychological harm) were used in only 10.3% (3-year-olds) and 8.9% (4-year-olds) of conflicts. Among 6- and 7-year-olds, Shantz and Shantz (1982) found that 5% of strategies involved physical and 4% verbal aggression. And finally, among college students, observer ratings of coercive tactics were 73% *smaller* than ratings of compromise (Graziano *et al.* 1996).

Laursen *et al.* (2001) conducted a meta-analysis to examine the relative prevalence of *negotiation* (e.g., compromise, sharing, turn taking, and talking things out), *coercion* (e.g., submitting to demands, commands, and denials, and physical or verbal aggression), and *disengagement* (e.g., discontinuing discussion and leaving the field) in children's conflict management. The results indicated that, across age groups, peers are more likely to resolve conflicts with negotiation than coercion and disengagement, and they are also more likely to resolve conflicts with disengagement than coercion. The results also showed that younger children were more likely to resolve conflicts with coercion than negotiation and disengagement, while older children and adolescents were more likely to resolve conflicts with negotiation than coercion and disengagement. While these findings have been widely interpreted to indicate developmental changes in conflict resolution strategies, I argue for a more cautious interpretation, because the primary studies' methodology likely biased the meta-analytic findings (e.g., observational studies of young children versus survey studies of adolescents).

Missing from research on conflict strategies is a more detailed account of the sequence of strategies and the degree to which strategies or combinations thereof contribute to ending or extending conflicts. Simply classifying conflict strategies (e.g., as coercion, negotiation, or disengagement) also ignores the way one strategy may interact with another. For example, the potentially negative effects of one strategy (e.g., aggression) may be offset by combining it with another (e.g., friendly affiliation) (de Waal 1996; Pellegrini 2008).

My colleagues and I observed the way conflict strategies interact in two preschool samples (Pellegrini, Roseth, *et al.* 2007; Roseth *et al.* 2007, 2011). In one sample, we found that rates of coercive resource control (using physical or verbal aggression) declined across the school year, while rates of prosociality (interacting positively or neutrally with peers) simultaneously increased (Roseth *et al.* 2011). This suggests that coercive resource control must be understood within the behavioral context of peaceful interaction. This also suggests that preschoolers may manage peer conflict using a combination of strategies – for example, using coercive strategies to negotiate social dominance relationships at the start of the school year, but then switching to peaceful resource control in order to maintain social dominance while maintaining positive peer relationships (Pellegrini 2008).

In a third sample of preschoolers, we used focal sampling and video coding to conduct a micro-analytic study of the relative frequency of different conflict strategies within individual conflict events (Roseth *et al.* 2015). To our surprise, initial findings suggest that individual preschoolers only employ *both* peaceful and coercive resource control strategies during 16.2% of conflicts. This is surprising because theory (e.g., resource control theory; Hawley 1999; see also de Waal 1986; Charlesworth 1996) and prior laboratory research (e.g., La Freniere & Charlesworth 1983; Charlesworth & Dzur 1987; Hawley 2002) have highlighted the efficacy of using both strategies to control resources while maintaining positive peer relations. Our naturalistic data, however, suggest that bi-strategic resource control occurs infrequently during a given conflict. It may be that preschoolers' strategy use is highly structured in naturalistic settings, and that the effects of using different conflict strategies on status are only realized over time.

Children's Peacemaking

Having reviewed the developmental and behavioral processes by which children avoid the obstacles of peace, in this section I review research examining the processes by which children make peace. Topics include together outcomes and postconflict reconciliation.

Together Outcomes

Historically, researchers have only documented the immediate consequences of children's conflicts, focusing on whether children stay together or separate after a conflict ends (e.g., Sackin & Thelen 1984; Hartup *et al.* 1988; Laursen & Hartup 1989; Pellegrini, Roseth *et al.* 2007; Roseth *et al.* 2007, 2011; but see Verbeek & de Waal 2001). The focus here was on the immediate and long-term consequences of aggression rather than on the peaceful processes by which the negative effects of aggressive conflict might be avoided or repaired (e.g., Dodge *et al.* 2006). Indeed, dispersal (i.e., postconflict separation) has been used as a criterion for differentiating between aggressive and nonaggressive behavior (e.g., Smith & Connolly 1972; Humphreys & Smith 1987; Savin-Williams 1987; Pellegrini 1988).

For preschoolers, most naturalistic studies suggest that together outcomes are less likely than separate outcomes (Roseth *et al.* 2011: 30% together; Sackin & Thelen 1984: 35% together; Verbeek & de Waal 2001: 22% together). Various factors influence together outcomes. For example, Sackin and Thelen (1984) found that conciliatory behaviors (e.g., cooperative propositions, apologies, symbolic offers, object offers, and grooming) were associated with together outcomes, while insistence strategies resulting in subordination (e.g., crying, screaming, flinching, withdrawal, and request cessation) were associated with separation. Similar results have since been documented by four additional naturalistic studies (Laursen & Hartup 1989; Verbeek & de Waal 2001; Ljungberg *et al.* 2005; Roseth 2006), although the relative frequency of behaviors varied by sample.

The use of conciliatory gestures likely reflects the increased sophistication of preschoolers' socio-cognitive and self-regulatory abilities, as these behaviors require both perspective taking and redirecting oneself from the issue that gave rise to the conflict. More distally, conciliatory gestures may also help to reduce stress associated with conflict and preserve valued relationships. The use of friendly gestures to signal peaceful intentions is well documented in the comparative literature (e.g., Preuschoft & van Schaik 2000), as is the use of grooming to alleviate stress and repair social relationships (de Waal 2000). Future research is needed to document whether the use of conciliatory gestures is also associated with status.

Preschoolers' relationships also moderate together outcomes. Friends' and nonfriends' conflicts do not differ in frequency, length, or the instigating issues, but together outcomes are more likely between friends than nonfriends (Hartup *et al.* 1988; Vespo & Caplan 1993; Verbeek & de Waal 2001). Friendships provide an important context for learning about reciprocity and commitment, not only because they are valued, but also because they are voluntary (Hartup 1989). Conflicts with friends likely prioritize relationship maintenance and favor the development of "softer" modes of conflict management to preserve those relationships (Verbeek *et al.* 2000). Functionally, the fact that preschooler friends are more likely to remain together after conflict is also consistent with the valuable relations hypothesis (de Waal & Aureli 1997), which predicts that social conflicts are more likely to be resolved constructively when opponents have a vested interest in their relationship.

Finally, for older children, research indicates that social interaction is more likely to continue after a conflict ends among friends and romantic partners than after conflicts with parents, siblings, and others (Rafaelli 1990; Laursen 1993). Research also suggests that adolescents' conflicts are equally likely to strengthen or have no effect on the quality of relationships among friends and romantic partners, while the opposite is true of conflicts with other peers and nonfamilial adults (Laursen 1993). Longitudinal research indicates that continued conflict is negatively associated with friendship (Berndt &

Keefe 1992). Broadly speaking, these findings are also in accord with the valuable relationship hypothesis and highlight that conflicts can strengthen relationships, presumably by clarifying interests and building trust through constructive conflict strategies (Deutsch 1973; de Waal 2000).

Postconflict Reconciliation

Developmental psychologists' traditional focus on immediate outcomes was motivated, in part, by the assumption that separation has a destructive effect on relationships. As detailed in this section, this assumption has largely been refuted by evidence that postconflict separation can also *increase* affiliation by way of postconflict *reconciliation*, defined as friendly interaction among former opponents after conflict-induced separation (Aureli & de Waal 2000).

To date, research using the postconflict–matched control (PC-MC) method (de Waal & Yoshihara 1983) has documented reconciliation among children ranging in age from three to 11 years old in 12 different samples from six different countries: Japan (Fujisawa *et al.* 2005, 2006), the Republic of Kalmykia (Butovskaya 2001), the Netherlands (Kempes *et al.* 2008), Russia (Butovskaya & Kozintsev 1999; Butovskaya *et al.* 2005; Butovskaya 2008), Sweden (Ljungberg *et al.* 1999, 2005; Horowitz *et al.* 2007; Westlund *et al.* 2008), and the United States (Verbeek & de Waal 2001; Roseth *et al.* 2008, 2011). This evidence suggests that children's conflict can actually enhance relations between former opponents when it is followed by reconciliation. In fact, and in keeping with research on nonhuman primates, this growing body of work has demonstrated that children's reconciliation promotes tolerance around resources (Butovskaya & Kozintsev 1999; Fujisawa *et al.* 2005; Ljungberg *et al.* 2005), reduces continuing aggression (Ljungberg *et al.* 1999), and decreases redirected aggression toward a third party (Butovskaya & Kozintsev 1999). In short, not only does children's reconciliation result in increased postconflict affiliation, but it also appears to strengthen peer relationships more than if conflict had never occurred.

The diversity of countries in which this early research has been conducted represents a unique strength of this small body of research, and it strongly suggests that reconciliation is part of the behavioral repertoire of typically developing children. These cross-cultural findings are also in accord with research on more than 30 species of social primates (Aureli *et al.* 2002; Verbeek 2008), suggesting that children's reconciliation has deep evolutionary roots. Unfortunately, this cross-cultural diversity also limits conclusions about what causes children's reconciliation and how it develops. Thus, until more evidence is available, the data on what factors moderate reconciliation's expression and development should be interpreted with caution.

In line with evidence in the comparative literature, one of the causes of reconciliation may be uncertainty about the status of the relationship between conflicting peers. Insofar as conflicting peers experience anxiety or stress following postconflict separation, the "uncertainty reduction hypothesis" (Aureli & van Schaik 1991) posits that they will seek relief from it through reconciliation. Fujisawa *et al.* (2005) found that reconciliation reduced Japanese preschoolers' conflict-induced self-directed behavior, a behavioral manifestation of stress (Maestripieri *et al.* 1992). And in Russia, Butovskaya (2008) found that reconciliation reduced seven- to 11-year-old boys' salivary cortisol, a hormone released in response to stress (Gunnar & Donzella 2002). These are promising findings that also align with the idea that postconflict separation may provide conflicting peers with a "cooling-off" period that improves future interaction (cf. Selman 1980; Shantz 1987). Future research should incorporate interviews to help clarify whether reconciliation is indeed motivated by a desire to reduce postconflict stress and anxiety (Verbeek 2008). Indeed, interview data can also reveal other factors, as one can imagine that one's own stress and anxiety might motivate separation, but so too might the goal of *inducing* stress and anxiety in the conflicting peer.

According to the valuable relationship hypothesis (de Waal 1996), mutually valued social relationships increase the likelihood of reconciliation (see also Deutsch 1973). The basic premise is that conflict-induced damage to valuable relationships incurs a large cost, and individuals are therefore motivated to restore harmonious relations (see also Adang *et al.*, this volume, Chapter 10). At the ultimate level, the assumption is that social relationships enhance survival and reproductive success (e.g., Kummer 1978). The valuable relationship hypothesis is generally supported among nonhuman primates (Arnold & Aureli 2006), but the evidence is mixed for human children. Two studies – one in the United States (Verbeek & de Waal 2001) and one in the Republic of Kalmykia (Butovskaya 2001) – found *no differences* in the likelihood of reconciliation between friends and nonfriends. The opposite pattern was found in two other studies (Butovskaya 2001; Fujisawa *et al.* 2005: among four-year-olds but not three-year-olds). Finally, two other studies – one in the United States and the other in Russia – found that reconciliation was *less* likely between friends than nonfriends (Butovskaya & Kozintsev 1999; Roseth 2006). These inconsistent findings contrast with meta-analytic findings showing that other aspects of conflict management (e.g., termination strategy and conciliatory behaviors) differ between friends and nonfriends (Newcomb & Bagwell 1995). As already noted, however, differences in country, culture, context, and methodology suggest that these differences should be interpreted cautiously.

Turning to the question of how reconciliation develops, the cross-cultural diversity of existing research and lack of longitudinal designs also limit conclusions. To date, reconciliation has been documented in four different samples of three- to six-year-olds (Verbeek & de Waal 2001; Fujisawa *et al.* 2005; Roseth *et al.* 2008, 2011; Westlund *et al.* 2008), two different samples of six- to eight-year-olds (Butovskaya & Kozintsev 1999; Kempes *et al.* 2008), one sample of seven- to 11-year-old boys (Butovskaya *et al.* 2005), and one sample of seven- to 15-year-old boys (Butovskaya 2008). Reconciliation has also been documented among four- to seven-year-old Swedish boys with language impairment (Horowitz *et al.* 2007), but not in six- to eight-year-old boys exhibiting clinical levels of aggression (Kempes *et al.* 2008). Taken together, this suggests that children's reconciliation emerges as early as three years of age and continues thereafter. The only evidence to date of age-related differences comes from a two-year study in Japan (Fujisawa *et al.* 2005). In year 1, proximity increased the likelihood of reconciliation among four-year-olds but not three-year-olds. And in year 2, the four- to five-year-old cohort exhibited age-related changes, but the three- to four-year-old cohort did not. That is, in year 2, five-year-olds reconciled more frequently and earlier than in year 1, while four-year-olds showed no such differences. This suggests that the frequency of reconciliation may increase around five years of age, but additional cross-cultural and longitudinal research is needed to clearly document age-related changes.

Some comparative insight on ontogeny might be gained from the few developmental studies on nonhuman primates. Pointing to the importance of social learning, Ljungberg and Westlund (2000) found that hand-reared, single-housed rhesus macaques (*Macaca mulatta*) showed no evidence of reconciliation, while those reared in a peer group did. Likewise, de Waal and Johanowicz (1993) found that the rate of reconciliation among juvenile rhesus monkeys increased dramatically when co-housed with juveniles of the more tolerant stumptail macaque (*Macaca arctoides*). Together, these two studies support the view that reconciliation emerges naturally out of peer-related social learning processes (cf. Piaget 1932; Killen & de Waal 2000; Verbeek *et al.* 2000).

Children's emotional reactions to conflict may also play a role in the development of reconciliation. Peer conflict typically provokes negative emotions (Arsenio *et al.* 1993; Whitesell & Harter 1996), and there is evidence that reconciliation reduces postconflict stress and anxiety (Butovskaya *et al.* 2005; Fujisawa *et al.* 2005). Emotional reactions may thus underlie children's conflict-related goals and behaviors. Supporting this view, Murphy and Eisenberg (2002) found that children's negative emotional reactions (e.g., anger and sadness) to self-reported peer conflicts were associated with unfriendly goals (e.g., hurt and annoy) and unconstructive behavior (e.g., threat, coercion, deception, and power assertion), even after controlling for age, gender, the conflict issue, and friendship.

Further supporting the link between emotions and reconciliation, a recent study in China on 4–6-year-old preschoolers (Liao *et al.* 2014) found that children's ability to recognize emotions was positively linked to their reconciliation tendency in conflict scenarios involving overt and relational aggression. Children's affective perspective taking and teacher ratings of peaceful behavior were also both found to be associated with reconciliation in the overt but not relational aggression scenarios. Taken together, these findings suggest that emotional reactions and the ability to detect others' emotional reactions may play key roles in the development of conflict management. Future research should continue this line of inquiry, as the regulatory processes involved in monitoring, evaluating, and modifying emotional reactions would all seem relevant to the causal mechanisms and development of children's reconciliation.

Given the strong evidence of reconciliation's evolutionary roots (Aureli *et al.* 2002), what would constitute normative reconciliatory processes in children, and what would constitute individual differences? The lack of within- and between-culture research makes it difficult to make inferences about normative developmental change (Butovskaya *et al.* 2000). At present, however, there is evidence of different rates of reconciliation, as evidenced by the corrected conciliatory tendency (CCT; Veenema *et al.* 1994) in studies using the PC-MC method. Across all samples, the overall conciliatory tendency ranges from 27% in one US sample (Roseth *et al.* 2011) to 35% in Japan (Fujisawa *et al.* 2005), 47% in a second US sample (Verbeek & de Waal 2001), 60% in Sweden (Ljungberg *et al.* 2005), and 70% in a Kalmyk sample (Butovskaya 2001). There is also evidence of different forms of reconciliation. Across all samples, children used both explicit (e.g., apology, object offer, and hugs) and implicit (e.g., simply resuming play) conciliatory behaviors. But only Kalmyk children used a ritual in which conflicting peers joined hands and recited peacemaking rhymes. Such differences point to a broad capacity to change the frequency and form of reconciliation, and additional research is needed to determine if variability also exists in timing, who initiates reconciliation, the extent to which others accept conciliatory gestures, as well as the goals motivating reconciliation (for a review, see Butovskaya *et al.* 2000). To the extent that these variables differ within and between cultures, the psychological mechanisms underlying reconciliation may also differ. After all, the advantage that conflict management confers on survival and reproductive success favors multiple developmental pathways to reconciliation, and relevant factors such as the issue causing the conflict, the strategies used during conflict, and the value of the peer relationship likely operate in different combinations to produce similar proximal and distal results.

Third-Party Intervention

While postconflict reconciliation focuses on the way opponents make peace, it is important to remember that many peer conflicts occur within the presence of others. Thus, conflict's social context involves more than a particular dyad's prior interactions and relationship; it also involves third parties (e.g., teachers and peers), who may also influence the course of conflict and its resolution.

Teacher Intervention

Teacher intervention moderates conflict outcomes, albeit not in the way that many developmental psychologists might expect (e.g., DeVries & Zan 1994; Grusec & Goodnow 1994; Katz & McClellan 1997). A consistent finding is that the majority of preschoolers' conflicts end without adult intervention (e.g., Dawe 1934; Houseman 1972; Bakeman & Brownlee 1982). In fact, research links the *absence* of teacher proximity to preschoolers' together outcomes (Laursen & Hartup 1989; Roseth *et al.* 2008), continued negotiation (Killen & Turiel 1991), and use of conciliatory behaviors (Verbeek & de Waal 2001; Fujisawa *et al.* 2005). Research also suggests that teacher intervention moderates reconciliation, but discrepant results limit clear conclusions. To date, one US study has shown that preschoolers' reconciliation is *more* likely after teacher intervention (Roseth *et al.* 2008), while another US study showed that the likelihood of reconciliation was unrelated to teacher intervention (Verbeek & de Waal 2001), and one Japanese study showed that reconciliation was *less* likely after teacher intervention (Fujisawa *et al.* 2005). These findings suggest that, in some cases, teacher intervention may inhibit the peaceful behavioral processes that children employ to manage conflict and restore harmonious relations. Future research is needed as differences in culture and context prevent clear conclusions.

Bystander Intervention

Bystanders, or third-party peers not involved in a conflict, may also intervene and influence the course of conflict and its outcomes. Research on this topic comes from three different cultures: Japan, Russia, and the United States.

Japan

In a two-year observational study, Fujisawa *et al.* (2005, 2006) examined peer intervention among three- ($n = 15$) and four-year-old ($n = 22$) Japanese preschoolers in two same-age classrooms. In year 1, Fujisawa *et al.* (2005) reported that four-year-old peers intervened in 26% (24 of 92) of aggressive conflicts, and that both reconciliatory attempts and acceptance were more likely when this occurred than when it did not. In year 2, Fujisawa *et al.* (2006) did not report the frequency of peer intervention but did find that five-year-olds' conciliatory tendency (CT) was higher when a peer did *not* intervene (50.6%) compared to when they did (12.6%).

United States
In the United States, peer intervention has been reported in three different observational studies of preschoolers. First, in a study of 120 children (three- to five-years-old) in a private preschool, Verbeek (1996) reported that peers intervened in ~1% (3 of 200) of conflicts that ended in separation. Second, in a study of 91 children (43 girls) enrolled in a university preschool, Roseth *et al.* (2008) found that peers intervened in 5% (10 of 195) of conflicts, and the majority of these (80%) resulted in separate rather than together outcomes (20%). Third, in a study of 111 children enrolled in seven Head Start preschool classrooms, Fedor *et al.* (2012) found that peers intervened in only 3% (9 of 304) of conflicts, and that these were associated with similar rates of together (55%) and separate outcomes (45%). In the Head Start sample, more girls ($n = 8$) than boys ($n = 2$) intervened, and older children were also more likely to intervene than younger children.

Russia
In an observational and interview study of 25 Russian children aged six to seven years old on the school playground, Butovskaya and Kozintsev (1999) found that bystanders intervened in 13% (21 of 159) of conflicts, and in approximately equal frequencies on behalf of the aggressor ($n = 9$) and victim ($n = 12$, n.s.). Interview data suggested that the vast majority of boys and girls believed that "one should always defend the target of aggression" (Butovskaya & Kozintsev 1999, p. 134), and that friends would not only console other friends who were the victim of aggression but also encourage the aggressor to apologize.

In a survey study of 212 Russian adolescents (11–15 years old), Butovskaya *et al.* (2007) found that both sexes tended to exaggerate third-party intervention, as self-rated scores exceeded peer-rated scores. For both sexes, they found a positive correlation between peer-rated popularity and self-rated intervention. On average, girls engaged in significantly more intervention than boys on both self- and peer ratings.

These studies provide preliminary evidence that peer bystander intervention emerges in early childhood (e.g., by four years of age), but various factors may affect its frequency and efficacy for peacemaking. The tendency to intervene in peer conflict likely depends on many of the same socio-cognitive skills as discussed for reconciliation, such as perspective taking and empathy (Eisenberg *et al.* 2006). To the extent that these skills become more sophisticated with age, one might expect that peer intervention would be more likely to occur as children become older. This same developmental mechanism may underlie sex differences in preschoolers, as girls typically develop such socio-cognitive skills earlier than boys (Eisenberg *et al.* 2006).

The extent to which peers intervene on behalf of one individual over another also likely depends on the development of socio-moral skills (Killen & de Waal 2000) as well as in- and outgroup biases (Rutland *et al.* 2010). For example, it may be that moral concerns direct intervention among ingroup peers, but outgroup biases direct it when conflict occurs between in- and outgroup individuals (see Otten *et al.* this volume).

Compared to teacher intervention, it may also be that bystanders' direct and indirect intervention strategies are equally effective at making peace, as peer intervention does connote the same power differences that characterize adult intervention. For example, Butovskaya *et al.* (2000) report that Kalmyk children employ a range of behaviors to intervene in peer conflicts, including physically pushing opponents apart, verbally persuading them to end a conflict, and even taking one of the conflicting peers by the hand and bringing them back to the former opponent to make peace. Peers may resist being told to make peace by a teacher, but be motivated to do so when a peer (and perhaps especially a high-status peer) encourages them to do so. Future research is needed to understand these dynamics.

Bystander Consolation

Besides intervention, another way that bystanders can influence conflict outcomes is by providing *consolation*, defined as a postconflict comforting behavior directed at a distressed party, such as a recent victim of aggressive conflict (de Waal 2008; see also Romero, this volume, Chapter 4). For example, a bystander may go over to a peer who has just lost a toy and offer them another one to play with. To date, however, only one study has reported specific data on children's consolation.

In their two-year observational study, Fujisawa *et al.* (2005) did not find any age-related differences in the frequency of consolation during year 1, as the triadic conciliatory tendency (TCT; Call *et al.* 2002) was equally small among three- and four-year-olds (9.9 vs. 7.7%, respectively). In year 2, however, the TCT was significantly higher among five-year-olds (28.7%) compared to one year earlier, while there was still no age-related difference between four-year-olds' TCT (13.3%) compared to one year earlier (Fujisawa *et al.* 2006). This suggests that the frequency of consolation increased rapidly between four and five years of age, most likely as a function of developing socio-cognitive skills, such as perspective taking and empathy (Eisenberg *et al.* 2006). Presumably, peers console distressed peers because they have developed the ability to differentiate their own feelings from others as well as accurately understand another individual's reaction to an event (Hoffman 1982). To the extent that these skills become more sophisticated with age, one would expect that consolation would also become more frequent with age.

Interestingly, Fujisawa *et al.* (2006) also found that consolation occurred more often when no reconciliation occurred than when it did among all four- and five-year-olds, and that consolation occurred more often *before* reconciliation than after among all but the three-year-olds. These findings are consistent with the view that consolation may substitute for reconciliation for distressed parties (de Waal & Aureli 1997; Palagi *et al.* 2004; Romero, this volume, Chapter 4). Before reconciliation, when peers are distressed by the conflict's outcome and subsequent social separation, bystanders may recognize the need for consolation. But this need may no longer be manifest after reconciliation, and as a result bystanders may infer that the distressed peer already feels better and would no longer benefit from consolation. Among older children especially, future research should examine these issues using methodologies (e.g., interviews, surveys, and hypothetical scenarios) that allow us to better understand the psychological mechanisms underlying consolation.

Future research should also examine whether bystanders' postconflict affiliation varies as a function of peer relationships and status. As discussed, Butovskaya and Kozintsev (1999) found that 6–7-year-olds believe that friends would intervene on their behalf if they are the victim of aggression, and the same inclination may motivate higher rates of consolation among friends than nonfriends. Bystanders may also be more motivated to console higher status peers than subordinate peers, as postconflict affiliation may not only provide solace to a distressed peer but also serve to appease and gain favor with a high-status peer (cf. Romero, this volume, Chapter 4).

Conclusion

Conflict is inherent to peer relationships, and managing peer conflict plays a key role in social development. Past conflict experiences inform future conflict experiences, and these experiences in turn shape peer relationships. In this way, conflict represents a formative experience for children's social development, and early conflict experiences likely promote the development of the peaceful behavioral processes that help children to avoid the obstacles of peace (peacekeeping) and restore harmonious relations (peacemaking).

The research reviewed in this chapter highlights current understanding of developmental changes in the frequency of children's conflict, the issues that give rise to conflict, the strategies used, and the immediate and delayed outcomes. Compared with research in younger children, existing data on older children's conflict management are much more limited, especially in terms of naturalistic observations. Thus, uncertainty remains about the scope and motivational significance of the issues that give rise to conflict and the strategies used during conflict. Uncertainty also remains about developmental continuity and the way early conflict experiences relate to later ones. What is more certain is that peaceful processes involved in children's conflict share many parallels in the comparative literature (cf. Verbeek 2008). This supports the view that children share a natural tendency to keep and make peace.

Looking forward, one of the major challenges for researchers taking an ethological approach to children's peacekeeping and peacemaking is how to integrate careful observation and description with a broader appreciation of the way objective stimuli may convey different psychological meanings. This is a challenge, because ethology has long cautioned against anthropomorphism, anthropocentrism, and unnecessarily complex psychological explanations. This makes sense, of course, insofar as we should not view animals as humans, nor should we see animals from a human perspective. But children are humans, and while they have their own problems to solve and their own age-related behaviors to solve them, they also think about those problems in human ways. Thus, children's goals, values, and beliefs matter for conflict management, as do their perceptions of the conflicting peer's goals, values, and beliefs, and their perceptions of an audience and how it will react to the conflict and its outcomes. In short, children's conflict is not completely and inevitably shaped by objective circumstances, and whether it takes a constructive or destructive course is subject to psychological mechanisms that may not be observable given traditional ethological methods. Looking forward, our challenge is therefore to integrate careful observation with psychological methods that allow us to ask questions about the mechanisms underlying children's peacekeeping and peacemaking.

References

Arnold, K., & Aureli, F. (2006). Postconflict reconciliation. In C.J. Campbell, A. Fuentes, K.C. MacKinnon, M. Panger, & S.K. Bearder (Eds.), *Primates in perspective* (pp. 592–608). New York: Oxford University Press.

Arsenio, W., Lover, A., & Gumora, G. (1993, March). Emotions, conflicts, and aggression during preschoolers' freeplay. Poster presented at the biennial meeting of the Society for Research in Child Development, New Orleans, LA.

Astington, J. (1993). *The child's discovery of the mind.* Cambridge, MA: Harvard University Press.

Aureli, F., Cords, M., & van Schaik, C. (2002). Conflict resolution following aggression in gregarious animals: A predictive framework. *Animal Behaviour, 63,* 1–19.

Aureli, F., & de Waal, F.B.M. (2000). *Natural conflict resolution.* London: University of California Press.

Aureli, F., & van Schaik, C. (1991). Post-conflict behaviour in long-tailed macaques (*Macaca fascicularis*): II. Coping with the uncertainty. *Ethology, 89,* 101–114.

Bakeman, R., & Brownlee, J.R. (1982). Social rules governing object conflicts in toddlers and preschoolers. In K.H. Rubin & H.S. Ross (Eds.), *Peer relations and social skills in childhood* (pp. 99–111). New York: Springer.

Bateson, P. (2015). Ethology and human development. In W.F. Overton & P.C. Molenaar (Eds.), *Theory and method: Vol. 1. Handbook of child psychology and developmental science* (7th ed., pp. 208–243). Hoboken, NJ: John Wiley & Sons.

Berndt, T.J., & Keefe, K. (1992). Friends' influence on adolescents' perceptions of themselves at school. In D.H. Schunk & J.L. Meece (Eds.), *Student perceptions in the classroom* (pp. 51–73). Hillsdale, NJ: Erlbaum.

Bernstein, I. (1981). Dominance: The baby and the bathwater. *The Behavioral and Brain Sciences, 4,* 419–457.

Broidy, L.M., Nagin, D.S., Tremblay, R.E., Bates, J.E., Brame, B., Dodge, K.E., et al. (2003). Developmental trajectories of childhood disruptive behaviors and adolescent delinquency: A six site, cross-national study. *Developmental Psychology, 39,* 222–245.

Butovskaya, M.L. (2001). Reconciliation after conflicts: Ethological analysis of post-conflict interactions in Kalmyk children. In J.M. Ramirez & D.S. Richardson (Eds.), *Cross-cultural approaches to aggression and reconciliation* (pp. 167–190). Huntington, NY: Nova Science.

Butovskaya, M.L. (2008). Reconciliation, dominance and cortisol levels in children and adolescents (7–15-year-old boys). *Behaviour, 145,* 1557–1576.

Butovskaya, M.L., Boyko, E.Y., Serverova, N.B., & Ermakova, I.V. (2005). The hormonal basis of reconciliation in humans. *Journal of Physiological Anthropology and Applied Human Science, 24,* 333–337.

Butovskaya, M.L., & Kozintsev, A. (1999). Aggression, friendship, and reconciliation in Russian primary schoolchildren. *Aggressive Behavior, 25,* 125–139.

Butovskaya, M.L., Timentschik, V.M., & Burkova, V.N. (2007). Aggression, conflict resolution, popularity, and attitude to school in Russian adolescents. *Aggressive Behavior, 33,* 170–183.

Butovskaya, M.L., Verbeek, P., Ljungberg, T., & Lunardini, A. (2000). A multicultural view of peacemaking among young children. In F. Aureli & F.B.M. de Waal (Eds.), *Natural conflict resolution* (pp. 243–258). Berkeley: University of California Press.

Call, J., Aureli, F., & de Waal, F.B.M. (2002). Postconflict third-party affiliation in stump-tailed macaques. *Animal Behaviour, 63,* 209–216.

Charlesworth, W.R. (1996). Co-operation and competition: Contributions to an evolutionary and developmental model. *International Journal of Behavioral Development, 19,* 25–38.

Charlesworth, W.R., & Dzur, C. (1987). Gender comparisons of preschoolers' behavior and resource utilization in group problem solving. *Child Development, 58,* 191–200.

Chen, D.W., Fein, G.G., Killen, M., & Tam, H. (2001). Peer conflicts of preschool children: Issues, resolution, incidence, and age related patterns. *Early Education and Development 12,* 523–544.

Chung, T., & Asher, S.R. (1995). Children's goals are related to their strategies for responding to peer conflict. Paper presented at the meeting of the American Psychological Society, New York.

Chung, T., & Asher, S.R. (1996). Children's goals and strategies in peer conflict situation. *Merrill-Palmer Quarterly, 42,* 125–147.

Dawe, H. (1934). An analysis of two hundred quarrels of preschool children. *Child Development, 5,* 139–157.

Delveaux, K.D., & Daniels, T. (2000). Children's social cognitions: Physically and relationally aggressive strategies and children's goals in peer conflict situations. *Merrill-Palmer Quarterly, 46,* 672–692.

Deutsch, M. (1973). *The resolution of conflict: Constructive and destructive processes.* New Haven, CT: Yale University Press.

DeVries, R., & Zan, B. (1994). *Moral classroom, moral children: Creating a constructivist atmosphere in early education.* New York: Teacher College Press.

de Waal, F.B.M. (1986). The integration of dominance and social bonding in primates. *Quarterly Review of Biology, 61,* 459–479.

de Waal, F.B.M. (1996). Conflict as negotiation. In W.C. McGrew, L.F. Marchant, & T. Nishida (Eds.), *Great ape societies* (pp. 159–172). Cambridge: Cambridge University Press.

de Waal, F.B.M. (2000). The first kiss. In F. Aureli & F.B.M. de Waal (Eds.), *Natural conflict resolution* (pp. 13–33). Berkeley: University of California Press.

de Waal, F.B.M. (2008). Putting the altruism back in altruism: The evolution of empathy. *Annual Review of Psychology, 59,* 279–300.

de Waal, F.B.M., & Aureli, F. (1997). Conflict resolution and distress alleviation in monkeys and apes. *Annals New York Academy of Sciences, 807,* 317–328.

de Waal, F.B.M., & Johanowicz, D.L. (1993). Modification of reconciliation behavior through social experience: An experiment with two macaque species. *Child Development, 64,* 897–908.

de Waal, F.B.M., & Yoshihara, D. (1983). Reconciliation and redirected affection in rhesus monkeys. *Behaviour, 85,* 224–241.

Dodge, K.A., Coie, J.D., & Lynam, D. (2006). Aggression and antisocial behavior in youth. In W. Damon & N. Eisenberg (Eds.), *Handbook of child psychology: Vol. 3. Social, emotional, and personality development* (pp. 719–788). New York: John Wiley & Sons.

Dunn, J., & Slomkowski, C. (1992). Conflict and the development of social understanding. In C. Shantz & W. Hartup (Eds.), *Conflict in child and adolescent development* (pp. 70–91). New York: Cambridge University Press.

Eisenberg, A.R., & Garvey, C. (1981). Children's use of verbal strategies in resolving conflicts. *Discourse Processes, 4*, 149–170.

Eisenberg, N., Fabes, R.A., & Spinrad, T.L. (2006). Prosocial development. In N. Eisenberg (Ed.), *Handbook of child psychology: Vol. 3. Social, emotional, and personality development* (6th ed., pp. 646–718). New York: John Wiley & Sons.

Erdley, C.A., & Asher, S.R. (1996). Children's social goals and self-efficacy perceptions as influences on their responses to ambiguous provocation. *Child Development, 67*, 1329–1344.

Fedor, M., Thelamour, B., & Roseth, C.J. (2012, April). "Stop fighting!" Third-party peer intervention and preschoolers' conflict resolution. Paper presented at AERA Annual Meeting, Vancouver, BC, Canada.

Fujisawa, K., Kutsukake, N., & Hasegawa, T. (2005). Reconciliation pattern after aggression among Japanese preschool children. *Aggressive Behavior, 31*, 138–152.

Fujisawa, K., Kutsukake, N., & Hasegawa, T. (2006). Peacemaking and consolation in Japanese preschoolers witnessing peer aggression. *Journal of Comparative Psychology, 120*, 48–57.

Galen, B.R., & Underwood, M.K. (1997). A developmental investigation of social aggression among children. *Developmental Psychology, 33*, 589–600.

Graziano, W.G., Jensen-Campbell, L.A., & Hair, E.C. (1996). Perceiving interpersonal conflict and reacting to it: The case for agreeableness. *Journal of Personality and Social Psychology, 70*, 820–835.

Grusec, J., & Goodnow, J. (1994). Impact of parental discipline methods on the child's internalization of values: A reconceptualization of current points of view. *Developmental Psychology, 30*, 4–19.

Gunnar, M.R., & Donzella, B. (2002). Social regulation of the cortisol levels in early human development. *Psychoneuroendocrinology, 27*, 199–220.

Hartup, W.W. (1989). Social relationships and their developmental significance. *American Psychologist, 44*, 120–126.

Hartup, W.W., Laursen, B., Stewart, M.I., & Eastenson, A. (1988). Conflict and the friendship relations of young children. *Child Development, 59*, 1590–1600.

Hawley, P.H. (1999). The ontogenesis of social dominance: A strategy-based evolutionary perspective. *Developmental Review, 19*, 97–132.

Hawley, P.H. (2002). Social dominance and prosocial and coercive strategies of resource control in preschoolers. *International Journal of Behavioral Development, 26*, 167–176.

Hawley, P.H. (2014). The duality of human nature: Coercion and prosociality in youth's hierarchy ascension and social success. *Current Directions in Psychological Science, 23*, 433–438.

Hay, D.F. (1984). Social conflict in early childhood. In G. Whitehurst (Ed.), *Annals of child development* (Vol. *1*, pp. 1–44). Greenwich, CT: JAI.

Hay, D.F., & Ross, H. (1982). The social nature of early conflict. *Child Development, 53*, 105–113.

Hinde, R.A. (1980). *Ethology*. London: Fontana.

Hoffman, M.L. (1982). Development of prosocial motivation: Empathy and guilt. In N. Eisenberg-Berg (Ed.), *Development of prosocial behavior* (pp. 281–313). New York: Academic Press.

Horowitz, L., Westlund, K., & Ljungberg, T. (2007). Aggression and withdrawal related behavior within conflict management progression in preschool boys with language impairment. *Child Psychiatry and Human Development, 38*, 237–253.

Houseman, J. (1972). *An ecological study of interpersonal conflicts among preschool children*. Unpublished doctoral dissertation, Wayne State University, Detroit (University Microfilms No. 73-12533).

Humphreys, A., & Smith, P.K. (1987). Rough-and-tumble play, friendship and dominance in school children: Evidence for continuity and change with age. *Child Development, 58*, 201–212.

Katz, L., & McClellan, D.E. (1997). *Fostering children's social competence: The role of the teacher*. Washington, DC: National Association for the Education of Young Children.

Kempes, M.M., Orobio de Castro, B., & Sterck, E.H.M. (2008). Conflict management in 6–8- year-old aggressive Dutch boys: Do they reconcile? *Behaviour, 145*, 1701–1722.

Killen, M., & de Waal, F.B.M. (2000). The evolution and development of morality. In F. Aureli & F.B.M. de Waal (Eds.), *Natural conflict resolution* (pp. 352–372). Berkeley: University of California Press.

Killen, M., & Turiel, E. (1991). Conflict resolution in preschool social interactions. *Early Education and Development, 2*, 240–255.

Kummer, H. (1978). On the value of social relationships to nonhuman primates: A heuristic scheme. *Social Science Information, 17*, 687–705.

La Freniere, P., & Charlesworth, W.R. (1983). Dominance, attention, and affiliation in a preschool group: A nine-month longitudinal study. *Ethology and Sociobiology, 4*, 55–67.

Laursen, B. (1993). The perceived impact of conflict on adolescent relationships. *Merrill-Palmer Quarterly, 39*, 535–550.

Laursen, B., Finkelstein, B.D., & Townsend-Betts, N. (2001). A developmental meta-analysis of peer conflict resolution. *Developmental Review, 21*, 423–449.

Laursen, B., & Hartup, W.W. (1989). The dynamics of preschool children's conflicts. *Merrill- Palmer Quarterly, 35*, 281–297.

Liao, Z., Li, Y., & Su, Y. (2014). Emotion understanding and reconciliation in overt and relational conflict scenarios among preschoolers. *International Journal of Behavioral Development, 38*, 111–117.

Ljungberg, T., Horowitz, L., Jansson, L., Westlund, K., & Clarke, C. (2005). Communicative factors, conflict progression and use of reconciliatory strategies in preschool boys – a series of random events or a sequential process? *Aggressive Behavior, 31*, 303–323.

Ljungberg, T., & Westlund, K. (2000). Impaired reconciliation in rhesus macaques with a history of early weaning and disturbed socialization. *Primates, 41*, 79–88.

Ljungberg, T., Westlund, K., & Forsberg, A.J.L. (1999). Conflict resolution in 5-year-old boys: Does post-conflict affiliative behaviour have a reconciliatory role? *Animal Behaviour, 5*, 1007–1016.

Maestripieri, D., Schino, G., Aureli, F., & Troisi, A. (1992). A modest proposal: Displacement activities as indicators of emotions in primates. *Animal Behaviour, 44*, 967–979.

Maynard, D.W. (1985). How children start arguments. *Language and Society, 14*, 1–29.

Murphy, B.C., & Eisenberg, N. (1996). Provoked by a peer: Children's anger-related responses and their relations to social functioning. *Merrill-Palmer Quarterly, 42*, 103–124.

Murphy, B.C., & Eisenberg, N. (2002). An integrative examination of peer conflict: Children's reported goals, emotions, and behaviors. *Social Development, 11*, 534–557.

Newcomb, A.F., & Bagwell, C.L. (1995). Children's friendship relations: A meta-analytic review. *Psychological Bulletin, 117*, 306–347.

Palagi, E., Paoli, T., & Tarli, S.B. (2004). Reconciliation and consolation in captive bonobos (*Pan paniscus*). *American Journal of Primatology*, *62*, 15–30.

Parke, R.D., & Slaby, R. (1983). The development of aggression. In E.M. Hetherington (Ed.), *Handbook of child psychology: Vol. 4. Socialization, personality, and social development* (pp. 547–641). New York: Wiley.

Pellegrini, A.D. (1988). Elementary school children's rough-and-rumble play and social competence. *Developmental Psychology*, *24*, 802–806.

Pellegrini, A.D. (2008). The roles of aggressive and affiliative behaviors in resource control: A behavioral ecological perspective. *Developmental Review*, *28*, 461–487.

Pellegrini, A.D., & Long, J.D. (2002). A longitudinal study of bullying, dominance, and victimization during the transition from primary to secondary school. *British Journal of Developmental Psychology*, *20*, 259–280.

Pellegrini, A.D., Long, J.D., Roseth, C.J., Bohn, K., & Van Ryzin, M. (2007). A short-term longitudinal study of preschool children's sex segregation: The role of physical activity, sex, and time. *Journal of Comparative Psychology*, *121*, 282–289.

Pellegrini, A.D., Roseth, C.J., Mliner, S., Bohn, C.M., Van Ryzin, M., Vance, N., ... Tarullo, A. (2007). Social dominance in preschool classrooms. *Journal of Comparative Psychology*, *121*, 54–64.

Pellegrini, A.D., Van Ryzin, M.J., Roseth, C.J., Bohn-Gettler, C.M., Dupuis, D.N., Hickey, M.C., & Peshkam, A. (2011). Behavioral and social cognitive processes in preschool children's social dominance. *Aggressive Behavior*, *35*, 1–10.

Piaget, J. (1932). *The moral judgment of the child*. New York: Free Press.

Piaget, J. (1966). *Judgment and reasoning in the child*. London: Routledge & Kegan Paul.

Preuschoft, S., & van Schaik, C.P. (2000). Dominance and communication: Conflict management in various social settings. In F. Aureli & F.B.M. de Waal (Eds.), *Natural conflict resolution* (pp. 77–105). Berkeley: University of California Press.

Roseth, C.J. (2006). *Effects of peacekeeping and peacemaking on preschoolers' conflict: A multi-method longitudinal study*. Unpublished doctoral dissertation, University of Minnesota, Twin Cities.

Roseth, C.J. (2016). Character education, moral education, and moral-character education. In L. Corno & E.M. Anderman (Eds.), *Handbook of educational psychology* (3rd ed.). New York: Routledge/Taylor-Francis.

Roseth, C.J., Chen, I-C., & Thelamour, B. (2015, March). Resource control strategies and conflict outcomes: A dyadic analysis. Paper presented at the Society for Research in Child Development (SRCD) biennial meeting, Philadelphia, PA.

Roseth, C.J., Pellegrini, A.D., Bohn, C.M., Van Ryzin, M., & Vance, N. (2007). An observational, longitudinal study of preschool dominance and rates of social behavior. *Journal of School Psychology*, *45*, 479–497.

Roseth, C.J., Pellegrini, A.D., Dupuis, D.N., Bohn, C.M., Hickey, M.C., *et al.* (2008). Teacher intervention and U.S. preschoolers' natural conflict resolution after aggressive competition. *Behaviour*, *145*, 1601–1626.

Roseth, C.J., Pellegrini, A.D., Dupuis, D.N., Bohn, C.M., Hickey, M.C., Hilk, C.L., & Peshkam, A. (2011). Preschoolers' bistrategic resource control, reconciliation, and peer regard. *Social Development*, *20*, 185–211.

Ross, H., Ross, M., Stein, N., & Trabasso, T. (2006). How siblings resolve their conflicts: The importance of first offers, planning and limited opposition. *Child Development*, *77*, 730–1745.

Rutland, A., Killen, M., & Abrams, D. (2010). A new social-cognitive developmental perspective on prejudice: The interplay between morality and group identity. *Perspectives on Psychological Science, 5,* 279–291.

Sackin, S., & Thelen, E. (1984). An ethological study of peaceful associative outcomes to conflict in preschool children. *Child Development, 55,* 1098–1102.

Savin-Williams, R.C. (1987). *Adolescence: An ethological perspective.* New York: Springer.

Selman, R. (1980). *The growth of interpersonal understanding.* New York: Academic Press.

Shantz, C.U. (1987). Conflicts between children. *Child Development, 58,* 283–305.

Shantz, C.U., & Hartup, W.W. (1992). *Conflict in child and adolescent development.* New York: Cambridge University Press.

Shantz, C.U., & Shantz, D.W. (1985). Conflict between children: Social-cognitive and sociometric correlates. In M.W. Berkowitz (Ed.), *Peer conflict and psychological growth.* San Francisco: Jossey-Bass.

Shantz, D.W., & Shantz, C.U. (1982, August). Conflicts between children and social-cognitive development. Paper presented at the annual meeting of the American Psychological Association, Washington, DC.

Shonkoff, J., & Phillips, D. (2000). Rethinking nature and nurture. In J.P. Shonkoff & D.A. Phillips (Eds.), *From neurons to neighborhoods* (pp. 39–56). Washington, DC: National Academy Press.

Sluckin, A., & Smith, P.K. (1977). Two approaches to the concept of dominance in preschool children. *Child Development, 4,* 917–923.

Smith, P.K., & Connolly, K. (1972). Patterns of play and social interaction in pre-school children. In N. Blurton Jones (Ed.), *Ethological studies in child behaviour* (pp. 65–96). Cambridge: Cambridge University Press.

Strayer, F.F., & Santos, A.J. (1996). Affiliative structures in preschool peer groups. *Social Development, 5,* 117–129.

Strayer, F.F., & Strayer, J. (1976). An ethological analysis of social agonism and dominance relations among preschool children. *Child Development, 47,* 980–989.

Tinbergen, N. (1951). *The study of instinct.* Oxford: Clarendon Press.

Vaughn, B.E., & Santos, A.J. (2007). An evolutionary-ecological account of aggressive behavior and trait aggression in human children and adolescents. In P.H. Hawley, T.D. Little, & P.C. Rodkin (Eds.), *Aggression and adaptation: The bright side of bad behavior* (pp. 31–64). Mahwah, NJ: Erlbaum.

Veenema, H.C., Das, M., & Aureli, F. (1994). Methodological improvements for the study of reconciliation. *Behavioural Processes, 31,* 29–38.

Verbeek, P. (1996). *Peacemaking in young children.* Unpublished doctoral dissertation, Emory University.

Verbeek, P. (2008). Peace ethology. *Behaviour, 145,* 1497–1524.

Verbeek, P., & de Waal, F.B.M. (2001). Peacemaking among preschool children. *Journal of Peace Psychology, 7,* 5–28.

Verbeek, P., Hartup, W.W., & Collins, W.A. (2000). Conflict management in children and adolescents. In F. Aureli & F.B.M. de Waal (Eds.), *Natural conflict resolution* (pp. 34–53). Berkeley: University of California Press.

Vespo, J.E., & Caplan, M.Z. (1993). Preschoolers' differential conflict behavior with friends and acquaintances. *Early Education and Development, 4,* 45–53.

Westlund, K., Horowitz, L., Jansson, L., & Ljungberg, T. (2008). Age effects and gender differences on post-conflict reconciliation in preschool children. *Behaviour, 145,* 1525–1556.

Whitesell, N.R., & Harter, S. (1996). The interpersonal consequences of emotion: Anger with close friends and classmates. *Child Development, 67,* 1345–1359.

8

The Role of Relationships in the Emergence of Peace

Ellen Furnari

Significant research has been done in the last two decades addressing the basic questions of whether peacekeeping interventions are successful and the conditions that contribute to success (Druckman *et al.* 1997; Bellamy and Williams 2005; Fortna 2008; Fortna and Howard 2008; Hegre *et al.* 2010). Peacekeeping, as used in this chapter, refers to organized action by third parties to prevent violence, protect civilians, and support local problem solving by controlling or influencing belligerents and/or their proxies as well as local people. This body of research addresses the following questions: is peacekeeping successful at decreasing the likelihood of return to war? If undertaken by the UN or other multilateral institutions, is it equally successful? And is the success impacted by national or regional contexts such as levels of poverty, geography, availability of natural resources, or involvement of other nations in the conflict? Much of this research uses statistical methods to examine the relationship between variables. There is variation in this research as to how success is defined or understood (Druckman *et al.* 1997). Success is defined by some as no return to war within five years (or some other time period), measured by 1000 or more battle-related deaths in a given year (Fortna 2008; Hegre *et al.* 2010). Others consider success to require fulfillment of a mandate or other broader criteria (Pushkina 2006; Howard 2008; Martin-Brûlé 2012). Recent research has also addressed the question of whether peacekeeping interventions succeed at protecting civilians; under what conditions; and, if so, whether the UN and other institutions are equally successful (Kreps and Wallace 2009; Hultman 2010; Hultman *et al.* 2013). In general, this research finds peacekeeping to make a significant contribution to preventing a return to war but less successful at meeting other criteria for success. The research generally suggests that UN peacekeeping is somewhat more effective than that undertaken by other organizations. Lastly, the research finds that peacekeeping has a poor record of protecting civilians, although, again, UN peacekeeping may do better at this than other organizations, and it may be improving.

Little of this research considers how actual practices of peacekeepers in the field contribute to effective peacekeeping, although Howard (2008) considers integrative learning, referring to learning that occurs during the intervention and contributes to

Peace Ethology: Behavioral Processes and Systems of Peace, First Edition.
Edited by Peter Verbeek and Benjamin A. Peters.
© 2018 John Wiley & Sons Ltd. Published 2018 by John Wiley & Sons Ltd.

improved practices, one of the critical factors for success. This chapter discusses findings from research with frontline peacekeepers that elicited their analysis of what contributed to effective peacekeeping (Furnari 2014). The research was guided by a desire to contribute to strengthening interventions by third parties into armed conflict within communities, regions, and nations. It was based on an assumption that those who did the work of peacekeeping would have perspectives and insights different from those of their organizational superiors or academics. In other words, the research investigated effective peacekeeping from a ground-level view.

The research used constructivist grounded theory methods. This approach assumes that meaning is constructed through interactions. In other words, how people understand each other and the surrounding world is socially constructed. In the case of this research, meaning is constructed between the researcher and participants in the research reflecting on their experiences. Furthermore, it assumes that meaning depends on who is doing the constructing, is contingent, and changes over time. This is in contrast with an approach that assumes there is truth that would be the same for all and that can be discovered or uncovered through specific methods (Moses and Knutsen 2007). Specifically, the research was carried out by interviewing former, as well as a few current, peacekeepers in person or via computer (Skype) technology. The interviews were semistructured – there was a general plan, but each interview was different – and generally lasted a bit over an hour. Initial interviews were carried out with people reached through personal contacts and the contacts of colleagues, sometimes referred to as a convenience sample. Most of the interviews were conducted with participants referred by people who had previously participated, referred to as a snowball sample.

The research included interviews with 57 former or current peacekeepers between June 2011 and November 2012. Participants served as military, police, civilians, or unarmed civilian peacekeepers and included 39 men and 18 women ranging in age from their mid-20s to late 60s at the time of the interviews. They came from 19 different countries: Australia, Canada, Egypt, Finland, Germany, Ghana, India, Ireland, Kenya, Nepal, New Zealand, Poland, South Africa, South Korea, Sri Lanka, Sweden, Switzerland, the United Kingdom, and the United States. They served in UN, NATO, and African Union missions as well as in unarmed civilian peacekeeping projects fielded by Peace Brigades International (PBI) and the Nonviolent Peaceforce (NP). Some of them served in only one intervention, but 34 served in two or more missions or the same mission at least twice. References in this chapter to what peacekeepers think or understand refer specifically to findings from these interviews and do not imply generalization to most or all peacekeepers.

A number of factors understood to impact effectiveness were identified in this research, including organizational issues such as leadership, adequate logistical equipment and support, as well as concerns about internal organizational politics. Local people's perception that the intervention was nonpartisan (not supporting one political faction or another) was also frequently mentioned as an important factor. Factors related to their actual practices included a concern that the intervention was accepted by local people; that the intervention supported local ownership, also referred to as local problem solving; and that the intervention developed "good" relationships across many social sectors. A relationship here refers to recurring interactions between two or more people through which, or during which, they are able to influence each other (Reis *et al.* 2000). Importantly, this highlights interactions where people influence each other:

these interactions vary over time, and thus, the character of the relationships can vary as they develop. These factors, that were understood to contribute to effective peacekeeping, are interrelated, and many believed good relationships are key.

The analysis discussed in this chapter, based on the research described here, suggests that peacekeeping relationships tend to be task oriented and move within a range from mistrustful, coercive, and disparate interests to relationships characterized by trust, shared interests, and cooperation, as well as everything in between (Hardin 2002; Deutsch 2011a, 2011b). They tend to encompass both formal and informal aspects and are multidirectional. They function in contexts of conflict-affected communities where violence has challenged trust and connections. This chapter elaborates on the findings and discusses what is meant by good relationships, how violence decreases and peace increases in part through developing good relationships that contribute to effective peacekeeping, and some of the implications of this analysis for peacekeeping practice.

Peacekeeping Takes Place in Relationships at the Local Level, on the Ground in Communities

While earlier peacekeeping tended to focus on observing a ceasefire at a physical border, today's peacekeepers primarily address intrastate conflict and work in communities in direct contact with others. Thus, peacekeepers working on the ground in communities experience the agency of local people, whether they are members of the national military, police, other armed actors, government officials, local civic and religious leaders, or just local people in their everyday activities in their home areas or internally displaced persons (IDP) camps. As Kalyvas (2003) and Leonard (2013) point out, local conflicts, or their absence, impact the overall possibility of maintaining peace, and the potential for local conflict is complex. Johnstone (2011) suggests that peacekeeping can be understood through relational contract theory, which addresses the need for flexibility within formalized, contractual relationships when there are ongoing interactions over a length of time rather than a short-term, single incidence. In these contexts, it is critical to maintain a generally cooperative relationship while carrying out contractual obligations. Applied to peacekeeping, Johnstone emphasizes that while initial consent to an intervention may have been given, peacekeeping interventions need to maintain relationships with civilians as well as the government for the duration of the intervention through many changing circumstances. Given the need for effective peacekeeping practices in the community over time, it seems important to understand what frontline peacekeepers themselves find effective in their day-to-day work.

Peacekeepers interviewed indicated that developing and maintaining good relationships comprise one of the critical factors for being effective in these local contexts. Like most human activity, peacekeeping takes place in relational fields. Relationships are the context in which peacekeeping activities take place, be they coercive or cooperative, as well as the vehicle by which peacekeeping has effect. In other words, peacekeeping takes place in a complex network of interacting relationships, and it is through "good" relationships that peacekeepers are able to exert influence to prevent violence and protect people. Peacekeepers also noted that it is through changes in behavior and what people said to them (implying in relationship) that they knew what was effective. Although discussing peacebuilding, Lederach (2005) uses the phrase "centrality of relationships" and describes relationships as both the context of peacebuilding and the source of generative energy to build peace in communities that have suffered great violence. He notes, "At the cutting edge of fields from nuclear physics and biology to systems theory and organizational development, relationships are seen as the central organizing concept of theory and practice" (p. 34). Lederach uses the imagery of webs of relationships as the field in which peacebuilding occurs and that it must affect.

Reflecting on the importance of relationships, one former military peacekeeper interviewed by the author stated, "Relationships are more important than any body armor or mine-resistant vehicle the government could purchase.... It is a human terrain, and in order to operate effectively on that human terrain, you have to have relationships." Peacekeepers described good relationships as being characterized by trust, mutual benefit, and cooperation (the working definition of good relationships used in this chapter). There was no significant difference between men and women interviewed, although the only three who felt that weapons were more important than relationships were men. This seemed related to their shared experience of being military peacekeepers in Bosnia during the time the mission shifted from operating under the UN to operating under NATO guidelines, which altered the rules of engagement. They believed that the more permissive use of military power made a huge difference that could not have been achieved in any other way. Although friendships may develop, peacekeepers described relationships that were primarily task oriented, focused on preventing violence, protecting people, and supporting local ownership of the process (the purposes of peacekeeping most often described in interviews). It takes time to build relationships; they are not onetime occurrences. Mac Ginty (2008) makes the point that in many societies, ongoing relationships are critical in that decisions are made through ongoing processes, not at onetime meetings.

Peacekeepers felt that good relationships were interrelated with both acceptance and supporting local ownership in effective peacekeeping. Local community acceptance of the mission and the specific local intervention created a context more conducive to good relationships. Some peacekeepers felt their missions were partisan, imposing solutions or undermining local initiatives, and this, in turn, undermined acceptance and made it more difficult to develop good relationships. Others noted the positive contribution that supporting local problem solving contributed to acceptance and good relationships.

Peacekeepers view relationships as core processes for keeping peace. Relationships contribute to preventing violence in a number of ways, including: with good relationships, local people warn peacekeepers of impending violence; local people approach peacekeepers with information and appropriate requests; local government, civil society leaders, and others cooperate with peacekeepers; and armed actors are influenced within relationships and respond to peacekeepers' concerns. Peacekeepers shared many explanations for why relationships were more powerful at preventing violence, in the long run, than weapons. While weapons might provide immediate influence in a situation, deterring violence or protecting someone, they do not change the underlying context. And, as the news demonstrates daily, having the biggest and most weapons does not automatically produce the desired results. Building relationships characterized by trust and cooperation provides a context in which peacekeepers can influence those who are themselves armed actors. Peacekeepers can gain valuable information, revise their own understanding, and even derive protection in and through these relationships. The unarmed civilian peacekeepers (UCPs) described relying solely on relationships to deter violence. For example, a former UCP explains how developing a complex network of relationships protected people through the influence they could exert in the area where she worked:

> But the big change was in dynamics, leveraging influence, establishing relationships that can influence actors, change dynamics. I do think we protected people via complex relationships, networks, coordination. Over time, we influenced decision makers, as dynamics changed. We had vertical and horizontal influence, people and events came together, overtime, a complex matrix.

This quote highlights how violence decreases and peace develops in and through relationships. Similarly, a military peacekeeper who had served in Africa suggested that in the beginning, you need to use military power, but said in the long run, relationships are critical to accomplishing peacekeeping goals:

> Obviously, if you have a good relationship with the specific rebel leader, he is much more open to you, much more positively disposed toward you. It means you'll be able to operate in that specific area, you will be able to approach them without being shot at.... The deeper the relationship develops, the more keen they will be to listen to what you want to accomplish, to listen to what you want to say. But it is obvious the better that the relationships are with the people that you work with, the better progress you will make for sure.

One of the UCPs interviewed explained the power of having a broad network of relationships, as key to effective peacekeeping:

> UCP strength is the relationships. And I think why it works, is the ability of people [UCPs] to live in the communities, where threats are happening, where there is violence, and to build relationships with all the key stakeholders, civil society, security, there on the ground … they also create a lot of respect for them, not just as internationals but as human beings who have decided to do this kind of work.

Protecting civilians is a key task and expectation for most peacekeeping missions today, and peacekeepers believed civilian protection to be a key purpose or goal of their work. As already mentioned in this chapter, good relationships help provide early warning and support preventive responses. This was understood to result from increased sharing of information, and accurate information, between peacekeepers and civilians and between peacekeepers and armed actors. Additionally, having a good network of relationships meant peacekeepers could understand the situation from various perspectives rather than relying on one particular bias. This supported more effective analysis and interventions. Peacekeepers also felt that with good relationships, protection was reciprocal. People warned of impending violence and areas to avoid when there were good relationships.

Peacekeepers from all backgrounds also acknowledged there were times when coercion was necessary. When cooperation and good relationships did not yet exist or circumstances had changed, peacekeepers attempted to coerce armed actors, pressuring them to refrain from attacking civilians and/or each other. At times, they also attempted to restrain government actions that, while not physically violent, infringed on civilians' human rights. While military peacekeepers used the threat or actual use of weapons as part of their coercive activities, UCPs relied on the coercion available through a network of relationships and the pressure that could be exerted through these networks. Peacekeeping exists within a wider network of relationships regionally, nationally, and internationally. Interventions can draw on the potential ramifications of drawing attention to violence against civilians or the breaking of a peace agreement (Mahony 2006; Mahony and Nash 2012). Most governments and other armed actors want to be seen as legitimate in the eyes of their constituency and the wider world. In addition to the threat of losing this perception of legitimacy, there may be a threat of sanctions, including the possibility of charges in the International Criminal Court.

The Dilemma of Carrying Weapons

The carrying and potential use of weapons posed a dilemma for a number of the military and police peacekeepers interviewed. They believed being armed was essential at times to protect themselves and be able to uphold peace and protect others. Here, a military peacekeeper speaks to this dilemma, valuing the benefits of relationship and engagement but worrying about self-protection:

> The security dilemma – the need to be out in the community versus security … the physical risk of staff versus being close to the local people … the view [that] having close relationships with local people, the more secure you are, in the sense you will be warned of anything that might happen, that they think anyone is meaning you harm, and that is certainly how it worked when I was there, because we were out in the community, we relied on the community to help us with our security, to help protect us…. But that was the conflict, the balance of the risk to ourselves, our own security, [versus] the need to be among the people and have good relationships and understanding what was going on.

As he and other military and police peacekeepers noted, they believed that being armed made it more difficult to establish and maintain good relationships, and yet they also believed they needed to be armed for self-protection, even though this undermined being effective. Some noted that people were more afraid or avoidant of them and that it made it more likely to be seen as dominating rather than helping. For example, a military peacekeeper who has served in several missions, sometimes as a soldier and sometimes as an unarmed observer, noted the differences in his relational experiences when he was armed or unarmed:

> When I used to go to the villages, [when] I was in the military, people do not prefer to see a stranger approaching to you, an armed stranger approaching to you, a stranger with a weapon. When I had no weapons, I could access people, people had confidence at the first sight…. People used to welcome us more, but if you are there with a weapon, you are looked at in a different way. People used

to think about it before talking with you. But without a weapon we have more, military observers have more access. Which is very true, that is what I have found myself.

Several peacekeepers pointed out that soldiers are trained to kill and win, and much of the relational work of peacekeeping might be better done by others with different training. Last (1995), in his research on the practice of peacekeeping, identifies the tension between the need to use military fire power versus the use of negotiations and other kinds of contact interventions (interventions that depend on contact with belligerents and civilians) as one of the key problems in peacekeeping. The NORDCAPS *Tactical Manual* (2007), while defending the use of the minimal force necessary, nonetheless notes that the use of force may undermine the legitimacy of the mission, leading to the withdrawal of consent and the failure of a mission.

UCPs believed they were safer because they were unarmed; carrying weapons could be seen as a threat and might draw violence toward them. At the same time, several addressed the limitations of being unarmed, acknowledging there were situations in which they could not operate if their own safety was at imminent risk and/or their presence was likely to increase harm to civilians nearby. Fundamentally, unarmed civilian peacekeeping by its very definition is committed to nonviolence and using nonviolent tactics to promote peace. One unarmed peacekeeper explained, for instance, that using weapons to keep peace did not lead to long-term peace, but rather maintained a focus on violence. She said,

> [I]f the only way people get from point A to point B is just because we have guns, then you get bigger guns, you haven't changed anything. Protection still comes from the same source; you haven't changed the dynamics. If it is about guns, it will escalate. But if you're trying to have a transformative impact, you are trying to work with people who work nonviolently … you aren't actually changing the situation. You want to change the source of power.

Peacekeeping can be understood, then, as moving within a system or spectrum of relationships that encompasses good relationships characterized by trust, mutuality, and cooperation, as well as coercive relationships characterized by threats and, at times, physical violence. Good relationships are understood to be more effective at creating sustainable peace, causing dilemmas for armed peacekeepers in particular.

Understanding Peacekeeping Relationships

As one peacekeeper said, "These are not normal relationships." Peacekeeping relationships have a number of peculiar characteristics. When the intervention begins, peacekeepers arrive as strangers and have to build relationships across many social sectors in multiple directions. While the intervention may take place over many years, specific peacekeepers come and go, often being in the field for only six months or a year (although UCPs tend to stay longer). So, in a sense, local people are often relating to strangers. Peacekeepers may not share similar cultures among themselves or with the local population. As noted in this chapter, these are primarily task-oriented relationships and so move between coercion and resistance to cooperation and back again. This process is based primarily on analysis of the context rather than emotion or personal conflict, and the stakes are high for all those involved. Good relationships can save lives (both civilian and peacekeeper) and/or strengthen local capacities and infrastructures for sustainably dealing with conflict nonviolently. Conversely, the absence of good relationships may contribute to continued conflicts and even death. These relationships take place in a broader context impacted by dynamics of struggle for power and dominance with related struggles over economic, cultural, and religious inclusion and exclusion, and they are impacted by international agendas and perceptions of the legitimacy of the intervention itself. The potential for building relationships is impacted when interventions are perceived as undermining local power structures, culturally inappropriate, and any number of other problematic systemic issues. Nonetheless, there are some sensitizing concepts that point to aspects of these relationships that deepen our understanding and highlight areas for further research.

Social capital is one theoretical framework that offers useful insights to thinking about these relationships (Nan 2009; Paffenholz 2009). Social capital refers to the strengths and capacities that develop when people are connected to each other in networks that operate cooperatively and for mutual benefit to some degree. One can think of peacekeeping as engaged in building networks between peacekeepers and others but also (re)building networks between various local social sectors directly. For example, peacekeepers can help bridge connections between conflicted groups by providing increased security, rumor control, and what some call shuttle diplomacy. This may simultaneously support increased social capital while preventing violence. One peacekeeper noted that while they needed good relationships in the communities where they worked, it was more important to support local people in building good relationships with government, aid agencies, and other critical actors. Another described a period of work to build a network that included diverse ethnic groups. When lethal violence flared in a nearby community, this network mobilized members and helped prevent the spread of violence to their area. For networks to be of mutual benefit and for members to cooperate, there must be a degree of trust. It may be a limited trust around specific issues or a more inclusive trust in each other's reliability.

There is a large body of work on trust (Nannestad 2008; Cox 2009; Newton and Zmerli 2011), which was the most frequently used word to describe good relationships. Trust was described as indicating a belief that the other encapsulates your interests (Hardin 2002), or, as peacekeepers said, having shared interests and mutual benefit. How peacekeepers and local people understand each other and the potential for mutual benefit seems both critical and complex. Murphy (2006, p. 429) conceptualizes trust as "a sociospacial process enacted by agents through relations mediated by structural factors, power differentials, emotions, meaning systems and material intermediaries." In other words, trust takes place in specific places and within particular social contexts, involving particular people. It is impacted by power, emotions, how those involved understand what they are about, and the actual physical objects involved in their interactions – be it money, guns, books, food, and so on. This highlights that these relationships occur over time and are not static. There are power differentials, both obvious and subtle. And the sense of trust emerges from the meaning local people and peacekeepers make of their interactions over time and in conflict-affected contexts.

A concept in the trust literature of particular interest to peacekeeping is termed the control dilemma (Miller 2004). Briefly stated, there is a dilemma between trying to control the other and trusting them in order for a job to be done well. This arises in part because it is not possible to specify and control everything in most jobs, and the attempts to do so may actually undermine performance. On the other hand, inadequate control may also contribute to poor performance, if people do not do what they should.

Jagd (2010) describes trust and control as related processes, not static attributes or concepts, and analyzes the circumstances in which they can complement or undermine each other. While his examples of control are related to typical businesses, and thus without discussion of nonviolent or force-based coercion, there are a number of useful parallels with peacekeeping. In the business world, control is thought to be needed when there is a high level of risk and uncertainty about another's intentions or actions. Trust is thought to be needed in complex environments when there is a need for flexibility and when excessive control might limit responsiveness to change. Thus, control is seen to be relevant when the larger risk relates to the (mis)behavior of others, and trust is seen to be needed in dynamic environments and where tasks cannot be easily circumscribed. He notes that both trust and control are emergent and in flux and are processes better described as trusting and controlling. Trusting is needed because of the uncertainty and risk in situations where the achievement of the goals or interests of actors depend on each other to some extent. He suggests that the expectations people have of each other in the beginning of a relationship affect the development of trust.

This body of work offers some fruitful lines of questioning for peacekeeping. It would be worthwhile to explore when trust and control complement or undermine each other in peacekeeping contexts. Control or coercion in peacekeeping occurs when peacekeepers perceive others as untrustworthy in some important ways, such as belligerents threatening to attack civilians or local authorities violating human rights, but many peacekeepers say coercion and the use of force may undermine trust and good relationships. There are significant risks in peacekeeping. Inaccurate information and analysis can lead to death for peacekeepers and local people at worst and, at the very least, may undermine political aspects of a peace process. It is not possible to exert complete control over belligerents, local authorities, or people in communities, nor over peacekeepers themselves.

Further investigation of this tension between trust and control in peacekeeping might prove fruitful as peacekeepers, military doctrine, and peacekeeping institutions themselves all refer to this control dilemma when discussing the need for a credible military threat while also recognizing that the use of military firepower may undermine acceptance and cooperation.

Relationships can also be understood according to various typologies and categorizations. Deutsch (2011b) describes a typology of social relationships based on interdependence constructed across a number of dimensions. By interdependence, he means either the degree to which people mutually need each other to achieve their goals or the degree to which the achievement of one person's or group's goals depends on the failure to achieve the goals of the other. In other words, he considers how much the actions of one affect the other. Interdependence does not necessarily imply cooperative or beneficial interdependence. Deutsch uses the term "negative interdependence" for competitive relationships in which the success of one depends on the failure of the other.

It may be helpful to think of peacekeeping with these concepts to better understand what kind of relationships are implied by peacekeepers' desire for "good" relationships. Yet the structure and language of categories and typologies imply dichotomies and distinctions that are often blurred or unclear and perhaps obscure the ways in which peacekeeping relationships, even between the same two people, change and evolve over time and as needed.

Deutsch (2011b) suggests that relationships can be thought of in two overarching categories: socio-emotional or task oriented. While Deutsch notes there is a continuum, he nonetheless tends to classify relationships as one or the other. One would generally consider peacekeeping as more task oriented. Listening to peacekeepers, however, it appears that in some circumstances, it is important to pay attention to the emotional, more informal, and even amicable aspects of a relationship. This was discussed in particular in terms of mentoring relationships where peacekeepers mentor other police or military, with coworkers, and in terms of building more personal connections with many different kinds of people in communities. Cross-cultural interaction further complicates the experience of relationship in peacekeeping, as what are considered appropriate task-oriented behaviors in one culture may not be in another. In the author's own experience working in Sri Lanka, the first part of most meetings started with asking each other about families and health, and talking about generalities. To try to address a task too soon was considered very rude, and it was generally ignored. In another example, a police peacekeeper, who has worked in several different missions, talks about how mentoring depends on close relationships with the police he is mentoring, that he can only work with a few at a time, and that his interactions must support mentees in developing confidence and skills. To do so, in one context, he has to go out drinking with those he is mentoring, even though it is not an allowed activity:

> [Effective mentoring] is about giving the local police confidence in what they do.... You have to go and sit beside the people you are working with.... You have to basically work alongside them and have important discussions and basically giving the local police confidence to actually go out into their community and actively talk and actively discuss, and by you being there, having that presence, well first of all you empower them ... [and culturally you have to engage in friendship activities too:] I'll have some beers with you, though it is a no-no ... because everyone in his service did it, it's the norm. And in the end, I can remember having some very beery afternoons sitting in a grass hut with some of the other police leaders, getting horribly drunk, and after that you were their best buddy.

Deutsch (2011b) describes a number of other relational dimensions along dichotomous lines, which combine to produce specific categories. These include cooperation versus competition, equal versus asymmetrical power, formal versus informal, and intensity of importance. He notes that a number of other dimensions appear in the literature, including temporary versus long term, voluntary versus involuntary, public versus private, as well as the number of people involved in a relationship (Deutsch 2011b).

While these are useful concepts for exploring peacekeeping, it is important to keep in mind that the process of relationship building in peacekeeping is in flux. Peacekeeping relationships are initiated and maintained with people from governments, local military and police personnel, other belligerent groups, and all sorts of groups and individuals. Peacekeepers report believing they have more power in some situations but less power in others. They may have both formal and informal aspects of a relationship with people, and whatever dimension is under consideration, it evolves over time. Significantly, perceptions of the degree of cooperation may be different between peacekeepers and the other people and groups with whom they are building and maintaining relationships.

As discussed in this chapter, peacekeepers describe striving to create and maintain cooperative relationships. Cooperative relationships are thought to have a preponderance of common goals, while conflictual or competitive relationships give disparate goals more salience or weight. Coleman *et al.* (2012) elaborate a model of conflictual or cooperative relationships, noting that the degree of cooperation or conflict will fluctuate over time and depends significantly on context. The experience of cooperation in peacekeeping may relate to the degree to which the intervention supports local ownership and the degree to which local people accept the intervention. It would appear that when peacekeepers are supporting local problem solving through local efforts, there is likely to be a higher degree of shared goals. Similarly, it is possible that the perception of shared goals is a significant component of local acceptance.

Two other dimensions highlighted by Deutsch (2011b) and Coleman *et al.* (2012) that also seem important in peacekeeping relationships are equality or power and the importance of the goals and the relationship's ability to impact them. Some relationships have shared or oppositional goals regarding issues that are of little importance. In other relationships, the goals are of greater importance, but the relationship may have little ability to influence achieving these goals. Applying this to peacekeeping, it would seem likely that peacekeepers' goals to prevent physical harm and stop armed conflict would be shared by many in the community and would be of greater importance. For instance, a peacekeeper here suggests that through building good relationships with the government and different groups of nonstate actors, they can both allay fears and misperceptions and work together on issues of concern, even reaching a point where criticism can be shared with positive results:

> We have to constantly engage the government and non-state actors which are there. This is an issue of trying to [address] any misunderstanding that we come in as an organization that wants to create trouble for the government or non-state actors.... We have regular meetings with different government ministries that are of great importance to our work here, to allay the fears of the government that we might do some subversive work … when we do peacekeeping, it is about building confidence and relationships: the more comfortable I am with you, the more I will buy your ideas, will support what you do, be interested in what you do.... If you really engage them well, it really plays well with our work on the ground … you find that your presence will be so much welcome by all parties, so the issues that might arise concerning our own intervention, concerning our own understanding of what is happening on the ground, will be so appreciated, and if there is a criticism to either party, it will be taken with some positiveness, rather than negativity.

Presumably, however, these peaceful goals would not be shared by all, as some groups may wish to continue to use physical violence to achieve their desired outcomes. Other goals currently involved in many peace support missions, such as government and security sector reform or developing free market economies, may have less or no shared interest with local people or may be of a lower priority than their physical safety or basic welfare. Conversely, some local groups may have goals that are not shared by peacekeepers such as dominating through the use of armed violence, forcing particular groups to move, or profiting from a no-war/no-peace situation. Furthermore, it may be that while shared, some aspects or goals are of little importance to particular people and groups, such as building relationships through playing sports together. Lastly, whether shared or not, peacekeepers may or may not be significantly able to influence the achievement of these goals.

Bringing together all of the above dimensions, it seems that peacekeepers perceive effective peacekeeping as more likely within relationships that share the following characteristics: primarily task oriented; able to identify, articulate, and work on shared goals cooperatively; a shared understanding of who has the power to affect what; and with shared assessments of what is more important to accomplish. This literature does not, however, address how peacekeepers move between cooperative and coercive interactions and how that affects their ability to maintain effective relationships. Nor does the research reviewed here, as it was an issue that emerged after the research was completed.

Higate and Henry (2009) discuss peacekeeping as embodied performances with props, by which they mean that real people bodily perform peacekeeping acts with specific materials. They emphasize that the props, which are weapons, other equipment, transportation, and the like, must be convincingly menacing for the performance to have a coercive influence. Deutsch (2011a) theorizes that acting cooperatively (i.e., acting as if one shares goals, trying to solve problems together, fostering mutual respect, and acting benevolently if one has more power) creates the conditions for cooperation. On the other hand, behaviors that indicate competition create conditions for further competition and negative interdependence. The so-called weapons effect theorizes not only that people carrying weapons are more likely to act aggressively but also that weapons may bring out anxiety and fear in others (Turner *et al.* 1977), responses not conducive to building positive relationships. Thus, military peacekeepers and their weapons may symbolize an embodiment (as well as enactment) of coercion upon arrival in a community, a negative interdependence that may trigger anxiety and even resistance and that must then be overcome in the process of creating "good" relationships. UCP peacekeepers arrive with actions and props that may symbolize and embody their intention for and reliance on cooperative interdependence. They must work to establish sufficient understanding of their capacities, however, to be able to be effectively coercive with those threatening violence. The performances of military and UCP peacekeepers are at least symbolically, and in many cases actually, quite different in terms of weapons, living conditions, and the like, yet both use coercive methods at times, which may undermine trust and cooperation. Still, these differences in performance may account for the perceptions many peacekeepers shared (both armed and unarmed) of the advantage of being unarmed. One military peacekeeper reflected on his experience that when his mission shifted to being unarmed, it was more fully accepted, and further discussed how being accepted led to information and cooperation, which was essential for their operation. He said,

> Better if you are unarmed and going to the people ... and going without the weapon. If you patrol out in the streets, with heavy equipment and APC [armed personnel carriers] ... they just got away from these things in the war ... they say – we just got away from all these things, now you people come with the same things ... what people told me, why do you come, what are you afraid of, why do you come in a big vehicle ... it was difficult to convince them it was for their security, later we started going without weapons ... and the people welcomed that.

As peacekeepers say, they are perceived differently whether armed or unarmed, which leads to different potentials for relationship building.

The relational model, based in ethology, provides another set of sensitizing concepts. In general, the model suggests that peace-promoting or reconciliation-promoting behaviors often follow conflict behaviors, and this seems to reflect the value of the relationship, as well as the confluence of shared interests to those involved (de Waal 2000). This research directs attention to the value of the relationship to each individual involved, the importance of the conflict versus the presence of shared goals, conflict as related to negotiating hierarchy, and the risks associated with the conflict such as physical harm or losing the relationship. The research on which this model is based (see de Waal [2000] for a historical summary) resulted from the study initially of animals and then of humans in stable social groups with shared history, culture, behavioral patterns, and so on, and it most frequently studied dyads, families, or small social groups. In general, the kinds of human conflicts studied did not involve lethal or potentially lethal physical violence. This is obviously quite different from the peacekeeping relational context described in this chapter. Still, there are some findings that point to areas for further exploration in understanding how peacekeeping relationships contribute to peace. The preference for cooperative relationships described by peacekeepers may reflect the context in which the need for each other is high for all who are genuinely working for peace and the risks associated with noncooperation are also high. The use of coercion by peacekeepers and by various armed actors may be part of a process to clarify power and dominance. Coercion reflects the existence of conflicting goals between peacekeepers and others, be they government officials, rebels, militias, criminal elements, or the like. At core, these conflicts concern peacekeepers' goals of preventing violence, protecting civilians, and supporting local nonviolent political struggle whether through social movements, elections, or other forms of contestation. Peacekeepers engage in coercive acts or processes to attempt to impose their goals, which tests their power or ability to dominate. So, while this chapter uses the word *coercion*, reflecting the language of peacekeepers, it clearly reflects the existence of conflicts in peacekeeping contexts.

Perhaps most interestingly, the relational model understands conflict as part of the natural process of negotiating connection and cooperation (de Waal 2000). The processes of reconciliation after conflict hold potential to develop deeper connection and cooperation. Unfortunately, this author's research does not address the processes by which peacekeepers, after a coercive incident or period, return to more cooperative periods. For instance, several UCPs described periods of time or incidents in which they understood their presence to have a coercive, violence-preventing impact. In other words, their presence coercively restrained potentially violent actors. Yet later they engaged cooperatively with the same actors. In an example of this dynamic, a UCP described a subtle coercive interaction with police, which resulted in a sort of cooperation. He shared,

> There was a [human rights] researcher we worked with, who went out to the field and interviewed people on human rights. Dark cars started to follow him around, there were threats to his family and friends about his life. We … made a plan that was acceptable to him. A couple of [UCP] folks went to the police office and mentioned our client, we're worried about our client, and we wanted you to know he is being threatened. We know you are in charge of safety, and we know you can keep him safe. And the threats stopped. We know the threats were coming from the police…. We took a non-confrontational, nonviolent approach. We knew we could bring more political clout to bear but this was effective, asking the head of the police to do his job and complimenting him.

In this case, the UCPs, in a sense, forced the police to stop threatening the researcher by expecting them to visibly do their job. The ability to apply pressure built on their previous relationship. The resulting cessation of threats was, in a sense, cooperation. They both were then working for a shared goal of maintaining police control of violence, which would lead to keeping the person safe. And they worked in a way that presumably made it possible to continue to work cooperatively in the future, despite having coerced the police to stop threatening and possibly even murdering this researcher.

This process of shifting between coercion and cooperation seems a critical area for further exploration. How do peacekeepers signal their intentions to shift from coercion to cooperation to armed actors and others in the communities where they work? How is the relationship, if not reconciled or repaired, at least recalibrated toward cooperation and trust? Does the previous period of coercion contribute to increased cooperation later?

Deutsch describes a "crude law of social relations" (Deutsch 2011a), which helps to draw these various strands together. He theorizes that acting cooperatively produces cooperation and vice versa with competition, using the analogy of genotype and phenotype. In other words, the initial behavior can be thought of as the genotype, which is the internal genetic coding in biology. The initial behavior sets a pattern. Phenotype references how this patterning is expressed externally, in further behavior in this case. Thus, the initial patterning sets the context for what follows, although it does not determine it precisely. This suggests that how peacekeepers enter a community is a critical factor. Entering in ways that symbolize and signal trust and expectations of cooperation may be more likely when unarmed or lightly armed. As expressed in several of the research quotes in this chapter, many peacekeepers implied that heavy weapons and body armor signal mistrust or at least make a peacekeeper less accessible to people (Furnari 2014). One peacekeeper who had served as both a military and a UCP peacekeeper summed it up nicely, saying, "Once you arrive with a uniform and a gun … even in civilian engagement, you are still a soldier, and all that means to people, even if you don't pull a trigger … you don't inspire people's confidence. Even if you only have a pistol, it affects women and children."

Rubinstein (2005) theorizes that peacekeepers are seen as the embodiment of international order. This suggests that the way peacekeepers are perceived locally is potentially affected not only by their own actions but also by the ways in which the overall intervention is understood in the local communities where they work. Some interventions are reported to be welcomed at the start. The perceptions peacekeepers and local people have of each other may be affected by their ongoing embodied performances (Pouligny 2006; Higate and Henry 2009), which indicate symbolically and concretely to each other cooperative or competitive intent and the extent of their power to impose their goals, if needed, within interdependent relationships. The achievement of preventing violence, protecting people, and supporting local ownership is positively interdependent in that peacekeepers and local people need each other to achieve their shared goals. It should be noted, however, that this does not address those local people who act to continue the violence, nor peacekeepers who are unwilling to protect people. Alternatively, it is negatively interdependent in that peacekeepers and local people need to block or overcome each other when they do not have shared goals.

Robust Relationships

This discussion of the importance of relationships has significant implications for the practice of peacekeeping. Some of the current discussion in the field of peacekeeping calls for more robust efforts (Tardy 2011). In general, robust peacekeeping implies a greater use of military force, although Sartre (2011) suggests robust peacekeeping is more complex. However, as the discussion in this chapter suggests, there are dilemmas related to the use of force. Cooperative relationships are considered most useful for effective peacekeeping in the long run. The use of coercion – and, in particular, the use of weapons – is understood by many peacekeepers to challenge and possibly diminish this cooperation. Whether because carrying and using weapons signals mistrust and an intention to be coercive, or that people are wary of approaching armed soldiers, or that military peacekeepers tend to live in barracks and only spend limited time in communities, as described, military peacekeepers reported dilemmas related to the tension between the need for weapons and the challenges posed by having them. UCP interventions developed from a long history of nonviolent practice (Moser-Puangsuwan and Weber 2000; Schweitzer 2010) and consciously use relationship building as a core practice (Wallis 2010). In the long run, to support effective violence prevention, protection of civilians, and long-term sustainable local peace efforts, building and maintaining reciprocally cooperative relationships comprise an essential component of effectiveness for all peacekeepers.

Thus, one of the implications of this research is the need to develop what might be termed robust relationships as a core strategy and practice in peacekeeping. Robust relationships are relationships strong enough to withstand both the movement between coercion and cooperation and the many stresses and strains on relationships that communities suffer during armed conflicts. Based on a solid understanding of good relationships as a key to effective and sustainable peacekeeping, a commitment to robust relationships as a core strategy would require a major reconfiguration in military peacekeeping. As one peacekeeper said,

> If you are going into a lot of cultures and pretending you care, but you upset them because you are working against their culture, and you are using force, there is no way they will come on board, but if you can meet them at a level where you can communicate with them, and gain their trust and support, then they will come to you with their problems, and you can look at their problems and problem solve with them.... I can't see by using a more robust force that will happen. I can see in other places, in conflicts, in missions around the world, if you go in with that attitude you are going to feel like you are taking over a community, you are doing it because you think it is good, but you are not being sensitive to their cultural issues, and not listening to them, it isn't going to work.

This suggests the need for more research to understand the tensions in military peacekeeping between the use of military force for protection and violence prevention and the need for robust relationships. When does the risk outweigh the potential advantage of being unarmed, and what does this imply for decisions on when and how to intervene? What are the actual practices that support the movement between coercion and cooperation? How do peacekeepers create, restore, and deepen cooperative relationships? This also suggests the need to train peacekeepers to be sensitive to developing good relationships and the related skills of listening, cultural sensitivity, and a variety of good communication practices.

Conclusion

Peacekeeping today primarily addresses intrastate conflicts, and peacekeepers mostly operate in communities, experiencing the agency of local people across multiple social sectors. Peacekeeping can be thought of as working within a field of relationships influenced by multidirectional interactions that build or undermine trust, attempt to assert influence or control, and take place in dynamic contexts affected by external forces that may contribute to further conflict or peace. Peacekeepers believe that good relationships characterized by cooperation, mutual benefit, and trust are a critical and central factor of peacekeeping effectiveness. Good relationships contribute to protecting civilians and interrupting armed violence by supporting accurate information and analysis, early warning and mutual protection, and cooperation around shared goals. Peacekeepers found that cooperation was more effective than coercion at meeting their goals of preventing violence, protecting civilians, and supporting local ownership of peacebuilding processes. The stakes are high. Ineffectiveness can lead to disrupted peace processes, destruction of property and livelihoods, and possibly death for peacekeepers and/or the civilians they are there to protect.

These relationships are somewhat peculiar. Peacekeepers are strangers entering conflict-affected contexts. In an intervention, peacekeepers themselves will come from different countries with different cultures and languages. Peacekeepers will also likely be from a different culture and speak different languages from those in use locally. They will have some shared goals relating to peace with some and conflicting goals with others, whether relating to preventing violence itself or the particular vision of peace that peacekeeping is meant to promote. Ideally, the use of coercion to attempt to impose their goals stems from political and contextual analysis rather than from emotions or a personal desire to dominate.

Peacekeeping relationships are affected by how peacekeepers enter communities and encounter local people. For example, whether peacekeepers enter carrying weapons or unarmed has an impact on the quality and depth of these relationships. The relationships move between periods of cooperation and episodes or periods of coercion. This suggests the need for robust relationships able to withstand various stresses and obstacles.

While the initial research on which this chapter is based identified peacekeepers' beliefs that relationships are a central component for being effective, there is a need for further research. General questions raised by this research include what happens in the process of constructing relationships, perceptions of power and influence, and other relational dynamics that create perceptions of increased costs of violence and increased benefits of cooperation for belligerents. How do relationships contribute to the prevention of violence and protection of civilians? More specifically, how do peacekeepers move between coercion and cooperation; how do they understand when the risks of being unarmed outweigh the advantages to relationship building and maintenance that being unarmed signals; how do peacekeepers think about the contexts and criteria for making these decisions; what are the actual practices that peacekeepers use to build and maintain cooperation; and does the use of coercion, in the long run, ever strengthen cooperation, and if so how and when? The initial research for this study used in-depth interviews with former peacekeepers from a wide variety of backgrounds and experience, which ultimately led to these new questions. Further research using interviews with open-ended questions that focus on these questions would presumably shed light on how ground-level peacekeepers understand these issues. It would also be useful to identify other contexts that resemble peacekeeping in some ways, such as community policing in high-crime areas, and interview people in various roles to see what might be similar to or different from peacekeepers' understanding.

Another avenue of research would include not only the experiences of frontline peacekeepers but also the experiences of people in local communities who experience peacekeeping. While there is some literature on how local people perceive peacekeepers (e.g., Pouligny 2006), there appears to be no literature comparing local perceptions of peacekeepers with the perceptions that peacekeepers have of themselves and of local people. Do local people also believe that good relationships are a central factor in peacekeeping, and would they describe them similarly?

Understanding peace as developing within peacekeeping (and peacebuilding) relationships opens new avenues for research and for more effective interventions. In the long run, it is hoped this will contribute to a more peaceful world.

References

Bellamy, A., & Williams, P. (2005). Who's keeping the peace? Regionalization and contemporary peace operations. *International Security, 29*, 157–195.

Coleman, P.T., Kugler, K.G., Bui-Wrzosinska, L., Nowak, A., & Vallacher, R. (2012). Getting down to basics: A situated model of conflict in social relations. *Negotiation Journal, 28*, 7–43.

Cox, M. (Ed.). (2009). *Social capital and peace-building: Creating and resolving conflict with trust and social networks*. London: Routledge.

Deutsch, M. (2011a). Cooperation and competition. In P.T. Coleman (Ed.), *Conflict, interdependence, and justice: The intellectual legacy of Morton Deutsch* (pp. 23–40). New York: Srpinger.

Deutsch, M. (2011b). Interdependence and psychological orientation. In P.T. Coleman (Ed.), *Conflict, interdependence and justice:The intellectual legacy of Morton Deutsch* (pp. 247–271). New York: Springer.

de Waal, F.B.M. (2000). The first kiss: Foundations of conflict resolution research in animals. In F. Aureli & F.B.M. de Waal (Eds.), *Natural conflict resolution* (pp. 15–33). Berkeley: University of California Press.

Druckman, D., Stern, P.C., Diehl, P., Fetherston, A.B., Johansen, R., Durch, W., & Ratner, S. (1997). Evaluating peacekeeping missions. *Mershon International Studies Review, 41*, 151–165.

Fortna, V.P. (2008). *Does peacekeeping work? Shaping belligerents' choices after civil war*. Princeton, NJ: Princeton University Press.

Fortna, V.P., & Howard, L.M. (2008). Pitfalls and prospects in the peacekeeping literature. *Annual Review of Political Science, 11*, 283–301.

Furnari, E. (2014). *Understanding effectiveness in peacekeeping operations: Exploring the perspectives of frontline peacekeepers*. PhD dissertation, University of Otago. Available from http://hdl.handle.net/10523/4765 *or* http://otago.ourarchive.ac.nz/bitstream/handle/10523/4765/FurnariEllen2014PhD.pdf?sequence=1

Hardin, R. (2002). *Trust and trustworthiness*. New York: Russell Sage Foundation.

Hegre, H., Hultman, L., & Nygard, H.M. (2010, September). Evaluating the conflict-reducing effect of UN peace-keeping. Paper presented at the annual meeting of the American Political Science Association, Washington, DC.

Higate, P., & Henry, M. (2009). *Insecure spaces: Peacekeeping, power and performance in Haiti, Kosovo and Liberia*. London: Zed Books.

Howard, L.M. (2008). *UN peacekeeping in civil wars*. Cambridge: Cambridge University Press.

Hultman, L. (2010). Keeping peace or spurring violence? Unintended effects of peace operations on violence against civilians. *Civil Wars, 12*, 29–46.

Hultman, L., Kathman, J., & Shannon, M. (2013). United Nations peacekeeping and civilian protection in civil war. *American Journal of Political Science, 57*, 875–891.

Jagd, S. (2010). Balancing trust and control in organizations: Towards a process perspective. *Society and Business Review, 5*, 259–269.

Johnstone, I. (2011). Managing consent in contemporary peacekeeping operations. *International Peacekeeping, 18*, 168–182.

Kalyvas, S.N. (2003). The ontology of "political violence": Action and identity in civil wars. *Perspectives on Politics, 1*, 475–494.

Kreps, S.E., & Wallace, G.L. (2009, September). Just how humanitarian are interventions? Peacekeeping and the prevention of civilian killings during and after civil wars. Paper presented at the annual meeting of the American Political Science Association, Toronto.

Lederach, J.P. (2005). *The moral imagination: The art and soul of building peace*. New York: Oxford University Press.

Leonard, D.K. (2013). Social contracts, networks and security in tropical African conflict states: An overview. *IDS Bulletin, 44*, 1–14.

Mac Ginty, R. (2008). Indigenous peace-making versus the liberal peace. *Cooperation and Conflict, 43*, 139–163.

Mahony, L. (2006). *Protective presence: Field strategies for civilian protection.* Geneva: Centre for Humanitarian Dialogue.

Mahony, L., & Nash, R. (2012). *Influence on the ground: Understanding and strengtheing the protection impact of United Nations human rights field presences.* Brewster, NY: Fieldview Solutions.

Martin-Brûlé, S.-M. (2012). Assessing peace operations' mitigated outcomes. *International Peacekeeping, 19*, 235–250.

Miller, G. (2004). Monitoring, rules and the control paradox: Can the good soldier Svejk be trusted? In R.M. Kramer & K.S. Cook (Eds.), *Trust and distrust in organizations: Dilemmas and approaches* (pp. 99–126). New York: Russell Sage Foundation.

Moser-Puangsuwan, Y., & Weber, T. (Eds.). (2000). *Nonviolent intervention across borders.* Honolulu: University of Hawai'i Press.

Moses, J., & Knutsen, T. (2007). *Ways of knowing: Competing methodologies in social and political research.* New York: Palgrave Macmillan.

Murphy, J.T. (2006). Building trust in economic space. *Progress in Human Geography, 30*, 427–450.

Nan, S.A. (2009). Social capital in excluse and inclusive networks: Satisfying human needs through conflict and conflict resolution. In M. Cox (Ed.), *Social capital and peace-buidling: Creating and resolving conflict with trust and social networks* (pp. 172–185). London: Routledge.

Nannestad, P. (2008). What have we learned about generalized trust, if anything? *Annual Review of Political Science*, 413–436.

Newton, K., & Zmerli, S. (2011). Three forms of trust and their association. *European Political Science Review, 3*, 169–200.

NORDCAPS. (2007). *NORDCAPS PSO tactical manual 2007.* Helsinki: NORDCAPS.

Paffenholz, T. (2009). Exploring opportunities and obstacles for a constructive role of social capital in peacebuilding: A framework for analysis. In M. Cox (Ed.), *Social capital and peace-building: Creating and resolving conflict with trust and social networks.* London: Routledge.

Pouligny, B. (2006). *Peace operations seen from below.* Bloomfield, CT: Kumarian Press.

Pushkina, D. (2006). A recipe for success? Ingredients of a successful peacekeeping mission. *International Peacekeeping, 13*, 133–149.

Reis, H., Collins, W.A., & Berscheid, E. (2000). The relationship context of human behavior and development. *Psychological Bulletin, 126*, 844–872.

Rubinstein, R.A. (2005). Intervention and culture: An anthropological approach to peace operations. *Security Dialogue, 36*, 527–544.

Sartre, P. (2011). *Making UN peacekeeping more robust: Protecting the mission, persuading the actors.* New York: International Peace Institute.

Schweitzer, C. (2010). Introduction: Civilian peacekeeping a barely tapped resource. In C. Schweitzer (Ed.), *Civilian peacekeeping a barely tapped resource* (pp. 7–16). Belm-Vehrte, Germany: Sozio Publishing.

Tardy, T. (2011). A critique of robust peacekeeping in contemporary peace operations. *International Peacekeeping, 18*, 152–167.

Turner, C.W., Simons, L.S., Berkowitz, L., & Frodi, A. (1977). The stimulating and inhibiting effects of weapons on aggressive behavior. *Aggressive Behavior, 3*, 355–378.

Wallis, T. (2010). Best practices for unarmed civilian peacekeeping. In C. Schweitzer (Ed.), *Civilian peacekeeping: A barely tapped resource.* Belm-Vehrte, Germany: Sozio Publishing.

9

Reintegration of Former Child Soldiers: Communal Approaches to Healing the Wounds and Building Peace in Postconflict Societies

Michael Wessells and Kathleen Kostelny

In diverse countries worldwide, armed forces (national armies) and armed groups (nonstate actors) recruit children, who are defined under international law as people under 18 years of age (Brett & McCallin 1996; Cohn & Goodwin-Gill 1996; Wessells 2006; Coulter *et al.* 2008; Denov 2010). Depending on the context, children enter armed groups through forced recruitment, often at gunpoint, or decide to join for various reasons. In Colombia, girls and boys have joined armed groups such as the FARC because they saw the armed group as a surrogate family and as offering better conditions than their biological families had offered (Human Rights Watch 2003a). In other conflicts, children have joined to obtain money or a sense of power, or to support what they regarded as a liberation struggle that aimed to end the oppression of their people (Wessells 2006). The exact number of child soldiers is unknown because commanders cleverly hide their recruitment of children.

Girls and boys may play a variety of roles inside armed forces and groups, such as cooks, porters, spies, bodyguards, sex slaves, and soldiers. Not uncommonly, children performed multiple roles concurrently. For example, girl soldiers inside the Revolutionary United Front (RUF) in Sierra Leone often were "wives," cooks, and porters (McKay & Mazurana 2004). It is difficult, perhaps inappropriate, to create universalized images of child soldiers. For example, widespread images of girls in Africa who have been recruited at gunpoint and sexually violated or assigned to a soldier as his "wife" do not apply in all contexts. Indeed, armed groups such as the Liberation Tigers of Tamil Elan (LTTE) in Sri Lanka had strict codes against the sexual abuse or exploitation of girl and women recruits. Also, dominant images portray commanders as men and boys, but in multiple conflicts, girls have also served as commanders (Wessells 2006).

However they were recruited, child soldiers are exposed to severe dangers such as lack of water, food, and medical care and multiple sources of violence such as witnessing and perpetrating killings, sexual abuse, and severe physical punishment. For forcibly recruited children, the exposure to violence often begins with the act of recruitment itself. In Sierra Leone, for example, the RUF sometimes recruited children by taking them at gunpoint and then handing the recruited child a weapon and ordering him to kill a member of his own village while other villagers watched. Such horrific orders were

Peace Ethology: Behavioral Processes and Systems of Peace, First Edition.
Edited by Peter Verbeek and Benjamin A. Peters.
© 2018 John Wiley & Sons Ltd. Published 2018 by John Wiley & Sons Ltd.

intended to destroy the bonds between child and community, thereby decreasing the child's intent to escape and return home. What follows forced recruitment is often even more violent. For example, in the Lord's Resistance Army (LRA) in northern Uganda, children who could not keep up were killed, and those who attempted to escape but were caught faced a more brutal outcome (Human Rights Watch 2003b). The child who had attempted to run away was put in a circle, and other children were given sticks or other weapons and forced to beat the child to death. Anyone who refused or who did not beat with sufficient force became the next victim in the circle of violence. This process successfully introduced ordinary children to committing horrific acts of violence, preparing them for what was to come, and cleverly it removed the moral pangs of killing someone since no one knew which blow had actually been the death blow. In addition, recruits are also subjected to brutal training regimens in which they are exposed to violence and beaten severely for any disobedience or poor performance.

The net result of recruitment is that children are socialized into a system of violence in which killings, seeing dead people, looting, and other horrors become normalized. As many children become skilled actors in the violence, there is a very real danger that they will embrace norms of violence and will see it as their primary means of obtaining what they want. This situation poses multiple problems for peacebuilding following the establishment of a ceasefire or peace agreement. Like adult soldiers, child soldiers who remain armed could undermine a peace process by continuing an armed struggle. Also, child soldiers could contribute to ongoing criminal violence and insecurity that destabilize society and thwart peace. Indeed, violence is sometimes as pervasive following the ceasefire as it had been during the armed conflict. Young people who have no jobs or means of meeting their basic needs may use their weapons to obtain what they need. Furthermore, communities may fear or stigmatize the former child soldiers, whose groups may have attacked civilians and left searing memories of brutal killings. Seeing few options for living as civilians, child soldiers sometimes decide to become mercenaries or participants in other armed conflicts, often contributing to regional destabilization.

A high priority, then, is to enable former child soldiers to reintegrate into civilian life following armed conflict. The purpose of this chapter is to outline how the reintegration of former child soldiers can be achieved by taking a community-oriented, resilience-based approach that is holistic and emphasizes the importance of social relationships. A central theme of the chapter is that both the *how* and the *what* are important in reintegration. Having pointed out the problems associated with excessively individualized approaches, it presents a social ecological framework that conceptualizes reintegration as an inherently communal process. Next, it suggests a community resilience approach to reintegration that stimulates empowerment and builds on existing community and family resources. It then examines how eight key elements contribute to reintegration efforts and shows how former child soldiers are not a "Lost Generation" but rather a vital resource for building peace.

Social Ecologies of Violence and Peace

The use of excessively individualized approaches has frequently limited the success of efforts to reintegrate former child soldiers. Field experience indicates that a stronger approach is to focus on the social environment and social relationships that influence reintegration and peace processes.

The Limits of Individualized Approaches

Individualized approaches are frequently seen in psychologists' focus on children's trauma, particularly posttraumatic stress disorder (PTSD; see Marsella *et al.* 1994; Apfel & Simon 1996), a well-documented source of suffering and an enabler of ongoing violence. A trauma focus has frequently led to clinical treatments that aim to alleviate individual suffering. Although the reduction of individual suffering is important, former child soldiers face profound issues that are communal and that relate not to past violence but to risks in their current environment. For example, stigmatization and lack of community acceptance can make it very difficult to reintegrate into civilian life (Wessells 2006; Betancourt *et al.* 2010, 2013). Particularly in collectivist societies, former child soldiers may be concerned more about their social relations than their individual issues. Following a war that has shattered social relationships, high priorities are to reestablish or establish social trust and cohesion and to enable the social acceptance of former child soldiers.

Excessively individualized approaches are also visible in formal processes of disarmament, demobilization, and reintegration (DDR). Quite often, DDR programs are constructed by a government and a UN agency as a means of standing down opposing armies and reforming the security sector through the creation of unified national armed forces. In a group context, former child soldiers turn in their weapons, and then they are individually demobilized or discharged and receive documentation indicating their civilian status. To aid reintegration into civilian life, each individual may receive seeds and tools, a cash dispensation, or other items according to the context, and then be returned to or "reinserted" into his or her home community. This approach, however, gives too little attention to the community dimensions of reintegration (McCallin 1998; Wessells 2009a). Little is done, for example, to address child soldiers' stigma, reduce communities' fears of former child soldiers, or restore the damaged relations between the former child soldiers and community members. Individualized benefits may also have pernicious effects. In Liberia, the benefits given to former child soldiers were seen by community people as "blood money" that had been given to the perpetrators of violence, when in fact the community had suffered the most due to attacks by the former child soldiers. As this example illustrates, well-intended individual benefits can cause unintended harm by creating or amplifying jealousies and social divisions at the moment when social unity is most needed.

A more appropriate and effective approach to reintegration is one that conceptualizes reintegration in terms of person–environment interactions at multiple levels and that highlights the importance of supportive social relationships in the reintegration process.

A Social Ecological Framework

Social ecological frameworks emphasize the importance of the social environment in children's development and well-being (Bronfenbrenner 1979; Dawes & Donald 2000). They depict child development as occurring within nested levels such as the family, neighborhood, community, and societal levels, and they feature the importance of the social interactions that occur within and across micro, meso, and macro levels. At each level, there can be risks, protective factors that mitigate risk, and promotive factors that prevent exposure to risk. For example, an abusive family exposes the child to significant physical, psychological, and social risks, whereas a supportive family provides psychosocial support and protection from harms. Risk accumulation has profound consequences, as the probability of developmental harm or psychopathology increases exponentially as the number of risks increases (Rutter 1979, 1985). Yet the effects of risk accumulation can be offset by exposure to protective factors. For example, a child who has been sexually abused by a stranger may continue to do relatively well if he or she has a supportive family and peer group. Furthermore, harm to children may be prevented by promotive factors such as being in a safe, supportive school where teachers care for children and there are positive peer relationships. In addition to being valuable in its own right, being in a safe school can keep children from being exposed to harms, such as dangerous labor and sexual exploitation, that are suffered by many children who are out of school.

Social ecological frameworks also show the importance of the social environment in influencing violence and peacebuilding. For example, if children had spent several years in an armed group, they will likely have been socialized into a system of violence and learned to view violence as normal and necessary to achieve their objectives. Conversely, if they had grown up in families and communities that discouraged violence and modeled nonviolent methods of handling conflict, children would likely have developed peacebuilding skills and values (Narvaez, this volume, Chapter 6).

This framework has far-reaching implications for efforts to reintegrate former child soldiers and to build peace. It depicts child development as an intensely relational process in which children's well-being depends on the nature and quality of their relationships with parents, friends, teachers, religious leaders, and community members, among others. It follows that reintegration programs should focus not only on the individual child soldiers but also on the quality of their social relationships and environment. Also, reintegration efforts should support peaceful values and practices at each level of children's social environment, thereby helping to resocialize former child soldiers away from values and practices of violence toward values and practices that support peace. The importance of relationships and of working at multiple levels resonates well with the work presented in other chapters of this volume (see Ali and Walters, this volume, Chapter 5; Furnari, this volume, Chapter 8; and Shnabel, this volume, Chapter 2).

Also, social ecological frameworks suggest that reintegration programs should not focus only on deficits or children's problems such as trauma. A strict deficits focus can contribute to children's stigmatization and lead one to overlook the supports that are present in the children's social environment. To support children's well-being, it is vital to simultaneously reduce the risk factors and strengthen protective and promotive factors. Even refugee or internally displaced person (IDP) camps or extremely poor neighborhoods have a mixture of risk factors and protective and promotive factors. Strengthening the latter is essential for achieving children's well-being and sustained reintegration (Vindevogel *et al.* in press).

Social ecological frameworks also cast a critical eye on views of former child soldiers as damaged goods or a Lost Generation. Such a depiction likely thwarts healing since it leads people to write off former child soldiers as somehow beyond repair. In fact, research increasingly indicates that the majority of former child soldiers exhibit remarkable resilience (Wessells 2006, 2009a) and can achieve meaningful, productive lives if given appropriate supports. Their resilience owes not only to individual factors such as the ability to self-regulate but also to the presence of protective and promotive factors in their social environment that mitigate the effects of and reduce exposure to the risk factors that cause mental health problems and psychosocial distress. To be sure, some children do not exhibit resilience, but this owes to a preponderance of risk factors. If the balance could be shifted in favor of protective and promotive factors, the children who had at one moment been overwhelmed and dysfunctional may do much better. Thus, neither resilience nor dysfunctionality are carved in stone – both are dynamic and are products of children's social ecologies.

Furthermore, social ecological frameworks fit well with the idea that systemic peacebuilding and child protection efforts are an integral part of reintegration work at the grassroots level. As discussed further in this chapter, effective reintegration often requires the nonviolent management of conflict at multiple levels such as family and community levels. Similarly, reintegration programs may become revolving doors if children who are released from armed groups are subsequently re-recruited. What is needed is a protective environment that prevents re-recruitment – one's approach to reintegration must be as systemic and holistic as are the causes of violence and harm to children. The following section outlines such a holistic approach.

A Community Resilience Approach to Reintegration

As discussed in this chapter, social relationships and the social environment are of fundamental importance in reintegration. Although family and interpersonal relationships are important, it is the wider relationship between returning children and their community that matters most. For example, family acceptance and support would have limited value if children continued to be isolated or stigmatized by the community. Also, families are not islands but themselves require community support. Moreover, it is in the community that children achieve a meaningful social role and become valued as civilians.

Unfortunately, DDR programs have often devoted inadequate attention to community aspects of reintegration. In some cases, communities have been viewed as sites into which former child soldiers have been "reinserted." The fallacy of this approach is conspicuous in the stigma or rejection that many former child soldiers experience when inserted back into their communities of origin. At best, reinsertion is a potential starting point for reintegration, but is not in itself the reintegration process, which includes extensive relationship transformation and strengthening.

A related challenge is that nongovernmental organizations (NGOs) have sometimes created dependency by playing the role of experts, making the key decisions, and doing much of the work to reintegrate former child soldiers. When this happens, communities tend to regard the reintegration work as an NGO "project," which tends to collapse when the external funding has ended. To achieve sustainable reintegration, a better approach is to promote community ownership. That is, the communities themselves need to see that the problem of reintegrating particular former child soldiers is theirs and that it is their collective responsibility to use their own resources to promote and achieve reintegration. The promotion of community ownership requires that NGOs play facilitative roles, leaving the decision-making power to communities.

What is a Community Resilience Approach?

Community resilience may be defined as the collective ability to cope with and adapt to adversities in a constructive manner. Like individuals, communities are not passive in the face of adversities such as armed conflict, poverty, or natural disasters. Instead, they actively engage with various problems or risks by using their social networks, resources or assets, and decision-making processes (Norris *et al.* 2008). For example, if in Sub-Saharan Africa a child's parents had been killed during an armed conflict, the extended family would be expected to care for the child. If no extended family members were present in the area, community leaders and people might discuss the problem and find a solution such as placing the child with an appropriate foster family.

As this example illustrates, communities are actors who analyze problems and take local steps to address them. The heart of community resilience is collective planning and action to address local risks or problems – in this case, the problem of reintegrating former child soldiers. Outsiders may facilitate the collective planning and action, yet it is communities who empower themselves and mobilize their resources or assets to address the challenge. A community resilience approach, then, focuses more on what communities do to support reintegration than on what outsiders such as NGOs or the government do. To be sure, partnerships with outsiders are crucial, but only if communities lead the process and make the key decisions.

A community resilience approach contrasts with the deficits approach that is commonly seen in war zones, where communities that have been attacked are viewed as suffering victims. This focus on victims neither invites empowerment nor recognizes that, although communities may have been hit hard, they have diverse resources that they can use to get back on their feet and to use in addressing the risks children face. A strengths focus corrects this problem as it invites communities to take stock of their resources and what they can do to address their challenges. The strengths focus promotes sustainability since the communities' resources, such as leadership, may continue to support vulnerable children long after external funding has dried up.

Community Resources

Communities' resources may be cognitive, social, cultural, spiritual, physical, and economic, among others. Cognitive resources include the analytic and problem-solving activities by various people in the community. When difficult situations arise, communities often look for advice from particular people who are known to be good problem solvers who are guided by more than self-interest. Communities may also look to elders who are repositories of knowledge and can help bring forward the lessons of prior community experience and action. Since these and other cognitive resources are shared, they form an important part of the social capital of the community.

Diverse social resources can support communities' collective planning and action. To begin with, most communities have a regular community meeting such as the *barray* in West Africa or the *baraza* in East Africa at which key issues are discussed. People come to the meetings because they are notified through their social networks or because a Town Crier or its equivalent announces the meeting. A particularly important social resource is leadership. Communities frequently have elders or chiefs who can help to solve problems related to children. Also, people such as teachers, religious leaders, and grandmothers may be particularly sensitive to children's issues and have social networks within the community that enable them to learn about and advise how to address problems concerning children. In addition, groups such as women's groups, youth groups, health groups, or religious groups may be active in supporting vulnerable children. These and other actors may be sustainable sources of action on behalf of vulnerable children.

Richly intertwined with social resources are communities' cultural resources, which vary considerably according to the context. Traditional leadership structures – for example, the *shuura* or council of elders in Afghanistan – may play an important role in supporting children and building peace at a grassroots level. Particularly in rural parts of Sub-Saharan Africa, there may be traditional healers who address various spiritual issues, including those that children face in war zones (Wessells & Monteiro 2004). However, it is important to neither dismiss nor romanticize traditional beliefs and practices. Cultural practices are dynamic and change frequently as people bring in and try out new practices, as youth do globally with the aid of social media. Although some traditional practices are useful and support children's well-being, others such as female genital mutilation and cutting are not. Taking a critical stance, it is appropriate to use child rights standards such as the UN Convention on the Rights of the Child and the African Charter on the Rights and Welfare of the Child to decide which practices are supportable.

Communities' spiritual resources, which frequently overlap with cultural resources, include religious and spiritual beliefs and practices, leaders, and organizations (Honwana 1997). In Afghanistan, for example, many children have reported that their Islamic faith has been of great importance in helping them to cope with their many adversities. In communities worldwide, people turn to religion as a means of finding meaning and to religious leaders for comfort amidst hard times. Also, religious leaders and communities can often be voices of peace, although as witnessed in countries such as Central African Republic, religion can also be a fault line for armed conflict (Wessells & Strang 2006).

Primary among communities' physical resources are land, water, and sources of food and economic resources such as labor, markets, and marketable resources such as timber. For children, these resources are essential for healthy development and also for achieving social standing. For example, in many parts of Sub-Saharan Africa, a teenage male is able to marry and start his family only if he owns land and has a house. In postconflict zones, a key reintegration priority is to provide jobs for former child soldiers lest they return to fighting or use guns to meet their basic needs.

A Critical Perspective

In discussing community resources and action, however, it is important to take a critical perspective. There are contexts and moments in which a community resilience approach is ill advised. For example, if the postconflict setting were rife with fears, spies, and ongoing violence, the act of bringing people together for collective planning purposes could be seen as a political act that might lead to violence. Since the use of a community resilience approach requires a modicum of security and stability, it cannot be regarded as a "silver bullet" for all situations.

Critical perspective is also warranted in regard to the focus on communities. An important question is "Who is the community?" Within communities, large power differences exist among different subgroups, and some people may have little access to particular resources or ability to participate in decision making. Furthermore, when community planning occurs, some people such as former child soldiers or children with disabilities may be marginalized or excluded. The achievement of community resilience requires a transformational dimension that seeks to include the marginalized people and give voice to many different people and subgroups. In this respect, a community resilience approach should include a social justice lens, without which community activities may reproduce existing inequities.

In addition, reintegration sometimes occurs in situations in which little sense of community exists. For example, former child soldiers may migrate in the postconflict period to urban settings characterized by high competition for resources, frequent movement of people into and out of the area, low social cohesion, few traditional structures or practices, and limited social controls. Having little or no sense of "community," people may not engage in collective planning and action at a wide level. Nevertheless, it may be possible in such settings to strengthen social cohesion by bringing groups together to identify common goals and develop collective means of addressing them. Although this likely requires a slow, prolonged process that may not fit well with the typical desire for quick results, it may fit with the slower tempo of community development processes. Regardless of the pace of collective planning and action, critical thinking is a necessary part of a community resilience approach to reintegration.

Key Elements of Reintegration

Broadly, eight key elements need to be developed and coordinated to achieve reintegration at the community level. Ideally, their implementation would be guided by careful assessments and contextual tailoring. Although each element is discussed separately in this section for purposes of clarity, the elements overlap and interact extensively in practice.

Child and Youth Participation

The participation of formerly recruited children and youth is essential for success at all stages, beginning with the design of reintegration programs. It would make little sense, for example, to plan to send demobilized children back into their communities if they had no interest in returning there. Similarly, the idea of training formerly recruited girls to be tailors would have limited merit if the girls preferred to be trained for other income-generating activities (Brett & Specht 2004). At every stage, it is valuable to draw on young people's agency, perspectives, and insights, including both people under 18 years (children) and 18–24-year-olds (youth) who are still in transition developmentally and have useful views to offer on the reintegration process (Chrobok & Akutu 2008). Since participation is a cross-cutting issue, it will arise in the other elements discussed in the remainder of this section.

Community Preparation

Community acceptance of former child soldiers is one of the most important determinants of successful reintegration. Community acceptance is best thought of not as a single step or set of activities but as an ongoing process that can take several years.

This process, however, begins with efforts to prepare the community to receive the former child soldiers. This can be quite challenging in contexts in which the child soldiers had been part of armed groups that had attacked the villages into which they hope to reintegrate. Strong fears or desire for revenge may require special attention to peacebuilding early on. By contrast, in settings such as Afghanistan where child soldiers had fought alongside uncles and other family and community members and were viewed as having helped their community, communities may be willing to accept former child soldiers with open arms.

Recognizing government authority in the reintegration process, an important first step is to meet with authorities such as district or provincial staff in the Ministry of Social Welfare or its equivalent and with other authorities such as police. In such meetings, it is useful to learn their understanding of the reintegration process, their role, and the challenges to reintegration. Such meetings allow the sharing of information on the DDR process, discussion of the holistic approach and elements that contribute to successful reintegration, and the engagement of partners in problem-solving dialogue about how to address challenges to reintegration. This process of mutual learning and problem solving nurtures the collaborative process that assists reintegration.

Next, there are meetings and dialogues with community leaders such as Chiefs and elders. To demonstrate proper respect, one follows the appropriate cultural script in the initial entry. In Sierra Leone, for example, this involved bringing a bag of rice to the Paramount Chief, explaining one's purpose, and requesting time to talk with the Chief, who might decide to talk with elders present as well. It is useful to learn from such authorities their understanding of the reintegration process, their role, and any challenges that they see. Quite often, local leaders have had few discussions with government authorities and may not understand the reintegration process or their role in the wider peace process. Also, they may harbor doubts about the return of former child soldiers, whom they may see as behaving like "wild animals" and not respecting authority.

An important step, then, is to build empathy by helping the leaders understand that the child soldiers themselves have endured considerable suffering. By asking and discussing how children entered armed groups, leaders are reminded of how children had been recruited at gunpoint or in the face of other adversities. This can be followed by educating people about how child soldiers suffered inside armed groups. Empathy helps to reestablish the humanity of the children and also to decrease the image of the children as dangerous perpetrators. Also, it can be useful to engage religious leaders and teachers in enabling reflection among local leaders about what the community can do to fulfill its responsibility to support them.

Education and empathy building are also needed with the wider community. This can be done via meetings and dialogues convened by Chiefs and elders, with help from NGOs or others who are facilitating the reintegration work. Typically, some community members recall the horrors of the war and express strong doubts about the return of former child soldiers. However, ongoing education and dialogues that follow the approach described here help to increase community members' receptivity to reintegration. A large boost to receptivity often occurs if the Chief and elders say and show publicly that they support the children's return and affirm that reintegration is the community's responsibility. Lingering fears and doubts may be addressed by having public discussions with government authorities who urge acceptance or by conducting local celebrations of peace such as peace festivals in which people use local practices such as song and dance to collectively celebrate the arrival or the opportunity for peace.

Linguistic and cultural considerations can also play a valuable role. If, for example, community meetings develop a narrative of supporting "our children," community members may be more likely to see the returnees as children who are to be supported by the entire community. In contrast, continued discourse about "child soldiers" may cast the children in a negative light and reawaken war traumas. In addition, local communities or groups may use cultural means of expressing their solidarity and commitment to supporting children. Through proverbs, for example, they may convey the importance of not "throwing children away," or they may sing songs saying that they are all parents who love children, and now our children are coming home. Culturally appropriate media, such as song and dance, may play a valuable role in opening the community doors to the return of former child soldiers.

As these discussions unfold, it is important to seek the inputs of different subgroups within the community such as girls, boys, women, and men. Not all these subgroups feel comfortable speaking at public meetings, which tend to be dominated by men. By organizing small-group discussions, however, one can elicit the views of different subgroups and enable them to have a voice by, for example asking the Chief whether one member of a subgroup can report on its discussions in a public meeting.

As awareness increases of what the children had experienced and of the ongoing problems they likely have, one can invite collective reflection about what the community can do to support their reintegration. In many cases, this is better done by a community facilitator than by a Chief, with whom local people may not be able to disagree publicly. In addition to identifying what they expect returning children to do (e.g., show proper respect for and obedience to local authorities), community members may discuss the practicalities of greeting the returning children by having an open celebration, providing for their education, discouraging arguments and taunting, and helping the families of the returning children. Consistent with a community resilience approach, the emphasis should be on what communities themselves can and should do to enable reintegration.

Family Support

Family reunification is key because family care and support are vital for children's protection and well-being. Also, being part of a family is a core part of people's identity, particularly in collectivist societies, and many former child soldiers want to return home to be with their families. Family tracing and reunification begin with the identification and documentation of the child and his or her place of origin, followed by efforts to locate and contact his or her family. Particularly if the situation remains dangerous, it is essential to accompany and provide transport home for demobilized children to enable family reunification.

Next comes the more challenging task of achieving family acceptance and harmony. Former child soldiers, who may have drunk alcohol and made their own decisions inside the armed group, may find it challenging to submit to parental authority. If, on returning home, children drank and engaged in unruly behavior, this would likely cause conflict with their parents. Such conflict could have wider repercussions since community members expect former child soldiers to return to and fit in well with their families, and they watch carefully to see how the reunification process is going. If a former child soldier were rejected by his family or seen as causing problems for the family, the community would be less likely to accept that child fully as a community member. In many respects, family acceptance is a precondition for community acceptance.

For this reason, peacebuilding supports at the family level are frequently included in reintegration work. Such supports may include training on nonviolent means of handling conflict; discussions about roles, responsibilities, and expectations; and family reflection on how peace within the family can contribute to peace at community and even societal levels.

Education

Like most war-affected children, former child soldiers frequently regard education as their future, and they tend to see the loss of their education as one of the most negative impacts of the war (Women's Commission for Refugee Women and Children 2005; Annan *et al.* 2008). For many children, going to school provides a sense of normalcy and helps to restore a sense of confidence and hope that the war has ended. Education helps to build cognitive skills such as reading and writing that prepare children for life and enable them to use and participate in the global media that engage young people world-wide. Also, the cognitive competencies developed by education boost children's abilities to solve problems, navigate complex environments, and engage in self-protection. Adults, too, usually prioritize children's education and see going to school as the age-appropriate social role for a child. Overall, education can play a key role in helping children to transition out of their military role, construct a civilian identity and role, and gain social acceptance.

To be supportive, however, education should be tailored to the unique circumstances and background of former child soldiers. Whereas the traditional pace of education is slow, former child soldiers frequently want to catch up on their lost years of education. Also, children who have made life-and-death decisions inside an armed group may be reluctant to take orders from or show deference towards teachers who have not experienced combat. They may also be unwilling to sit in a classroom alongside children who are much younger or to endure being called "rebel" by other children. Accordingly, reintegration programs may provide accelerated or catch-up education that is specifically for the former child soldiers. This education typically moves at a rapid pace, extends the entire year, and is taught by teachers whom the former child soldiers respect. The teachers may use various means such as rule setting and empathy exercises to help reduce the stigma of former child soldiers. Numerous school-based interventions support the psychosocial well-being of war-affected children (Tol *et al.* 2013), and these can be used to support former child soldiers.

Psychosocial Support

Psychosocial support for former child soldiers is a pivotal part of a holistic approach to reintegration (Wessells 2009a; Betancourt *et al.* 2013). Without healing and coming to terms with their war experiences and dealing constructively with ongoing stresses such as stigmatization and inability to meet their basic needs, children are at risk of continuing cycles of violence and experiencing problems that limit their chances of gaining social acceptance and finding a place in civilian life.

In some conflicts, psychosocial support begins in interim care centers (ICCs) before children have rejoined their families and communities. Those supports, which typically involve peer support and also trauma counseling, can be valuable, but only if the focus on the ICCs does not distract from the community reintegration work. In northern Uganda during the war, a cottage industry grew up around reception centers that tended to become long-term supports that garnered significant attention and resources, while relatively little was done to enable community reintegration. As emphasized by the Paris Principles, which provide interagency technical guidance, ICCs should be short term, and the greater emphasis in supporting former child soldiers should be on community reintegration.

Psychosocial supports for former child soldiers should be multilayered, as in the intervention pyramid of the *IASC Guidelines on Mental Health and Psychosocial Support in Emergency Settings* (Inter-Agency Standing Committee [IASC] 2007). The bottom layer of the pyramid, which applies to the largest numbers of children and other people, consists of establishing security and meeting basic needs for food, water, shelter, healthcare, and other necessities. When these basic needs have been met, nonformal supports such as those provided by family, peers, religious leaders, women's groups, and youth groups come into play. Often, it is the support from these "natural helpers" that is most contextually appropriate and sustainable. Thus, the second layer of the pyramid consists of family and community supports. This could include processes such as family reunification, family training on nonviolent conflict resolution, and the provision of education, even in the nontraditional format described above.

The third layer, at the top of the pyramid, are the specialized mental health supports that are needed by people who suffer on an ongoing basis from problems such as PTSD, depression, anxiety, and other psychological disorders. Although this layer applies to the smallest number of people, it can be highly important for former child soldiers who show evidence of ongoing psychosocial difficulties and whose suffering might continue without professional supports such as interpersonal therapy (Bolton *et al.* 2007) or trauma-focused cognitive behavioral therapy (Bass *et al.* 2013). Unfortunately, many conflict and postconflict zones have only a handful of trained clinical psychologists or psychiatrists who are qualified to offer specialized support. Also, the provision of these multiple layers of support requires extensive coordination and referrals across layers. For example, a child who needs trauma therapy may also need family reunification, education, and access to healthcare. The construction of a system of upward and downward referrals, which requires strong coordination and access to diverse services, is often one of the greatest challenges.

Of particular value in the organization of psychosocial supports are participatory approaches and cultural approaches. Participatory approaches enable former child soldiers to choose which supports they want and also engage them in a process of self-help. For example, girls and boys can be given choices about whether they want vocational training and income-generating activities and also which kinds of activities they wish to participate in. The act of choosing helps to strengthen the sense of self-efficacy that contributes to psychosocial well-being (Bandura 1982; Hobfoll *et al.* 2007).

Cultural approaches to psychosocial support are the ones that relate to or build upon cultural beliefs and practices that enable well-being. In some rural areas of Sierra Leone, for example, former girl soldiers who had been around dead bodies or who had been sexually violated in the bush were seen as spiritually contaminated. Both they and community members believed that the girls carried angry spirits that could cause illness, crop destruction, and other problems for the community members if they returned home. To address this problem, traditional healers performed a cleansing ritual that was seen as driving away the angry spirits and restoring harmony with the ancestors (Kostelny 2004). After the girls had eaten a special diet and received advice on how to behave, they participated in a cleansing ceremony with the entire village present. The healer washed the girls with a black ash soap believed to pull out the girls' spiritual impurities, which were carried away by the river. In some cases, the girls drank a cleansing mixture of boiled herbs and inhaled an herbal vapor believed to help cleanse the girls. Following this ritual washing, the girls and healers ate a special meal together and danced all night to drumming. The girls were then wrapped in white cloth, which symbolized purity, and were presented as "new" to the community members, who welcomed them. This ritual produced a sharp increase in the girls' social acceptance, as they were seen as "able to eat off the same plate" with others. In addition, the girls said that their "minds were steady," meaning that they no longer feared the spirits and were in a position to engage in normal community activities. As this example illustrates, healing is very much communal as well as individual. Cultural approaches such as this may be particularly useful in reducing stigma and restoring the relationship between community members and former child soldiers.

Economic Support
Since severe poverty and very high levels of unemployment are prevalent in postconflict settings, economic support is vital for enabling former child soldiers to meet their basic needs and develop hope for the future. Also, economic supports can help former child soldiers to achieve a positive role and status in civilian life. Young people who have money can either start or support their families, thereby adhering to social norms and meeting expectations in ways that enable social acceptance.

A case in point comes from participatory action research to aid the reintegration of girl mothers, over two-thirds of whom had been formerly recruited in Sierra Leone, Liberia, and northern Uganda (McKay *et al.* 2011). After the ceasefire, they were badly stigmatized as "rebel girls" and were distressed over their inability to meet basic needs and to be good mothers. With facilitation by international NGOs, over 600 girls formed groups of approximately 20 people each to discuss what they saw as their key issues and develop social actions or interventions that would enable them to address those issues. Group discussions over a period of months provided peer-based psychosocial support (though they never used the term *psychosocial*) in the form of caring, supportive interaction, learning that they were not alone, and group problem solving about how to move forward. All of the groups identified stigma and lack of a livelihood as their biggest problems. Wanting to achieve a livelihood, the girls then developed, with the aid of small grants or loans, group or individual income-generating activities such as collective farming, animal husbandry, baking, selling soap, or engaging in petty business activities. They worked with self-selected community advisors, who advised the girls on appropriate behavior, gave them tips on doing business, advocated for them within the community, and helped them gain access to social and business networks. Many of the groups decided to do community service to give back to their communities.

As the girls earned money, they supported themselves and met their children's needs. Community members soon began to see them as good mothers and productive community people rather than as rebels. Their gains in community acceptance, social connectedness, self-esteem, and self-efficacy contributed to their psychosocial well-being. This example illustrates that psychosocial support does not always involve counseling and therapy. In some cases, a combination of peer-based, family, and economic supports is quite powerful.

Sustainability, however, is of pivotal importance in enabling economic activities to support reintegration. Too often, well-intentioned economic programs have overstretched local markets by, for example, training 50 people in a small, rural village to be bicycle mechanics. To succeed, economic activities need to be guided by a systematic market analysis and be connected with macrolevel supports to insure economic vitality.

Restorative Justice

Issues of justice frequently loom large in the postconflict environment. Community members typically see former child soldiers as bearing responsibility for some of the bad things that had happened during the war, particularly if they had been part of armed groups that had attacked their village. Although in Western societies people frequently tend to press for the imposition of stiff penalties that fit the severity of the crime, people in developing countries typically tend to want former child soldiers to make restitution to the community. In essence, this is a means of restoring damaged relationships and maintaining the delicate web of reciprocities that bind people together in a community. It also serves as a nonviolent means of handling the conflict between community members and returning child soldiers.

The work of ChildFund Sierra Leone illustrates a restorative justice approach that also included nonviolent conflict management between people who had fought on different sides during the war (Wessells & Jonah 2006). In the northern province, returning young men and children who had fought on different sides were returning to rural villages, thereby creating a tense situation that was ripe for ongoing violence. ChildFund used the social psychological method of superordinate goals (Sherif *et al.* 1961), which reduces intergroup tensions by having members of each group cooperate on the achievement of shared, superordinate goals. Following extensive community dialogue and preparation, groups of three or four neighboring villages identified a project that would help them to address the needs of vulnerable children. Some groups chose to rebuild a school or to build a health post, whereas others chose to repair a road or a bridge that enabled economic connections with other areas and brought income that helped families meet their children's needs.

The returning former child and youth soldiers from both sides formed teams that then worked together on the community-identified projects. To make this possible without fighting, considerable work was done to prepare the young people for working together. Without incident and receiving payment for their efforts over several months, the work teams then built the structures or repaired the bridge or road, often with the community members looking on. As a result, the returning children and youth came to see each other in a more positive light and set aside the war-related hostilities. The communities came to see the returning children and youth not as potential troublemakers but as village members who were giving back for what they had done. This simultaneously provided a measure of restorative justice and also helped communities to see the young people as having appropriate civilian roles and promise. Young people learned valuable skills of carpentry and construction in the process, and this learning continued subsequently in more structured vocational training and income-generating activities under the guidance of a business and behavioral mentor.

This work shows how the reintegration process entails both individual and communal transformation. Individual children and youth who participated underwent a transformation of their behavior and their role. Prosocial behavior replaced their tendencies to fight, and their role changed from warrior to civilian. At the same time, communities were transformed by becoming more accepting of the returning children and youth and by supporting them in civilian roles and lives. This interactive process of reciprocal change is at the heart of reintegration and also the wider processes of building peace with justice following armed conflict.

Child Protection

To help guard against re-recruitment and also risks such as participation in dangerous labor, NGOs typically work with communities to form Child Welfare Committees (CWCs) or Child Protection Committees in war and postconflict settings. Such committees typically consist of 10–15 people, including teenagers, who monitor and help to mitigate abuses against children; respond to minor violations by, for example, mentoring a child who has stolen a loaf of bread; and report criminal violations to authorities or refer children who need specialized mental health services as needed.

CWCs are potentially useful in addressing diverse child protection issues (Wessells 2009b; Eynon & Lilley 2010). They tend to be more effective when they are linked with formal structures and actors in the wider child protection system, such as district-level social workers and police. Community ownership, however, is the main determinant of their effectiveness and sustainability (Wessells 2009b). Unfortunately, communities frequently see CWCs as NGO projects rather than as their own means of protecting vulnerable children. The resulting low levels of community ownership reduce the use and the effectiveness of the CWCs (Wessells *et al.* 2012). This problem could likely be corrected by taking a community resilience approach that puts communities in charge and enables them to make the key decisions. It remains for the next generation of action research and practice to develop and test this approach more fully in the context of work to support the reintegration of former child soldiers.

Conclusion

Two key themes from this chapter warrant emphasis. First, reintegration of former child soldiers is not an individual process but a collective process of social transformation. The next generation of work on reintegration should be animated by this insight and driven by communities in significant respects. Second, reintegration work requires a holistic approach that cuts across multiple sectors within the humanitarian system. Sustained reintegration of former child soldiers requires work on peacebuilding, restorative justice, education, child protection, and mental health and psychosocial support, among others. At the end of the day, efforts to achieve reintegration and to build peace must be as systemic as are the causes of armed conflict and militarization.

References

Annan, J., Blattman, C., Carlson, K., & Mazurana, D. (2008). *The state of female youth in Northern Uganda: Findings from the survey of war-affected youth (SWAY), phase II.* Retrieved from http://www.sway-uganda.org

Apfel, R., & Simon, B. (Eds.). (1996). *Minefields in their hearts.* New Haven, CT: Yale University Press.

Bandura, A. (1982). Self-efficacy mechanism in human agency. *American Psychologist, 37,* 122–147.

Bass, J., Annan, J., Murray, S., Kaysen, D., Griffiths, S., Cetinoglu, T., Wachter, K., Murray, L., & Bolton, P. (2013). Controlled trial of psychotherapy for Congolese survivors of sexual violence. *New England Journal of Medicine, 368,* 2182–2191.

Betancourt, T., Borisova, I., Williams, T., Meyers-Ohki, S., Rubin-Smith, J., Annan, J., & Kohrt, B. (2013). Research review: Psychosocial adjustment and mental health in former child soldiers – a systematic review of the literature and recommendations for future research. *Journal of Child Psychology and Psychiatry, 54,* 17–36.

Betancourt, T., Brennan, R., Rubin-Smith, J., Fitzmaurice, C., & Gilman, S. (2010). Sierra Leone's former child soldiers: A longitudinal study of risk, protective factors, and mental health. *Journal of the American Academy of Child & Adolescent Psychiatry, 49*(6), 606–615.

Bolton, P., Bass, J., Betancourt, T., Speelman, L., Onyango, G., Clougherty, K., Nugebauer, R., Murray, L., & Verdeli, H. (2007). Interventions for depression symptoms among adolescent survivors of war and displacement in northern Uganda: A randomized control trial. *JAMA, 298*(5), 519–527.

Brett, R., & McCallin, M. (1996). *Children: The invisible soldiers.* Vaxjo, Sweden: Radda Barnen.

Brett, R., & Specht, I. (2004). *Young soldiers: Why they choose to fight.* Boulder, CO: Lynne Rienner.

Bronfenbrenner, U. (1979). *The ecology of human development: Experiments by nature and design.* Cambridge, MA: Harvard University Press.

Chrobok, V., & Akutu, A. (2008). *Returning home: Children's perspectives on reintegration.* London: Coalition to Stop the Use of Child Soldiers.

Cohn, I., & Goodwin-Gill, G. (1994). *Child soldiers.* Oxford: Clarendon.

Coulter, C., Persson, M., & Utas, M. (2008). *Young female fighters in African wars: Conflict and its consequences.* Uppsala, Sweden: Nordiska Afrikainstitutet.

Dawes, A., & Donald, D. (2000). Improving children's chances. In D. Donald, A. Dawes, & J. Louw (Eds.), *Addressing childhood adversity* (pp. 1–25). Cape Town: David Philip.

Denov, M. (2010). *Child soldiers: Sierra Leone's Revolutionary United Front.* Cambridge: Cambridge University Press.

Eynon, A., & Lilley, S. (2010). *Strengthening national child protection systems in emergencies through community-based mechanisms: A discussion paper.* London: Save the Children UK.

Hobfoll, S., Watson, P., Bell, C., *et al.* (2007). Five essential elements of immediate and mid-term mass trauma intervention: Empirical evidence. *Psychiatry, 70*(4), 283–315.

Honwana, A. (1997). Healing for peace: Traditional healers and post-war reconstruction in Southern Mozambique. *Peace and Conflict: Journal of Peace Psychology, 3*(3), 293–305.

Human Rights Watch. (2003a). *Stolen children*. New York: Author.

Human Rights Watch. (2003b). *Uganda: Child abductions skyrocket in north*. Geneva: Author.

Inter-Agency Standing Committee (IASC). (2007). *IASC guidelines on mental health and psychosocial support in emergency settings*. Geneva: Author.

Kostelny, K. (2004). What about the girls? *Cornell International Law Journal, 37*(3), 505–512.

Marsella, A.J., Bornemann, T., Ekblad, S., & Orley, J. (1994). *Amidst peril and pain: The mental health and well-being of the world's refugees*. Washington, DC: American Psychological Association.

McCallin, M. (1998). Community involvement in the social reintegration of former child soldiers. In P. Bracken & C. Petty (Eds.), *Rethinking the trauma of war* (pp. 60–75). London: Free Association Books.

McKay, S., & Mazurana, D. (2004). *Where are the girls? Girls in fighting forces in Northern Uganda, Sierra Leone, and Mozambique: Their lives during and after war*. Montreal: International Centre for Human Rights and Democratic Development.

McKay, S., Veale, A., Worthen, M., & Wessells, M. (2011). Building meaningful participation inreintegration among war-affected young mothers in Liberia, Sierra Leone and northern Uganda. *Intervention, 9*(2), 108–124.

Norris, K., Stevens, S., Pfefferbaum, B., Wyche, K., & Pfefferbaum, R. (2008). Community resilience as a metaphor, theory, set of capacities, and strategy for disaster readiness. *American Journal of Community Psychology, 41*, 127–150.

Rutter, M. (1979). Protective factors in children's response to stress and disadvantage. In M. Kint & J. Rolf (Eds.), *Primary prevention of psychopathology. Vol. 3: Social competence in children* (pp. 49–74). Hanover, NH: University Press of New England.

Rutter, M. (1985). Resilience in the face of adversity. *British Journal of Psychiatry, 147*, 598–611.

Sherif, M., Harvey, O. White, B. Hood, W., & Sherif, C. (1961). *Intergroup cooperation and competition: The Robbers Cave experiment*. Norman, OK: University Book Exchange.

Tol, W., Song, S., & Jordans, M. (2013). Annual research review: Resilience and mental health in children and adolescents living in areas of armed conflict – a systematic review of findings in low- and middle-income countries. *Journal of Child Psychology and Psychiatry, 54*, 445–460.

Vindevogel, S., Wessells, M., DeSchryver, M., Broekaert, E., & Derluyn, I. (In press). Dealing with the consequences of war: Resources of formerly recruited and non-recruited youth in northern Uganda. *Journal of Adolescent Health*.

Wessells, M.G. (2006). *Child soldiers: From violence to protection*. Cambridge, MA: Harvard University Press.

Wessells, M. (2009a). Supporting the mental health and psychosocial well-being of former child soldiers. *Journal of the American Academy of Child and Adolescent Psychiatry, 48*(6), 587–590.

Wessells, M. (2009b). *What are we learning about protecting children in the community? An inter-agency review of the evidence on community-based child protection mechanisms in humanitarian and development settings*. London: Save the Children UK.

Wessells, M.G., & Jonah, D. (2006). Recruitment and reintegration of former youth soldiers in Sierra Leone: Challenges of reconciliation and post-accord peacebuilding. In S.

McEvoy-Levy (Ed.), *Troublemakers or peacemakers? Youth and post-accord peacebuilding* (pp. 27–47). Notre Dame, IN: University of Notre Dame Press.

Wessells, M., Lamin, D., King, D., Kostelny, K., Stark, L., & Lilley, S. (2012). The disconnect between community-based child protection mechanisms and the formal child protection system in rural Sierra Leone: Challenges to building an effective national child protection system. *Vulnerable Children and Youth Studies, 7*(31), 211–227.

Wessells, M.G., & Monteiro, C. (2004). Healing the wounds following protracted conflict in Angola. In U.P. Gielen, J. Fish, & J.G. Draguns (Eds.), *Handbook of culture, therapy, and healing* (pp. 321–341). Mahwah, NJ: Erlbaum.

Wessells, M., & Strang, A. (2006). Religion as resource and risk: The double-edged sword for children in situations of armed conflict. In N. Boothby, A. Strang, & M. Wessells (Eds.), *A world turned upside down: Social ecological approaches to children in war zones* (pp. 199–222). West Hartford, CT: Kumarian Press.

Women's Commission for Refugee Women and Children. (2005). *Youth speak out.* New York: Author.

Part Three

Function

10

Keeping the Peace or Enforcing Order? Overcoming Social Tension between Police and Civilians

Otto Adang, Sara Stronks, Misja van de Klomp, and Gabriel van den Brink

Police in modern society can function either as "peace keepers" or as "enforcers of order" (Fassin 2013), stereotypically exemplified by the British bobby on the one hand, closely linked to the community and moved by a sense of civic duty, and the US cop on the other hand, doing the dirty work of society in a relationship of reciprocal hostility with the public (Fassin 2013). According to pioneering police researcher Reiner (2000, p. 3), policing is aimed at securing social order, and the essential concept of policing is the attempt to maintain security through surveillance and the threat of sanctioning (either immediately or in terms of the initiation of penal processes, or both). Not only professionals employed by the state do policing; many other types of police exist and many others, including citizens, engage in policing functions. This chapter concerns itself with the police as the state institution exercising the (delegated) monopoly on the legitimate use of force within the territory of the state. According to Bittner (1974, p. 30), police tasks involve "something that ought not to be happening and about which someone had better do something now!" As Reiner (2000, p. 6) indicates, the police may invoke legal powers to handle these situations, but more often they use a variety of ways and means to keep the peace without initiating legal proceedings. However, also for police forces that promote themselves as service providers and do not routinely carry firearms, underlying all attempts at peacekeeping is, ultimately, the potential or actual use of legitimate force. In many countries, the relationship between police and citizens is seen as crucial for effective and legitimate police work (e.g., Tyler and Huo 2002; Skogan 2006). A variety of policing strategies, such as Community-Oriented Policing (e.g., Friedmann 1992; Skogan 2006), Problem-Oriented Policing (e.g., Goldstein 1979; Reisig 2010), and Reassurance Policing (e.g., Fleming 2005; Fielding and Innes 2006), strongly advocate building, securing, and maintaining good police–citizen relationships for routine police work. Della Porta and Reiter (1998), Otten *et al.* (2001), Fielding (2005), Waddington *et al.* (2009), Adang *et al.* (2010), and King (2013) all stress the need of good police–civilian relationships to prevent community riots, escalating protests, or other escalating confrontations. Furnari (this volume, Chapter 8) points to the importance of a relational approach to policing with regard to international peacekeeping operations.

Peace Ethology: Behavioral Processes and Systems of Peace, First Edition.
Edited by Peter Verbeek and Benjamin A. Peters.
© 2018 John Wiley & Sons Ltd. Published 2018 by John Wiley & Sons Ltd.

This approach fits well with the theory of relational distance (Black 1980), where this distance is determined by the scope, frequency, and duration of interaction between people and by the nature and number of links between them in a social network. According to Black, the relational distance between conflicting parties determines their willingness to use the legal system and the preparedness of police officers to arrest individuals. Police officers switch between more repressive and more conciliatory styles, but where parties are known to each other a conciliatory approach is often preferred. Ayres and Braithwaite (1992) talk about responsive regulation by enforcers who adopt a more responsive approach and act more as facilitators than as controllers and investigators. The aim is to increase the sense of responsibility of citizens and businesses to convince them of the importance of playing by the rules. Cooperation leads to bonds of trust that increase spontaneous compliance. Several studies show that the (perceived) legitimacy of police action determines compliance (Tyler 1990; Mastrofski *et al.* 1996; Tyler & Huo 2002). Especially ethnic minorities distrust police and authorities when they feel treated with disrespect (van Stokkom 2004). This is all the more important since gaining and maintaining authority among citizens imply active and ongoing investment from the police (Tyler & Huo 2001). It is important to make a distinction between different dimensions of social organization here. On the one hand, the relationship between police and citizens has a vertical dimension: the police are the strong arm of the law. On the other hand, there is a horizontal dimension to the relationship as well, as both police officers and citizens form part of social networks and participate in social interactions on a daily basis.

There is by now an impressive literature on police and policing, including on the use of force by police. Countries differ in the way police are organized, and some police forces are more service providers than law enforcers. In some countries, police mainly exist for the benefit of the state; in others, for the benefit of civilians. In spite of these differences, conflict with citizens is built into the relationship between police and civilians everywhere. From time to time, the performance of police tasks will inevitably result in at times physical conflicts between citizens and police officers, such as when police officers use force against citizens or when citizens resist the police by violent means. On occasion, eruptions of collective violence follow police–citizen interactions gone wrong. Vivid examples are the 2005 riots in France, the August 2011 riots in London and other English cities, and the 2014 riots in Ferguson, Missouri (United States), all following the death of one or more citizens in police–citizen interactions. Almost invariably, after such high-profile events, politicians, experts, and others stress the need for improved police–community relationships (e.g., Kerner 1968; Scarman 1982; Fielding 2005; Waddington *et al.* 2009; see Adang *et al.* 2010 for earlier cases). Analyzing riots in a neighborhood in the Dutch city of Utrecht, van de Klomp *et al.* (2011) argued that a strategy of combining empathy and enforcement, building on the relationship that existed with the local population, prevented riots from spreading following the death of a citizen shot by a police officer. They showed that preexisting relations between the actors (police–citizens–local authorities) in question influenced the way in which the conflict unfolded. According to van de Klomp *et al.* (2011), there was a history of mistrust between the citizens and the police. However, responding to complaints about "loitering youths," police had made the control of such nuisance their primary priority in the neighborhood in an active attempt to win the trust of residents.

At some point, residents even made a formal request for a more permanent community police officer presence in the neighborhood. In the months prior to the riots, many of the officers reportedly felt like they had turned the corner in a positive way in that regard. They felt this contributed to the positive outcome.

The Relational Model

Although there is by now a lot of literature on the causes and prevention of community riots, there is much less research on how police–citizen conflicts are resolved and on how police–citizen relationships are negotiated in the aftermath of overt confrontation. For inspiration, we looked to ethological literature on reconciliation. In their book *Natural Conflict Resolution*, ethologists Aureli and de Waal (2000) introduce a way of looking at conflicts from a relational perspective. In their Relational Model, the analysis of postconflict interaction between parties is central to the understanding of conflict resolution but also highly reliant on the nature of the relationship. They argue that postconflict interaction, just like the confrontation itself, holds meaning for the relationship between two (or more) parties. As Aureli and de Waal state:

> The first implication of the Relational Model, then, is that it allows for the full integration of competition and cooperation. This integration is not just an alternation between the two, or an uneasy coexistence; conflict and its resolution may actually contribute to a fine-tuning of expectations between parties, a building of trust despite occasional disagreement, hence a more productive and closer relationship than would be possible if conflict were fully suppressed. The reparability of relationships permits aggression to have a testing quality. (2000, p. 29)

In line with this assertion, Aureli and de Waal (2000) predict that the tendency to initiate aggression increases with the number of opportunities for competition, the value of contested resources, and the reparability of the relationship, while it decreases the higher the risk of injury and the value of the relationship. When the mechanisms of conflict resolution between two opponents are more developed, individuals will be less reluctant to engage in open conflict. The most consistently confirmed hypothesis springing from this line of thinking is the Valuable Relationship Hypothesis, which states that those relationships that are most valuable to the partners are also most likely to be reconciled after conflict. As such, reconciliation allows for a continuation of cooperative behavior to the benefit of the community. Building on this hypothesis, Cords and Aureli (2000, p. 182) hypothesized that three independent qualities of the relationship between opponents are key to evaluating the costs and benefits of reconciliation:

- *Value*: What the subject gains from her or his relationship with a partner, which depends on what the partner has to offer, how willing she or he is to offer it, and how accessible she or he is
- *Security*: The perceived probability that the relationship with the partner will change, which relates to the consistency of the partner's behavioral responses

- *Compatibility*: The general tenor of social interactions in a dyad, which may result from both the temperament of the partners and their shared history of social exchanges.

So far, few studies have attempted to test and demonstrate the integrated relevance of value, compatibility, and security. However, the results to date suggest their significance (e.g., Cooper *et al.* 2005; Fraser *et al.* 2008; Bonaventura *et al.* 2009).

The ideas of reconciliation and the reparability of relationships are the result of several decades of research, replicated consistently among well over 30 different social species, including humans (e.g., Koyama 2001; Fujisawa & Kutsukake 2005; Wittig & Boesch 2005). This is where the added value of an ethological perspective lies in its approach of (post)conflict interaction as an integrated process rather than a set of variables, while retaining attention to its specific forms and functions. This resonates strongly with findings from the social sciences, in which general notions of equal-status contact, group cooperation, and institutional support are suggested in order to restore or build group relations (e.g., Allport 1954; Rusbult *et al.* 2005, p. 193; Hewstone *et al.* 2008, p. 203; Janoff-Bulman & Werther 2008, p. 161). As these authors also point out, however, it is not entirely clear under which circumstances intergroup contact improves the relationship and what types of behavior are most beneficial. Shnabel (this volume, Chapter 2) developed the Needs-Based Model of reconciliation to attempt to understand from a social-psychological perspective how the transformation of the relations between former adversaries takes place.

In this chapter, we report on our first attempts to apply the Relational Model to police–citizen conflicts. More specifically, we explore the applicability of value, compatibility, and security in human conflict. In different studies, we analyzed seven cases: confrontations following a shooting incident, where police shot at a citizen (three cases; in two of them, the citizen got killed); confrontations during a party or celebration (two cases); one confrontation during a protest; and one football riot. All cases occurred in the Netherlands. It should be noted that, although in practice variations may occur from one place to another, Dutch police are generally known for their relation-oriented, community policing approach. This was helpful for the exploratory research that we wanted to do. However, this means the results are not necessarily generalizable to other countries where police forces are more "enforcers of order" and may be less oriented toward maintaining good relationships with citizens.

Police and Citizens in Conflict

Van de Klomp *et al.* (2014) explored the explanatory potential of the relational model in a study of the reconciliation process between police and members of the minority Moluccan community in a Dutch town. The conflict between the groups became salient during the celebration of New Year's Eve 2007–2008. While the neighborhood did not stand out from any other throughout the year, New Year's Eve had become the exception to this rule. For several years, youths in the neighborhood had built celebratory bonfires in violation of municipal regulations. To enforce the zero-tolerance policy with regard to such bonfires, the deployment of the fire department, protected by the police, had become something of a yearly tradition. This time around, however, the youths wore protective padding and had prepared firebombs and other projectiles, which they threw at a manned police vehicle. The riot squad had to retreat from the neighborhood, and it took several hours before they could disperse the youths and restore order. They were unable to make any arrests. At least two officers and one civilian were injured. Damage to public and private property amounted to approximately 25,000 euros throughout the city, but the subjective impact was much greater for the police officers at the scene, who feared for their lives. The incident caused major tension in the relationship between the local police and municipality on the one hand, and the local Moluccan community on the other. One year later, however, Moluccans, police, and the municipality cooperated to organize a peaceful and festive New Year's Eve celebration. The question van de Klomp *et al.* (2014) addressed was: what had happened to change the situation so drastically?

Van de Klomp *et al.* (2014) analyzed interviews and contemporaneous documents to explore whether there was something going on that could be called reconciliation in the postconflict interaction process between the police and citizens. *Relationship nature* was measured by qualifying statements made in the interviews that referred to relationships between police and citizens as positive, negative, or ambivalent. Comparing accounts referring to the relation before the confrontation with accounts referring to the relation after cooperation, the study established that respondents of both parties indicated that relationships had changed for the better. Before the confrontation, Moluccans and local police had a somewhat distant and formal relationship: as long as they did not bother each other, they implicitly agreed to ignore each other. When interactions occurred, they tended to be conflictive, and respondents from both parties assessed the nature of their relationship in negative terms. Directly following the confrontation, interactions were mainly conflictive and the negative tenor continued. As the year passed, it became clear that despite the tense relationship, actors in both parties felt a need and expressed a desire for a peaceful celebration of the next New Year's Eve. Van de Klomp *et al.* (2014) described several phases in the reconciliation process: the first phase, lasting several months, was characterized by avoidance: the different actors had no direct contact with one another, and the police proceeded with their investigations into the riot that had taken place. The second phase was characterized by tentative cooperation at the institutional level, involving the mayor and the local police as well as representatives from the Moluccan community. Cooperation at the operational level between Moluccans and police officers was only achieved in the third phase, just a month before New Year's Eve. This phase started with a meeting that took place at the Moluccan community center. The officers involved were quite anxious for this meeting:

> We went to the lion's den, that's how we experienced it. We said to each other, "if only we can make it back home alive," so to speak. It's just to give an indication of how tense things got with regard to that meeting: will we score or will we be even further from home?

The meeting went well, and the Moluccans, police, and local authorities were looking toward the upcoming New Year's Eve with increased optimism. The optimism proved to be justified, and the successfully festive New Year's Eve celebration was an expression of the newly forged union between the police and the Moluccan representatives.

The peaceful nature of the New Year's Eve celebration in the Moluccan neighborhood seemed even greater due to the fact that (albeit smaller scale) disturbances did occur in other neighborhoods (overall, there was approximately 10,000 euros less damage than in the previous year). The Moluccan organizers performed their duties without a hitch, which served to illustrate that the trust that was placed in their cooperation was not misplaced. The mayor commended them, mentioning the cooperation with the Moluccan neighborhood as a shining example for the future of the town in his New Year's speech. From a relational perspective, the actual celebration was an affirmation of the trust that was built in the previous weeks.

The analysis by van de Klomp *et al.* (2014) suggests that confrontations do not necessarily weaken relationships between police and civilians. When the aftermath is handled with care, confrontations may even have a constructive effect. This adds weight to seeing a move from confrontational to cooperative relations as a process of reconciliation and is in line with the definition of the Relational Model. The case also shows that reconciliation may be a long-term process with ebbs and flows rather than beginnings and endings. Neufeld-Redekop and Paré (2010, p. 133), who report on a case study of crowd–police relations surrounding protests, similarly state that "reconciliation is not a linear process: rather it is cyclical and iterative."

How did the change in behavior from violent confrontation to peaceful cooperation come about, and how did individual and institutional forms of interaction between the parties affect this change? It is important to recall the other assertion of the Relational Model. It predicts that the nature of a relationship between parties should influence the social process that follows after a confrontation. A more positive relationship should make re-establishing cooperative relations easier. However, the model does not delineate exactly how the nature of a relationship is experienced by the actors and how it can or should be measured.

The months-long process that it took the actors to reconcile in this case is an indication of the facts that reconciliation may be difficult and there are barriers on the way. Further study is needed into how the barriers to reconciliation are best dealt with during postconflict interactions. How are the costs and benefits of reconciliation negotiated, and when can reconciliation be successfully initiated? Aureli and de Waal (2000) suggested that the tendency to initiate aggression increases with the resource value and decreases with the risk of injury (among other factors). Van de Klomp *et al.* (2014) argue that this same tendency may also apply to postconflict interactions; Moluccan youths and police were fighting over the same "resource": control over the progression of the celebration of New Year's Eve. Specifically, the confrontation was centered around the lighting of a bonfire. More generally, it was an issue of control and public order; who is in charge in the Moluccan neighborhood? This issue was not fully resolved in favor of either party, but it was circumvented by the alternative celebration, which also served as a bargaining tool between the parties. Furthermore, the risk of injury had become greater in the perception of the police, who had been surprised by the violence directed at them but were now fully aware of the danger. Nevertheless, they also believed that an increased use of force would only lead to escalation of the conflict. The Moluccan representatives wanted to prevent further damage to their reputation and feared legal and financial repercussions from authorities. However, it was only under a certain amount of pressure from the municipality and the certainty of a rapidly approaching New Year's Eve that led the parties to seek cooperation more rapidly and simultaneously. The timing of conflict resolution is more commonly discussed in Ripeness Theory (e.g., Zartman 2001), which suggests that conflicts can only be resolved when the timing is right. In the field of ethology, an alternative theory of Benign Intent (Silk 2002) has also been suggested to give more weight to the importance of short-term objectives, such as access to desirable resources, over the long-term motivation for relational repair. In Shnabel's (this volume, Chapter 2) Needs-Based Model of reconciliation, conflicts threaten victims' sense of agency and perpetrators' moral image. Consequently, when victims feel empowered and perpetrators feel morally accepted, they are more likely to reconcile. Neither Ripeness Theory nor Benign Intent or the Needs-Based Model contradicts the Relational Model, but they do offer incentive to develop the motivational aspects of reconciliation in more depth.

Distinguishing between individual, group-level, and institutional forms of interaction in the postconflict process, the case shows a clear difference between the roles that different forms play. The interaction at the institutional level starts as a formalized method of confrontation in an attempt to re-establish power relations from the side of the authorities. The first attempt at reconciliation by the Moluccan representatives was repelled by the insistence on the side of the authorities to find the culprits before considering any other form of cooperation. However, another form of formal interaction at the institutional level, the signing of a covenant between the municipality and the Moluccan representatives, did serve as a guarantee for cooperation in the absence of trust at the group level. It seems that the distance that is created between the parties by such formal proceedings helps to prevent further escalations, but it may also stand in the way of more informal means of reconciliation. It would be interesting to explore whether it thereby serves to preempt reconciliation or helps to facilitate it in the long run.

Looking at the individual level of the postconflict interaction, in this case trust between the parties was only formed and brought to fruition when individuals from both sides of the conflict interacted on a face-to-face basis. Two meetings served as key moments of interaction after which cooperation was established. Direct personal contact became the engine of cooperation between the parties. At the very start of these meetings, the individuals who were present took the time to vent their emotions and frustrations with regard to the events that had taken place. The meetings only slowly progressed toward more friendly forms of interaction. Was this an interindividual expression of the process that also took place on the group and institutional levels: a setting of boundaries to let each other know where the limits are and how interaction may proceed? Or was this more of an emotional discharge and a moment to express empathy? Future research should delve into this matter, possibly by way of other theoretical perspectives such as those offered by Social Identity Theory (Tajfel & Turner 1986).

Van de Klomp *et al.* (2014) conclude that the relational perspective offered important clues in understanding the interactional shift from confrontation to cooperation. Moreover, approaching the subject from this perspective also allowed for the possibility that different forms of interaction may be part of the process of reconciliation. In this case study, reconciliation was a lengthy process (rather than the minutes reported in animal studies), with specific moments of contact that marked the change from confrontation to cooperation. It is yet to be established if such moments follow set patterns and if factors like emotion and identity play a similar role in different cases.

Critical Moments

In a second study, Stronks and Adang (2015) used new data, and elaborated on these findings in two ways. First, in recalling the crucial function of the social relationship between opponents in understanding postconflict interaction, and in attempting to objectify the subjective connotation of the concept of relationship strength, they applied the aforementioned notions on relationship quality of Cords and Aureli (2000).

Second, in focusing on interactions at the individual and group (rather than the institutional) levels, Stronks and Adang (2015) recalled the crucial importance of the two face-to-face meetings in fostering the group reconciliation process between the police and Moluccans. In Negotiation Theory, such moments are referred to as Critical Moments (CMs; e.g., Putnam 2004) and are commonly agreed to alter the meaning of events and transform or redefine the relations and interactions of the actors involved.

Following van de Klomp *et al.* (2014), Stronks and Adang (2015) defined police–citizen group reconciliation as an objectively determinable process from conflictive to cooperative interaction by meaningful group representatives. They selected three cases of media-salient confrontations between police and citizen groups that were characterized by the occurrence of group reconciliation but differed with regard to their local context, type of confrontation, and actors.

The first case occurred in a quiet rural community, where a violent confrontation between revelers and riot police took place when the final evening of the annual Oranjefeesten festival drew to a close. The confrontation seriously disrupted the relationship between the police and the local community, which judged the violence used by the regional riot police as unnecessary and unseemly. The second case occurred in an infamous neighborhood of a big Dutch city after police officers fired at a suspect considered armed and dangerous and arrested him. The suspect turned out to be a minor with a fake gun. Within a few hours, there were clear signs of indignation and anger among the local minority community directed at the police. In the third case, conflict arose after a youth jumped over the service desk of a local police station one Sunday morning and stabbed two officers. Police subsequently fatally shot him. The incident shocked both parties; the minority community had lost a brother, and the police had been attacked on its own territory with two officers seriously injured. The relationship between the police and part of the minority community (especially a group of juvenile delinquents) had been tense for some time.

Across the three cases, a total of 25 in-depth, semistructured interviews with 26 respondents were conducted. The interviews lasted between one and three hours and were recorded and transcribed verbatim afterward. Respondents were asked to provide a description of events and (reasons for) interactions after the overt conflict. They were encouraged to reflect on relevant relationships and their attitudes and feelings. The interviewers did not specifically ask about value, compatibility, and security in those words. For coding, each CM-related quotation received codes specifying the actor-stakeholder and indicating the tenor (positive, negative, or neutral as assessed by the coder) and behavior (e.g., questioning, expressing emotions, or sharing information). CM-related quotations in which respondents reflected on the relationship at hand between police and citizens received a Value, Compatibility, Security, or Other code (Stronks & Adang 2015). During this process, it soon became clear that it was fairly easy to assign a relationship code to each quotation (i.e., the Other code was hardly needed), but that compatibility appeared visible through more than Cords and Aureli's (2000) assumed temperament and shared history of social exchanges. When reflecting on relationships with (members of) the other party, respondents clearly added an identification aspect; they described (interactions of) themselves and others and relationships in terms of similarities and differences to the other (group).

This corresponds with Social Identity Theory (e.g., Tajfel and Turner 1979; Burke and Stets 2009), which presupposes that the interaction between people is strongly influenced by the extent to which individuals see themselves as members of their group, how they characterize their group, and how they link this to the character of, and the relationship with, (members of) other groups. Consequently, all quotations referring to such processes of social identification were coded as reflections on compatibility (Stronks & Adang 2015).

As hypothesized, analysis of the cases supported van de Klomp *et al.*'s (2014) assumption that certain moments of direct contact were critical in fostering reconciliation between police and citizens. During the CMs, three chronological stages could be distinguished: tension, transformation, and dialog (Stronks & Adang 2015). In the tension stage, insecurity expressions were recorded (e.g., negative emotional expressions, and expressions of disappointment). These comments of the community officer attending a CM with members of the festive committee in case 1 (the Oranjefeesten festival) provide a good illustration:

> So, when we finally sat together, after eight months without contact, we [representatives of the police party] were confronted with quite a grumpy festive committee. It was rather emotional. In the sense of … that, ehm, the disbelief that their beloved Oranjefeesten festival had ended like that. Their anger that it had ended like that … naturally they were very frustrated. It was quite tense.

Also, the tension stages were characterized by attempts to reduce insecurity (e.g., via appeasement, asking questions, and sharing information). For instance, one of the neighborhood team chiefs who decided to speak to the minority community during the CM the day after the stabbing incident (case 3) said:

> And people called out: "Why did she have to kill him?" So I said: "Well, I understand if you put it like that … I understand you say that if you watch television and see what you sometimes see there, but…." And then I tried to explain to tell the story from our side.

Very few value- or compatibility-related statements were made. Feelings of distrust, hesitation, nervousness, and unpredictability were dominant.

In the transformation stage, CM attendants engaged in the actual transformation to more cooperative interactions. Expressions of positive compatibility (e.g., talking in terms of "we" instead of in terms of "us vs. them," and referring to humanity as a common denominator in explaining behavior of self or other) characterized respondents' reflections on interaction. The neighborhood team chief of the previous quotation continues:

> I said: "I understand that. But" – and I was walking through the place with a microphone and out of the blue I used my finger to poke someone and said – "what would you do if suddenly out of nowhere someone came and...?" – because it is a split second – "How would you react if you had a pistol and he had a knife? This is not something you see coming ... and no matter how well we are trained in situations and how you can have the feeling that you are in control ... you cannot see this coming."

Also reflections of behavior, through which the positive value of the relationship was emphasized, were important (e.g., verbal emphasizing of the importance of cooperative contact or the relationship with the other, listening, and a motivated attitude). A team chief attending a CM in case 2 (the fake gun shooting incident) reflects:

> So I asked them: please help us to keep peace in our neighbourhood. Keep your sons at home. Help us in preventing the tensions felt, from escalating. It is up to us ALL to ... – because this disturbs, it is grist to the mill for the familiar troublemakers. That, dear people, is why I do not want the media to attend this meeting. Because it is grist to the mill for troublemakers. Please understand that, respect that and help us.

The third stage, dialog, was recorded mostly near the end of all CMs. Respondents described the interactions during this stage as more friendly, cooperative, positive, and future oriented, in which the insecure, uneasy tension felt at the start of the CM was reduced (Stronks & Adang 2015).

During the CMs analyzed, Stronks and Adang (2015) concluded, exchanged notions on positive relationship value and compatibility during CMs repress and reduce feelings of insecurity and open space for dialog. Not surprisingly, and in accordance with the assumptions of negotiation theorists, the rebalancing of insecurity was an important factor in the successful development of such meetings and their aftermath. Thus, security seemed to be the quality that mainly needed remedying during CMs, and insecurity could function as an obstacle to reconciliation. Direct, explicit, face-to-face verbal and nonverbal communication, through which it is possible to assess the status of the relationships and to determine the other's intentions, appeared essential during the CMs. This quote from a team chief attending a CM following the fatal shooting incident (case 3) exemplifies the symbolic value of a particular type of greeting:

> I don't know if I said this already, but I have never been kissed by Moroccan men as after that meeting. You know? That is really special, because it is about taking the trouble to be honest with people.... Taking the trouble with everything that is going on at the time, that you go to to meet 150 people from the neighbourhood.

Value, Compatibility, and Security

In a third explorative study, the conflicts analyzed for the Critical Moments study were now explored for the relatedness between police–citizen value assessments, compatibility assessments, and security assessments on police–citizen conflict interaction and the function of reconciliation specifically (Stronks 2015). To add to the diversity and comparability of results, two more cases were included for analysis. The fourth case concerned football riots in a provincial capital city that followed after the home team lost a decisive football match. During the riots between the riot police and "supporters" of the home team, which is known for a specific group of violent football supporters, significant damage was caused to the neighborhood around the football stadium. The fifth conflict concerned a conflict in a small city, specifically the police's use of batons while dispersing a group of 500 protesting secondary school students. The students had organized a spontaneous disruptive protest march in the neighborhood of their high schools.

Following the same method as van de Klomp *et al.* (2014), for each case, accounts referring to the relation before the confrontation and accounts referring to the situation after the last interaction that could be objectively related to the conflict were selected and analyzed. Following the conceptual scheme applied by Stronks and Adang (2015), relationship accounts were also assigned a Value, Security, or Compatibility code. The period in between these points was also analyzed in terms of their interactions, specifically with regard to the occurrence of CMs and reconciliation.

Again, some interesting preliminary conclusions could be drawn. Assessed relationship value contributed to an atmosphere in which the opponents were at least open to change and were prepared to adopt a vulnerable stance. Only when the need for the other in resolving the conflict was acknowledged, could a meeting take place. And only when the importance of good contact and cooperation with the other was acknowledged, stressed, and confirmed during that meeting, did the critical shift toward positive interaction and cooperation seem possible. A citizen attending a CM in the third case reflects:

> It was smart what the police did. They invited me and explained things to me – that's smart. If they'd done nothing, I'd have thought "They're all the same".... That's why it was good. That's why it was organized so quickly, to prevent misunderstandings. The mayor could say a few words. That REALLY had an effect at that moment.

Other value-emphasizing signals, such as emphasizing mutual understanding, sharing information, the motivated attendance of an authority figure, dedication, and friendly behavior, also seemed to pave the way for successful cooperation. By contrast, behavior through which a lack of emphasized relationship value is communicated appears to have negative effects on the assessed relationships, and on interactions. See, for example, these reflections of the chief constable of the police with regard to the attendance of fan representatives in "football match meetings" between the local authorities in case 4 (football riots):

> What we should stop doing is involve them in decision-making. We have a football-triangle [mayor–public prosecutor–police] which our extra partner [the football club] attends, but as far as I am concerned, not anymore the fan representatives. They are a partner for the club, not for us.

Six months after the decision, fan representatives were no longer welcome at the safety meetings (and seven months after the football riots). The chairman of the fan representatives commented:

> There is a safety meeting that we no longer attend. It used to be on Friday morning, before the match. Police and municipality and club decided that to be inconvenient, because it is too late then to arrange things. So that had to be moved to Thursday afternoon. Now, all of us work, so that's not possible anymore, we cannot attend that any longer. We indicated this clearly: "this is not possible, do it in the evening so we can attend." They say: "it is our work." I think, well, OK. So we are no longer there, and that is prior to every home match. They do not trust us. It feels a bit like a hunt [on the supporters – *Ed.*]. Whether they are good ones or bad ones. It is just … why punishing us for the mess made by few[?] I have a meeting scheduled with the police chief, that she drops by to talk about it, so she does that…. Because she says: "Yes, you think we do not trust you and paint you in a bad light," but they have the same feeling about us.

Relationship compatibility can function as a catalyst when it is actively sought and emphasized during interaction (e.g., seeking/emphasizing common goals, features, limitations, or enemies; and appealing to universal human characteristics, such as emotions, which everyone shares and understands), as illustrated by the following excerpt:

> Says the chief constable attending a CM after the shooting incident: Well, you go there, you make sure you are on time so you can shake hands when people come in – not showing up last, but making sure you are first – and while everyone comes inside you shake hands, you connect briefly, you chat briefly.

Social scientific research on social identification, the cognitive process of categorizing self and others in groups through similar (ingroup) or different (outgroup) features, confirms the positive effect of such behaviors. Interactions that encourage identification with (members of) other groups can be used effectively in assuaging, preventing, or reducing conflicts because the psychologically perceived distance and differences between parties are experienced as less severe or are even neutralized (e.g., Kramer & Brewer 1984; Gaertner *et al.* 1993; Ashmore *et al.* 2001; Pruitt & Kim 2003). But the same cognitive process may also function as an obstacle to reconciliation. The identification of individuals with their own group can lead to stereotyping, exclusion, and all kinds of other harmful processes that stimulate the explication of differences (incompatibility) rather than similarities (Macrae *et al.* 1996; Pruitt & Kim 2003), and hence can be a potential source for numerous misunderstandings and conflict.

The need for security was well reflected in respondents' stories about the interactions, self, and other. The rise of conflict created feelings of distrust and unpredictability between opponents, which in turn caused varying degrees of stress, depending on the intensity of the conflict and the assessed damage to the relationship. The expression of emotions seemed to play an important role relative to security: when displayed along with value signals on the one hand, venting stress and negative emotions seemed to appeal to compatibility (everybody has feelings, and it is not wrong to show anger, frustration, and grief) and fostered a sense of relationship security. However, (continued) emotional outbursts combined with signals of lack of value (e.g., walking away, ignoring, refusing to listen, and harm) and/or incompatibility (e.g., talking in terms of we–them, and direct references to incompatibility) may also have the potential to injure the security of a relationship even more. In such instances, a CM fails and reconciliation seems unattainable. In such cases, a single confrontation between a minority of group representatives may lead to enduring and escalating conflict to the extent of group level. We can see this in case 4 (football riots). In this case, a CM could not be detected, and it had the most negatively scored quotations for all three relationship qualities, in particular compatibility. The last quotation was only one of the many examples that explicated the fact that group relations suffered from many nonreconciled confrontations and a lack of cooperation. This notion is confirmed in the Relational Model, which postulates that interactions that are mainly discordant may harm a relationship in the long run (Aureli & de Waal 2000).

Discussion: Methodological Issues

The data from ethologists who test the Relational Model come from observational research. Behavior is quantified and analyzed on the ratio measurement level, in which connections between variables (i.e., value, compatibility, and security) and their effects on behavior (i.e., postconflict behavior) are tested statistically. The qualitative data gathered in our studies well after the events have taken place were generated by means of interviews and document analysis through which we analyzed individual interpretations of (reflections on) behavior. All statements on relationship nature and quality, whether referring to the period before or after the conflict occurred, were made after cooperation had been (re)established. Our data can only be quantified to a certain extent. Thus, we are not able to measure, compare, and analyze our data on a ratio level of measurement.

Another limitation can be found in the fact that a two-opposing-party perspective (police–citizen) was adopted in our measurement of relationship nature. While this simplifies analysis, there is a risk of overgeneralization. In-depth analysis of the quotations indicates that a given police–citizen relationship is rarely defined by general evaluations of the groups involved. Large parts of the quotations refer to specific between-group dyads, subgroup relationships, or key persons or groups, which respondents identify as illustrative for the general police–citizen relationship. This presents a challenge for future research that we have not yet found a solution to.

Conclusion

The application of a relational model to police–citizen confrontations is still in its infancy. The intention of this chapter was to show that interactions and confrontations between police and citizens, at both the individual and institutional levels, are mediated not only by legal procedures and considerations but also by the nature of police–citizen relationships. Reconciliations after police–citizen conflicts do occur – at least in the cases studied so far. It remains to be seen how far the results are applicable across different cultures and different police models and philosophies. It is doubtful, for instance, whether they carry the same meaning in the order-enforcing policing of French urban neighborhoods as described by Fassin (2013) in his excellent ethnography of urban policing. Also, as Fry (2000) makes clear, there is considerable variation across cultures in how conflicts are managed. Having said that, Furnari (this volume, Chapter 8) emphasizes the importance of maintaining good relationships for effective peacekeeping operations in a variety of non-Western countries. For the cases dealt with in this chapter, however, with police forces that adopted a community policing approach, the relational perspective offered important clues in understanding the interactional shift from confrontation to cooperation and to keeping the peace. A first step was taken in identifying and demonstrating factors related to the social features of the relationship between police and citizens that influence this shift and the tendency to reconcile. No doubt, the analytical distinction between the value, compatibility, and security of relationships needs to be objectified and should be applied and analyzed more systematically. However, despite the practical and theoretical barriers that still need to be overcome, the possibilities and potential of a relational perspective are promising, especially in light of frequently recurring escalations, with those in Ferguson, United States (in 2014), just being a recent example. The following summary may give some indication as to what could be relevant for those involved in these types of altercations.

The analysis of the cases studied through a relational perspective indicates that notions of value, security, and compatibility influence the way in which postconflict interactions develop. Without a sense of between-group value of at least some stakeholders, reconciliation seems out of the question. Conflicts between police and citizens cannot be dealt with at only an institutional level; there must be face-to-face contact between individuals (including those in authority). The analysis also indicates the relevance of feelings of insecurity when actors actually have to meet their opponents after a confrontation. Given the energy it takes to overcome such feelings and to (re) establish cooperation, and the fact that the police are by duty and mandate unable to avoid conflict and contact with citizens, it is no wonder that the police prefer secure and trustworthy relationships. However, our results indicate that without a sense of value, the chances that actions of the police are assessed as trustworthy seem small. This leaves the police with an important task: (strategically important) relationships can be "used" and strengthened in times of conflict, but they must be built and maintained in times of peace.

Direct social interaction (i.e., direct contact) appears to be of crucial importance during Critical Moments, where reconciliation takes shape. Confrontation and conflict drive group representatives into opposing camps, both literally and figuratively, which evokes uncertainty regarding the quality of the relationship, the damage it has suffered, and the possibilities for reparation. As such, notions of value, compatibility, and security can be both a catalyst and an obstacle in the transition to cooperation. During Critical Moments, these different notions are exchanged and negotiated through both explicit communication as well as symbolic and nonverbal communication. The latter are difficult to communicate via less direct channels such as email or phone. Only through direct contact was it possible to assess the status of the relationships and determine the other's intentions. Emotional expression and information sharing were important signals in (re)defining and negotiating the relationship.

Our analysis showed that Critical Moments are crucial to improving or repairing relationships. This confirms the basic principle of the Relational Model – the quality of the relationship nature influences conflict interaction. Sometimes, the quality of the relationship is even improved compared to before there was a confrontation (as was the case with the Moluccan community), but this is not always the case. Even when an active attempt is made during Critical Moments to create a positive group relationship, it remains to be seen in the long term whether assessed security, value, and compatibility actually improve and whether this has effects at the group level. In other words, reconciliation is a process, and the transformation from confrontation to cooperation cannot be made without one or more Critical Moments of conflict resolution. Interestingly, during the Critical Moments we studied, we could not find clear indications of the apology–forgiveness cycle mentioned by Shnabel (this volume, Chapter 2): no apologies were issued, and in all cases both police and citizens felt themselves to be victims, a situation that is particularly vulnerable to escalation, as Shnabel indicates.

Taking a relationship perspective, many questions remain, at both the individual and group levels as well as the institutional level. To name just a few: how do police leaders and police authorities value the relationship between police and citizens: do they see citizens as customers, as subjects, as partners, or as potential criminals? How does the institutional role of police as peacekeepers or enforcers of order affect police–citizen interactions at the group and individual levels? How do police interact with (groups of) citizens? How do police gain and maintain the trust of citizens? To what extent do apologies and symbolic gestures play a role? How do police monitor relationships with (groups of) citizens? With regard to processes of reconciliation: what factors influence who takes the initiative and when? How can the analytical distinction between the value, compatibility, and security of police–citizen relationships (or other between-group relationships in humans) be improved and analyzed more systematically? Comparative empirical research will be very important in answering such questions. The possibilities and potential of a relational perspective are, in our opinion, beyond question.

References

Adang, O.M.J., van der Wal, R., & Quint, H. (2010). *Zijn wij anders? Waarom Nederland geen grootschalige etnische rellen heeft* [Are we different? Why the Netherlands have not had large-scale ethnic riots]. Apeldoorn: Stapel & de Koning/Politieacademie.

Allport, G.W. (1954). *The nature of prejudice*. Reading, MA: Addison-Wesley.

Ashmore, R.D., Jussim, L., & Wilder, D. (Eds.). (2001). *Social identity, intergroup conflict and conflict reduction*. New York: Oxford University Press.

Aureli, F., & F.B.M. de Waal (Eds.). (2000). *Natural conflict resolution*. Berkeley: University of California Press.

Ayres, I., & J. Braithwaite (1992). *Responsive regulation: Transcending the deregulation debate*. New York: Oxford University Press.

Bittner, E. (1974). Florence Nightingale in pursuit of Willie Sutton: A theory of the police. In H. Jacob (Ed.), *The potential for reform of criminal justice*. London: Sage.

Black, D. (1980). *The manners and customs of the police*. New York: Academic Press.

Bonaventura, M., Ventura, R., & Koyama, N.F. (2009). A statistical modelling approach to the occurrence and timing of reconciliation in wild Japanese macaques. *Ethology, 115*, 152–166.

Burke. P.J., & Stets, J.E. (2009). *Identity theory*. London: Oxford University Press.

Cooper, M.A., Bernstein, I.M., & Hemelrijk, C.K. (2005). Reconciliation and relationship quality in Assamese macaques (*Macaca assamensis*). *American Journal of Primatology, 65*, 269–282.

Cords, M., & Aureli, F. (2000). Reconciliation and relationship qualities. In F. Aureli & F.B.M. de Waal (Eds.), *Natural conflict resolution*. Berkeley: University of California Press.

Della Porta, D., & Reiter, H. (Eds.) (1998). *Policing protest: The control of mass demonstrations in western democracies*. Minneapolis: University of Minnesota Press.

Fassin, D. (2013). *Enforcing order: An ethnography of urban policing*. Cambridge: Polity Press.

Fielding, N.G. (2005). *The police and social conflict*. London: Glass House.

Fielding, N.G., & Innes, M. (2006). Reassurance policing, Community policing and measuring police performance. *Policing and Society, 16*(2), 127–145.

Fleming, J. (2005). Working together: Neighbourhood watch, reassurance policing and the potential of partnerships. *Trends and Issues in Crime and Criminal Justice, 303*, 1–6.

Fraser, O.N., Schino, G., & Aureli, F. (2008). Components of relationship quality in chimpanzees. *Ethology, 114*, 834–843.

Friedmann, R.R. (1992). *Community policing: Comparative perspectives and prospects*. New York: St. Martin's Press.

Fry, D.P. (2000). Conflict management in cross-cultural perspective. In F. Aureli & F.B.M. de Waal (Eds.), *Natural conflict resolution*. Berkeley: University of California Press.

Fujisawa, K.K., Kutsukake, N., & Hasegawa, T. (2005). Reconciliation pattern after aggression among Japanese preschool children. *Aggressive Behaviour, 31*, 138–152.

Gaertner, S.L., Dovidio, J.F., Anastasio P.A., Bachman, B.A., & Rust, M.C. (1993). The common ingroup identity model: Recategorization and the reduction of intergroup bias. *European Review of Social Psychology, 4*(1), 1–26.

Goldstein, H. (1979). Improving policing: A problem-oriented approach. *Crime and Delinquency, 25*(2), 236–258.

Hewstone, M., *et al.* (2008). Stepping stones to reconciliation in Northern Ireland: Intergroup contact, forgiveness, and trust. In A. Nadler, T.E. Malloy, & J.D. Fisher (Eds.), *The social psychology of intergroup reconciliation*. New York: Oxford University Press.

Janoff-Bulman, R., & Werther, A. (2008). The social psychology of respect: Implications for delegitimization and reconciliation. In A. Nadler, T.E. Malloy, & J.D. Fisher (Eds.), *The social psychology of intergroup reconciliation*. New York: Oxford University Press.

Kerner, O., *et al.* (1968). *Report of the National Advisory Commission on Civil Disorders*. New York: Bantam.

King, M. (2013). Birmingham revisited: Causal differences between the riots of 2011 and 2005? *Policing and Society: An International Journal of Research and Policy, 23*(1), 26–45.

Koyama, N.F. (2001). The long-term effects of reconciliation in Japanese macaques *Macaca fuscata*. *Ethology, 107*, 975–987.

Kramer, R.M., & Brewer, M.B. (1984). Effects of group identity on resource use in a simulated commons dilemma. *Journal of Personality and Social Psychology, 46*(5), 1044–1057.

Macrae, C.N., Stangor, C., & Hewstone, M. (1996). *Stereotypes and stereotyping.* New York: Guilford Press.

Mastrofski, S.D., Snipes, J.B., & Supina, A.E. (1996). Compliance on demand: The public's response to specific police requests. *Journal of Research in Crime and Delinquency, 33*(3), 269–305.

Neufeld-Redekop, V., & Paré, S. (2010). *Beyond control: A mutual respect approach to protest crowd-police relations.* London: Bloomsbury Academic.

Otten, M.H.P., Boin, R.A., & Van der Torre, E.J. (2001). *Dynamics of disorder: Lessons from two Dutch riots.* Amsterdam: Elsevier Science.

Pruitt, D.G., & Kim, S.H. (Eds.). (2003). *Social conflict: Escalation, stalemate, and settlement* (3rd ed.). London: McGraw-Hill.

Putnam, L.M. (2004). Transformations and critical moments in negotiation. *Negotiation Journal, 20*(2), 275–293.

Reiner, R. (2000). *The politics of the police.* Oxford: Oxford University Press.

Reisig, M.D. (2010). Community and problem-oriented policing. *Crime and Justice, 39*, 1–53.

Rusbult, C.E., Hannon, P.A., Stocker, S.L., & Finkel, E.J. (2005). Forgiveness and relational repair. In E.L. Worthington Jr. (Ed.), *Handbook of forgiveness.* New York: Taylor & Francis Group.

Scarman, L. (1982). *The Scarman report: The Brixton disorder, 10–12 April 1981.* Harmondsworth: Pelican.

Silk, J.B. (2002). The form and function of reconciliation in primates. *Annual Review of Anthropology, 31*, 21–44.

Skogan, W.G. (2006). Advocate: The promise of community policing. In D. Weisburd & A.A. Braga (Eds.), *Police innovation: contrasting perspectives.* Cambridge: Cambridge University Press.

Stronks, S. (2015). Exploring police-citizen conflict and reconciliation by through a relational model. *European Journal of Policing Studies, 3*(1), 342–366.

Stronks, S., & Adang, O.M.J. (2015). Critical moments in police-citizen reconciliation. *Policing: An International Journal of Police Strategies & Management, 38*(2), 366–380.

Tajfel, H., & Turner, J.C. (1979). An integrative theory of intergroup conflict. *Social Psychology of Intergroup Relations, 33*, 47.

Tajfel, H., & Turner, J.C. (1986). The social identity theory of inter-group behavior. In S. Worchel & L.W. Austin (Eds.), *Psychology of intergroup relations.* Chicago, IL: Nelson-Hall.

Tyler, T. (1990). *Why people obey the law.* New Haven, CT: Yale University Press.

Tyler, T., & Huo, Y.J. (2001). *Trust and the rule of law: A law abidingness model of social control.* New York: Russel Sage.

Tyler, T.R., & Huo, Y.J. (2002). *Trust in the law: Encouraging public cooperation with the police and courts.* New York: Russell Sage Foundation.

van de Klomp, M., Adang, O.M.J., & van den Brink, G.J.M. (2011). Riot management and community relations: Policing public disturbances in a Dutch neighbourhood. *Policing and Society: An International Journal of Research and Policy, 21*(3), 304–326.

van de Klomp, M., Stronks, S., Adang, O.M.J., & van den Brink, G.J.M. (2014). Police and citizens in conflict: Exploring post-confrontation interaction from a relational perspective. *Policing and Society: An International Journal of Research and Policy, 24*(4), 459–478.

van Stokkom, B. (2004). *Handhaven: eerst kiezen, dan doen. Sociaal wetenschappelijke mogelijkheden en beperkingen* [Maintaining public order: Acting after choosing. Social scientific opportunities and obstacles]. The Hague: Ministry of Justice Department.

Waddington, D., Fabien, J., & King, M. (Eds.). (2009). *Rioting in the UK and France: A comparative analysis*. Cullompton, UK: Willan Publishing.

Wittig, R.M., & Boesch, C. (2005). How to repair relationships: Reconciliation in wild chimpanzees (*Pan troglodytes*). *Ethology, 111*, 736–763.

Zartman, I.W. (2001). The timing of peace initiatives: Hurting stalemates and ripe moments. *The Global Review of Ethnopolitics, 1*(1), 8–18.

11

Constitutions as Peace Systems and the Function of the Costa Rican and Japanese Peace Constitutions

Benjamin A. Peters

Introduction

In this volume, we use the concept *peace system* to mean an institution or arrangement between individuals or groups that patterns their interactions toward peace (see Verbeek and Peters, this volume, Chapter 1). This chapter is an investigation into a particular kind of human peace system, peace constitutions. In states that are both liberal and democratic, constitutional frameworks institutionalize patterns of interactions between groups within the state and between the state and members of society. These constitutions exist either as a single document such as the US Constitution or within a liberal, "rule of law" framework adopted over time as in the United Kingdom. As a particular kind of constitutional framework, peace constitutions affirm peace as a foundational value, provide the legal grounds for people to secure the right to peace, and outlaw war and/or the standing army. Although several countries have adopted peace constitutions in recent years (Sasamoto 2010), the constitutions of Japan and Costa Rica are exemplary for the duration of their use and for their demonstrated functions of limiting war making by the state and fostering emergent cultures of peace.

While many concepts and terms used throughout the chapter derive from the study of politics, this chapter is also an attempt to study human political behavior (here in terms of "activities" or repertoires of behaviors) through the *peace ethology* approach (for an introduction to the field, see Verbeek 2008; Verbeek & Peters, this volume, Chapter 1). Therefore, this section aims to familiarize the reader with both our species-typical capacity for peaceful behavior and the development of modern states and constitutional frameworks. The chapter then continues with case study analyses of the function of peace constitutions in Japan and Costa Rica.

Our Species-Typical Capacity for Peaceful Behavior

Humans belong to the genus *Homo* that emerged some 2 million years ago. The particular kind of animal we are, *Homo sapiens*, is relatively new on the evolutionary scene. According to Fry, our species "dates from at least 200,000 years ago, while anatomically modern *Homo sapiens* (modern humans) appear in the archaeological record only in

Peace Ethology: Behavioral Processes and Systems of Peace, First Edition.
Edited by Peter Verbeek and Benjamin A. Peters.
© 2018 John Wiley & Sons Ltd. Published 2018 by John Wiley & Sons Ltd.

the last 40,000 to 50,000 years" (2007, p. 61). Of specific relevance to the study at hand is anthropological and archaeological evidence that:

> [W]ar is simply absent over the vast majority of human existence.... But with gradual worldwide population increase, the shift from universal nomadic foraging to settled communities, the development of agriculture, a transition from egalitarianism to hierarchical societies – and, very significantly, the rise of state-level civilizations five thousand to six thousand years ago – the archaeological record is clear and unambiguous: war developed, despots arose, violence proliferated, slavery flourished, and the social position of women deteriorated ... taking place in the last 10,000 years. (Fry 2013, p. 15)

Several of these scientific findings are germane to the study of politics and peaceful behavior. Specifically, for approximately 95% of the distinct genetic history of *Homo sapiens*, we lived without war through species-typical behavioral repertoires of small-band, nomadic foraging and egalitarian social relationships (on the species' evolved developmental niche and nomadic forager past, see the respective chapters in this volume by Narvaez [Chapter 6] and Evans Pim [Chapter 12]). It is only in the last 5% of our species' genetic history that the behavioral repertoires of state making and war making emerged. It follows, then, that these occurred first as species-atypical repertoires of behaviors, behaviors not frequently shown by members of the species (on species-typical and species-atypical behaviors, see Verbeek 2013). Furthermore, state making and war making are forms of aggression insofar as groups carry them out to pursue active control of resources and the social environment at the expense of others (cf. de Boer in Kruk & Kruk-de Bruin 2010, cited in Verbeek 2013).

Modern State Making and the Emergence of Constitutional Frameworks

As the anthropological evidence indicates, war making emerged in tandem with agriculture approximately 10,000 years ago, and state making followed approximately 5000–6000 years ago. Just as there was a link between war making and state making activities during the emergence of both, social scientists have made the relationship between them the focus of one of the leading theories of the development of the early modern state as it emerged in Renaissance Europe. This theory identifies the change in military technology that began at the end of the 14th century as a catalyst for the transition from feudal warfare to standing armies and mercenary forces made possible by efficiencies of scale and the expansion of taxation (Spruyt 2002).

Charles Tilly, a proponent of the war-makes-states theory, argues that the emergence of modern states resulted from the gradual institutionalization of four activities: war making, state making, protection, and extraction (Tilly 1985). Each activity had a particular aim: war making to eliminate the state makers' rivals outside their territory, state making to eliminate state makers' rivals inside their territory, protection to eliminate enemies of their clients, and extraction to acquire the means to carry out the other three activities. The four activities also had corresponding organizations to carry them out: armies and navies for war making, police and surveillance for state making, courts and representative assemblies for protection, and tax and accounting agencies for extraction.

The second part of Tilly's theory accounts for the emergence of constitutional frameworks as a popular reaction against the organized violence of the state. According to Tilly, the modern state's enhanced capacity for organizing and using violence, especially through the activity of war making, brought with it threats to ordinary people's lives and livelihoods such as conscription, death in battle or as a result of invasion by foreign armies, use of the army against the domestic populace, and higher taxes. Because of these threats and the fact that there was little to prevent war making for reasons of personal greed, ambition, or honor (including revenge), popular resistance arose both to the onerous taxation necessary to fund the state's violence and to the monopolistic authority that carried it out. In other words, "popular resistance to coercive exploitation forced would-be power holders to concede protection and constraints on their own action" (Tilly 1985, p. 170). Here, then, is the connection between the development of the modern state and the emergence and function of constitutional frameworks:

> [P]opular resistance to war making and state making made a difference. When ordinary people resisted vigorously, authorities made concessions: guarantees of rights, representative institutions, courts of appeal. Those concessions, in their turn, constrained the later paths of war making and state making. (Tilly 1985, p. 183)

Viewed within the context of our species' evolutionary history and from the standpoint of peace ethology, we can understand Tilly's theory in the following terms. First, war making and state making emerged as forms of species-atypical aggression. In other words, among the behaviors exhibited across the evolutionary timespan of our species, they occur infrequently – with no evidence of such activity during 95% of the time we have existed as a species. Likewise, individual members of the species infrequently exhibit behaviors that are part of the activities of state making and war making, and there is an innate psychological avoidance of the latter (Grossman 1996). Second, in the modern era, people resisted the species-atypical aggression of war making and state making by demanding what emerged as constitutional frameworks. The original function of these frameworks was to limit war making and state making. In other words, the result of popular resistance, when successful, was a recuperation of species-typical egalitarian social relations and limitations on species-atypical aggression. This recuperation was partial and tenuous – partial because it still existed in the context of life in states organized and socially reproduced through species-atypical aggression and tenuous because state-level actors might still use the organizations of the state to carry out species-atypical aggression beyond agreed-upon limits.

Constitutions as Peace Systems and the Function of Peace Constitutions

Every constitutional framework is a peace system insofar as it establishes arrangements between individuals or groups that pattern their interactions toward peaceful behaviors. This is evidenced by both their "negative peace" and "positive peace" functions (on negative peace and positive peace as used here, see Verbeek and Peters, this volume, Chapter 1). On one hand, the *negative peace function* of constitutions is to institutionalize historical concessions and constraints won through popular resistance to war making and state making. This is the ubiquitous model and function of "rule of law" politics, even if particular constitutional frameworks simply adopted that function from earlier, model constitutions. The political theorist Judith Shklar makes explicit this negative peace function, asserting that "the ultimate … political struggle is always between war and law" (1998, p. 25). By "war," she means the aggression of war making and unconstrained state making carried out by a state *against its own people*. Alternatively, the *positive peace function* of constitutions is to create affirmative obligations of the state to members of society. These began as general obligations such as insuring domestic tranquility and promoting the general welfare, as in the US model. Later, more specific obligations emerged with the introduction of affirmative public welfare rights such as provisions of financial assistance to the unemployed or aged, publicly financed education, and guaranteed health provisions (Shane 2006).

All peace constitutions place negative peace limitations and positive peace obligations on the state. The defining negative peace limitations are (1) renunciation or strict limitation of the state's right to make war, and (2) the prohibition against maintaining a permanent military establishment. Peace constitutions include either or both of these limitations, with the second being more effective for reasons that follow in this chapter. Examples of positive peace obligations include recognition of the right to live in peace; some level of guaranteed welfare or employment; education, health, and housing provisions; and even guarantees of a cultured life. When both the negative provisions and positive obligations are functional, they are mutually reinforcing. For example, a peace constitution may renounce the state's right to maintain an army and affirm citizens' right to live in peace. As long as the state does not violate the first provision, citizens have secured their right to live in peace to the extent that the state does not engage in activities such as conscripting citizens to kill and die in war, using military force against its own society, and extracting revenue from citizens for military purposes. Even if a peace constitution does not explicitly enumerate a right to live in peace, citizens may enjoy that right by effectively limiting the state's state making and war making activities.

When state-level actors govern within the limitations and act to fulfill the obligations, the peace constitution is functional and is conducive to maximizing benefits of the peace system. Reduced budgetary outlays for preparing for and making war, reduced military violence against members of society, reduced mortality in war, and long periods of peace with other states are examples of benefits that accrue from such limits (on the economic benefits of peace, see Hyslop and Morgan, this volume, Chapter 13). The benefits of these positive obligations result directly from their fulfillment: enjoying the right to live in peace, adequate employment, welfare, education, health, and so on. Taken together, the overall benefit of the limitations and obligations is life in an emergent culture of peace. The qualifying term *emergent*, indicates that there may still be other aspects of a culture of peace that remain unfulfilled. This may be due to problems such as systematic police brutality, high unemployment or income inequality, widespread organized crime activity, prejudice and discrimination, or environmental pollution. As long as the state both complies with the limitations and fulfills its obligations, however, we can say that the peace constitution is functional. If the state does not fully comply with or fulfill the negative and/or positive peace provisions of its peace constitution, then the peace constitution is semifunctional. Finally, when the state does not comply with, does not fulfill, or abrogates these provisions, the peace constitution is nonfunctional.

While there is not a definitive count of the number of countries with peace constitutions, Jun Sasamoto (2010), Secretary General of the Japan International Association of Lawyers, recognizes seven such cases (Bolivia, Costa Rica, Ecuador, Japan, Panama, the Philippines, and Venezuela). Additionally, Japanese legal scholar Akira Maeda (2008) recognizes 27 countries without armies, and Swiss lawyer Christophe Barbey (2015) recognizes 26, although not all of those countries' constitutions can be considered peace constitutions since most do not explicitly prohibit the maintenance of armed forces. The remainder of this chapter presents overview analyses of the functioning of the peace constitutions of Costa Rica and Japan, two cases on which the author has previously conducted research. These countries adopted their peace constitutions at the same historical moment and following the cessation of periods of military violence but under quite different circumstances. The Constitution of Costa Rica (1949) resulted from the settlement of a brief civil war as a solution to the problem of rival political groups vying for control of the state and, with it, the country's military establishment. The adoption of the Constitution of Japan (1947) resulted from that country's defeat in World War II and US-led Occupation officials' insistence on the provisions that make it a peace constitution. As I demonstrate in this chapter, Costa Rica has a functional peace constitution, while Japan's is semifunctional.

Costa Rica's Peace Constitution

Given the frequency and scale of state violence in 20th-century Central America, Costa Rica's abolition of its army in 1948 and codification of that limit in its 1949 peace constitution provide a remarkable case in the context of the region. Policies related to both positive and negative peace functions, however, characterize the early history of the country. Upon its independence in 1838, Costa Rica adopted democratic governance and quickly made institutional changes conducive to the development of peace. These included the establishment of free and mandatory elementary school education in 1869 and abolition of the death penalty in 1877. Apart from a two-year dictatorship following a *coup d'état* in 1917 and a four-year period of instability including one year of civil war in the 1940s, Costa Rica has had the most stable democracy in Central America until the mid-1940s and has remained so since then (Katz & Lackey 2010).

The adoption of a peace constitution was a breakthrough solution to the problem of state violence in the Costa Rican case, and it has served the country well in terms of the achievement of a functional peace system, particularly but not only in comparison to its neighbors in Central America. The fact that Costa Ricans maintained it even in the face of invasion and regional strife verifies the function of peace constitutions in limiting war making. In addition, the peace constitution provided the institutional foundation upon which Costa Rican citizens established peace as a fundamental value and secured the right to live in peace, as the analysis of key legal cases demonstrates further in this chapter. As this case study also shows, the peace constitution has played a central role in the emergence of a Costa Rican culture of peace.

Limits on War Making

High levels of state violence were a distinguishing feature of Central American politics in the 20th century. In the 1980s and 1990s alone, civil wars in the region "led to the killing of at least 300,000 people, the vast majority of whom were civilians" and "produced between 1.8 million and 2.8 million refugees" (Lehoucq 2012, p. 30). Costa Rica, adhering to its peace constitution, avoided such violence almost altogether. It may seem ironic and even unlikely, then, that this constitution emerged as a solution to Costa Rica's own experience with civil war.

The Costa Rican civil war of 1948 followed several years of political turmoil centered on disputes over the results of the 1944, 1946, and 1948 national elections. In addition, the state had amassed the largest army in its history, partly as a result of US policy to provide military equipment and training to the government to enhance protection of the Panama Canal (Bird 1984, p. 99). It was in this context that the civil war broke out following the February 1948 election.

When the Legislative Assembly annulled the election results to prevent the duly elected opposition leader, Otilio Ulate, from assuming the presidency, former private militia leader and pro-democracy figure José Figueres Ferrer formed an Army of National Liberation, defeated the armed forces of the Costa Rican government, and established martial law. The civil war, which resulted in 2000 deaths and Figueres' subsequent rule by *junta*, came to an end when he dissolved the army and held new elections in December 1948. An immediate invasion originating from Nicaragua and led by the deposed former president, Calderón Guardia, was the first test for the army-less state. Without resorting to retaliatory violence, the Costa Rican government invoked the Inter-American Treaty of Reciprocal Assistance, and, after an immediate investigation, an Organization of American States (OAS) commission ruled that Nicaragua had not taken adequate measures to prevent the invasion. The Costa Rican and Nicaraguan governments then signed a Pact of Unity, and, with no base of support, Guardia's forces ceased hostilities (Bird 1984, pp. 110–115).

Costa Rica institutionalized this negative peace function in the 1949 constitution, Article 12, which proclaims the permanent abolition of the standing army. Although the constitution permits organization of an army in limited circumstances, the government has never done so. This is remarkable given that it faced popular revolts in 1950 and 1951 and faced a second invasion by Guardia and his forces in 1955. In the case of the revolts, the government relied on the Civil Guard to maintain order, and in the case of the invasion, Costa Ricans voluntarily organized resistance to the invading forces while the government appealed for intervention from the OAS. When the OAS sent civilian monitors to the border with Nicaragua, the invasion collapsed, and the two countries pledged to prevent insurgents from organizing in their territories and from crossing their shared border (Bird 1984, p. 127).

Costa Rica's 1983 Declaration of Perpetual, Active, and Unarmed Neutrality provides another example of the benefits of the peace constitution. Under pressure from the United States to support Contra rebels in Nicaragua's civil war and in response to terrorist attacks in the early 1980s, President Luis Alberto Monge allowed the CIA and Contras to operate inside the country. This did not prevent Costa Rica from voting against the US invasion of Grenada in 1983, and, to avoid getting further drawn into Nicaragua's civil war, President Monge declared neutrality. It was not until 1986, under the leadership of Óscar Arias Sánchez, however, that the government finally shut down Contra bases and US paramilitary operations in Costa Rica. Punctuating the country's neutral stance, Arias publicly refused a private meeting with CIA Director William Casey (LeMoyne 1987). Again, Costa Rica's peace system functioned to avoid war making, thus sparing unnecessary civilian deaths, military spending, and the establishment of an army. Furthermore, it was this history of the institutionalization of peace processes made possible by the peace constitution that provided citizens with the legal grounds for securing their right to live in peace.

The Right to Peace

The United Nations General Assembly affirmed "the inherent right to life in peace" with the adoption of the Declaration on the Preparation of Societies for Life in Peace in 1978, and the African Charter on Human and People's Rights (1981) further affirmed the right. According to Roche, the significance of these instruments is that they place the onus for ensuring the right to peace on governments and emphasize the duty of citizens to work toward fully securing the right (2003, p. 124). The UN General Assembly again affirmed the right with its adoption of the Declaration on the Right of Peoples to Peace in 1984, which established the right to peace as "a fundamental prerequisite for the fulfillment of other basic rights." Further progress toward securing the right was made, however, with the UNESCO Declaration of a Human Right to Peace in 1997 and the Oslo Draft Declaration the same year (Roche 2003). At the time of this writing, progress toward a draft UN declaration on the right to peace continues under the leadership of an intergovernmental working group established for this purpose by the UN Human Rights Council (United Nations Office of the High Commissioner on Human Rights 2016).

While the Constitution of Costa Rica does not explicitly mention the right to peace, Costa Ricans have recognized their constitution's strict limitation on war making along with the Declaration of Perpetual, Active, and Unarmed Neutrality as evidence and partial fulfillment of the government's responsibility to ensure peace. Citizens have also acted to further secure the right. This is most evident in legal cases by citizens against the government for violating the right to peace and by the state's ultimate affirmation of that right by compliance with high court rulings to that effect (Peters 2013).

The first legal test of the right to peace occurred in 2003 when Luis Roberto Zamora, a law student, brought suit against the government for agreeing to add Costa Rica's name to the Coalition of the Willing, the US-led alliance that invaded Iraq in 2003. Zamora's case reflected widespread public opposition to the government's policy, and when the Costa Rican Bar Association and the national Ombudsman filed similar suits, the three were combined into one suit (Benjamin 2014). Arguing before the Constitutional Chamber of the Supreme Court, Zamora charged that participation in the war, even in name only, violated the constitution and the declaration of neutrality. The Court unanimously ruled in favor of the claimants, finding that the government's support for the coalition violated the declaration of neutrality and the UN Charter since the UN Security Council had not approved the war. Additionally, the Court ruled that the government's action violated "a fundamental principle of 'the Costa Rican identity', which is peace as a fundamental value" (Zamora 2010). The government complied by requesting that the United States remove Costa Rica's name from the coalition.

In a second case brought by Zamora against the government in 2006, the plaintiff successfully challenged the constitutionality of an Executive Order by President Arias, a Nobel Peace Prize laureate, regulating the importation and manufacture of weapons as well as permitting extraction of uranium and thorium, nuclear fuel development, and manufacturing nuclear reactors. The Court again ruled unanimously in Zamora's favor, finding that the Executive Order violated the right of Costa Ricans to a healthy environment, threatened human health, violated international law and the Nuclear Non-Proliferation Treaty, and violated the value of peace and the UN Declaration on the Right of Peoples to Peace (Katz & Lackey 2010).

In a third ruling that affirmed Costa Ricans' right to peace, the Constitutional Chamber of the Supreme Court ruled unanimously in 2011 that a 2008 Executive Order to arm police with military-grade weapons was illegal. Again, the plaintiff was Zamora, and the Court agreed with his charge that Costa Rican law prohibits the use of weapons such as Uzis, M-16s, and AK-47s except in the case of invasion, a state of siege, or during a state of emergency (Zamora n.d.). In this case, as in the 2006 case, the peace constitution functioned to limit the potential violence of state making as well as the potential capacity for war making.

The rulings in the three cases definitively established peace as both a fundamental value of Costa Rica and a right in line with the constitutional limits on war making, the declaration of unarmed neutrality, and its obligations under international law, especially the UN Charter, the UN Declaration on the Right of Peoples to Peace, and the Nuclear Non-Proliferation Treaty. In addition, the second and third cases limited the potential violence of state making by preventing the militarization of the police. In summary, the peace constitution of Costa Rica functions to ensure a fairly robust peace system for Costa Ricans by providing the institutional framework upon which the process of claims making about the right to peace effectively limits the state's potential for war making and violent state making. Finally, the peace constitution provides the foundation upon which Costa Rica's emergent culture of peace has developed.

Emergent Culture of Peace

Costa Rica ranked 34th out of 162 countries on the 2015 Global Peace Index (Vision for Humanity 2015). While the country lags on indicators such as access to weapons, incarceration, and homicide, it leads on those related to state and war making: intensity of internal conflict, weapons imports, terrorism impact, deaths from internal conflicts, internal conflicts fought, armed services personnel, nuclear and heavy weapons, weapons exports, external conflicts fought, and deaths from external conflicts. According to the New Economics Foundation (NEF), Costa Rica also ranked as the happiest country in the world in the 2015 Happy Planet Index, which reflects a country's life expectancy, experienced well-being, and ecological footprint (NEF 2015). These rankings, especially the latter, have brought international attention to Costa Rica as they highlight its peace system.

As the oldest and most politically stable democracy in Central America and one that allocates minimal outlays in terms of military and security spending, Costa Rica has accrued many positive peace benefits. For example, the share of Costa Rica's population in poverty was lower than both the Latin American and Central American averages from 1980 until 2008. In 2008, for example, Costa Rica's rate was 16% compared to the Latin American average of 33% and the Central American average of 46%. It was also much lower than that of its neighbors to the north and south, Nicaragua (62%) and Panama (28%). In terms of the top quintile's share of national income in the same year, Costa Rica's rate of 53% was just below the Latin American and Central American averages (57% for both) as well as those of its neighbors (58% for Nicaragua and 56% for Panama). Finally, Costa Rica's social expenditures per capita as a share of the Latin American average have remained in the range of 125–175% since 1990. This compares to a range of 50–100% for Panama and 0–25% for Nicaragua (Lehoucq 2012).

Domestically, the promotion of the culture of peace has generally had state support toward institutionalizing peace processes. These have been both symbolic and substantive. When Figueres abolished the army in 1948, for example, the Minister of Public Security handed over the keys to the main army barracks to the Minister of Education in a public ceremony, and the government then converted them into a museum of art (Bird 1984, p. 89). In another show of its commitment to the reduction of state-directed violence, during the 1950 and 1951 revolts mentioned above, the president disbanded the Ministry of Public Security despite the crisis conditions. As Shutts (2009) points out, more recently the government has promoted peace processes by passing a law requiring every school to offer peace education (1997), implementing changes to the legal system to utilize mediation and peaceful conflict resolution (1997), creating a National Directorate of Alternative Conflict Resolution (2004) and a National Commission for the Prevention of Violence and Promotion of Social Peace (2006), and establishing Latin America's first Ministry of Justice and Peace (2009).

Costa Rica's emergent culture of peace has also had positive effects in terms of Central American politics. Costa Rican President Arias played a leading role in resolving civil wars in El Salvador, Nicaragua, and Guatemala through the Esquipulas Peace Agreement, for which he won the Nobel Peace Prize in 1987. In addition, he is credited with having urged the government of Panama to abolish its army, which it did in 1994 (Brysk 2009) and as stipulated in Article 310 of its amended constitution. Costa Rica has also been a signatory to all of the international treaties and protocols on death penalty abolition. For these reasons, Costa Rica has the potential to become the world's first nonkilling state (Peters 2013). On the other hand, there are state-level actors who want to upend Costa Rica's peace constitution. According to Marujo (2011), in recent years some members of the government have suggested amending Article 12 of the constitution to loosen the restriction on the army, and then–Foreign Minister Rene Castro suggested that Costa Rica spend 2 to 4% of its Gross Domestic Product (GDP) on a national defense force. Still, compared to most countries, Costa Rica has already benefited from its functional peace constitution and will continue to do so as long as there is consensus between members of the state and society.

Japan's Peace Constitution

The Constitution of Japan is a comprehensive peace constitution in that it affirms peace as a founding value, "recognize[s] that all peoples of the world have the right to live in peace" (Preamble), "renounce[s] war as a sovereign right of the nation and the threat or use of force as means of settling international disputes," and prohibits "land, sea, and air forces, as well as other war potential" for that purpose (Chapter II, Article 9). Because of its fundamental principles of peace and strict limits on war making, it is known in Japan as *heiwa kenpō*, "the peace constitution." Adopted during the US-led Occupation, the question of the constitution's origins has been a topic of much interest (see McNelly 2000; Hellegers 2001; Schlichtmann 2009). Another prominent line of research centers on the degree to which Japan has actually achieved and/or maintained its pacifist principles (see Hook 1996; McCormack 2007; Hughes 2009).

As with the Costa Rican case, Japan's adoption of a peace constitution was a breakthrough solution to the problem of state violence. Japan's belligerent actions before and during World War II resulted in approximately 6,000,000 civilian deaths throughout the Asia-Pacific region (Rummel 1994). However, since 1947 Japan has not waged war, and not one death in war since 1947 is attributable to the direct action of the Japanese state. While Japan has participated in a security alliance with the United States since the end of the Occupation and maintains Japan Self-Defense Forces (JSDF) that rank near the top of defense establishments in the world in terms of expenditures and armaments, the peace constitution continues to limit the war making capacity of the state. For example, the JSDF is not legally a military, there is no system of military law parallel to the civilian judiciary, and the Japanese state has no authority to conscript citizens into its service. Until 2014, the Japanese government considered even the universally sanctioned right of collective self-defense unconstitutional. Because state-level actors have successfully narrowed and eroded these limits, however, Japan's peace system is semifunctional. The analysis in this section, however, focuses on the viable aspects of the Japanese peace system: its limits on war making and state making, the right to peace, and facets of Japan's emergent culture of peace.

Limits on War Making

The adoption of the 1947 Constitution of Japan ended a period of total war. Because the Japanese state had been both authoritarian and militaristic, two of the goals of the US-led Occupation were democratization and demilitarization. The Constitution of Japan reflects both. In addition to the explicit peace provisions in the Preamble and Chapter II, Article 9, it designates the Emperor as a symbol rather than the head of the state (Chapter I, Article 1), guarantees citizen sovereignty and fundamental human rights (Preamble and Chapter III), establishes a representative government and independent judiciary (Chapters IV, V, and VI), and provides for local self-government (Chapter VIII). Considered in whole, *constitutional democratic antimilitarism* best describes the Japanese peace system. What follows in this section is an accounting of some of the moves taken by Japanese state-level actors, especially members of the long-ruling Liberal Democratic Party (LDP), that have rendered this peace system semifunctional, but it is important to remember that the Constitution of Japan remains unamended and continues to put relatively strict limitations on war making.

The outbreak of the Cold War presented the first tests of the negative peace aspects of the constitution. US tensions with the Soviet Union in the early postwar years resulted in a "reverse course" in Occupation policy away from the goals of democratization and demilitarization. With the establishment of the People's Republic of China and the beginning of the Korean War, Occupation and Japanese government officials de-purged 20,000 former military men (including accused war criminals), established a 75,000-man National Police Reserve led in part by former imperial military officers, and increasingly centralized control over local government (Duke 1973, p. 101). The Korean War in particular tested the peace constitution, as former militarists urged US General Douglas MacArthur to send them to fight in the former colony under US command and members of parliament urged him to establish a one-million-strong Japanese Volunteer

Defense Force (Sodei 2001, pp. 230–232). Although MacArthur did not oblige these requests, he did order more than 40 Japanese minesweepers to Korean waters in 1950, two of which sank resulting in one death.

Coinciding with the end of the Occupation in 1952, the National Police Reserve became the National Security Force. Two years later, that became the current JSDF. To justify the existence of the JSDF under the constitution's limits, the government pledged to achieve only the minimum troop level necessary to defend Japan in accordance with its international right of self-defense, utilize the JSDF for natural disaster relief, and maintain it as an apolitical, civilian organization (Buck 1967). While the establishment of such a force re-established the state's war making capacity organizationally, state-level actors continued to adhere to certain constitutional limits.

Despite these moves by the governing LDP, the public generally opposed remilitarization. As an example, when the United States and Japan began renegotiating their security treaty in 1959, 50% of the public favored a declaration of Japanese neutrality while only 26% favored a continued alliance with the United States (Packard 1966, p. 150). As further proof of popular opposition, Sasaki-Uemura (2001, p. 16) notes that "from the spring of 1959 to the fall of 1960, an estimated sixteen million Japanese engaged in protests against the Security Treaty." Although LDP Prime Minister Nobusuke Kishi forced approval of the treaty through parliament against strong opposition and advocated for amendment of the constitution, the mobilization of the public on such a large scale resulted in his resignation and a period during which the government avoided public calls for constitutional revision or expansion of the state's war making authority.

Following the events of 1959–1960, LDP governments focused on economic growth and even created new self-restraints on war making, for example the three nonnuclear principles prohibiting the possession and manufacture of nuclear weapons and the introduction of such weapons into Japanese territory. The government also capped defense spending at 1% of Gross National Product (GNP), reaffirmed its defense-only policy, banned acquisition of offensive weapons, and restricted weapons exports. This post-1960 consensus continued until the 1980s, when officials began testing some of these limits. However, the constitutional framework was still strong enough by the 1991 Gulf War that the Japanese government had to give up its desire to participate due to parliamentary opposition and an overwhelmingly skeptical public. Giving in to US pressure, however, Japan contributed $13 billion to the war effort and sent six minesweepers to the Persian Gulf. Failure to participate in the war elicited ridicule of Japan's "checkbook diplomacy" from US officials, who insisted that Japan would have to put "boots on the ground" in future wars. As analyzed below, it would not be until the 2001 terror attacks in the United States that Japanese militarists would find an opening to further challenge the limits of the peace constitution.

The Right to Peace

The preamble of the Constitution of Japan mentions peace as a foundational value and proclaims, "We recognize that all peoples of the world have the right to live in peace, free from fear and want." While lower courts have affirmed that Japanese citizens do, indeed, have a constitutional right to peace, higher courts have failed to uphold this right. For example, in 1968 when the Agriculture Ministry tried to reclassify local land so that the JSDF could install an antimissile base, residents of nearby Naganuma, Hokkaido, challenged the Ministry in the Sapporo District Court. Although the lower court found that the villagers had a constitutional right to live in peace and that the reclassification of the land violated that right, the Sapporo High Court overturned the ruling, and the Supreme Court affirmed the Sapporo High Court's decision (Matsui 2011, p. 241–242).

Despite the state's failure to recognize the right to peace as an actionable right, an unexpected indicator of Japanese citizens' potential for securing that right is their relative unwillingness to fight in war. Eide (1980) interprets the right to live in peace as one that includes the right of individual refusal of military service on pacifist grounds alone but also when one's state makes war aggressively or illegally, makes military preparations that might potentially be used for aggressive purposes, or engages in military preparations for defensive purposes that exceed stipulated and proper limits. Lacking systematic observations of such refusal since the Japanese state has no constitutional basis to require military service, we can instead examine Japanese attitudes toward fighting in war in comparative perspective using data from the World Values Survey (WVS 2015). Unfortunately, the WVS does not include Costa Rica, so it is difficult to make a direct comparison of the two.

As the data in Figure 11.1 indicate, affirmative responses by Japanese to the statement "Of course, we all hope that there will not be another war, but if it were to come to that, would you be willing to fight for your country?" are markedly lower than responses for South Korea, the United States, and Germany (West Germany through 1990). For example, in the most recent wave of the WVS, only 15.2% of Japanese expressed a willingness to fight in war as compared to 40.9% of Germans, 57.7% of respondents in the United States, and 63% of South Koreans. It is notable that the highest rate for

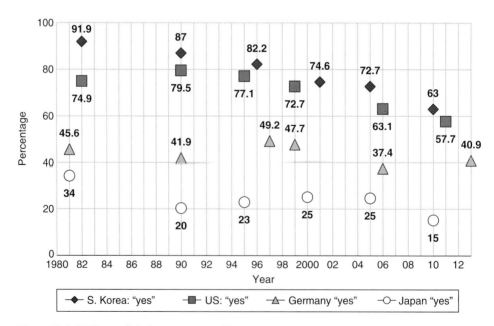

Figure 11.1 "Willing to fight for your country?"

Japanese (34.2% in 1981) was lower than the lowest rate for Germans (37.4% in 2006) over the past 35 years. Comparison of Japan and these countries helps to determine the function and benefits of the peace constitution attitudinally as a popular expression of the right to peace.

Looking at the "within case" data for Figure 11.1, the affirmative Japanese response rate declined from 25 to 15% between 2005 and 2010 after remaining fairly steady from 1995 until 2005. This decline may have partially resulted from widespread pro–Article 9 activism during this period, in particular activism by the nationwide Article 9 Association (*Kyūjōnokai*, hereafter A9A) described in more detail in this chapter. An examination of the cases of South Korea and the United States, however, also shows a decline in the willingness to fight during this period, but in those cases, the decline began earlier. While a full analysis of those cases is outside the scope of this study, a decrease in international tensions following the end of the Cold War may help to explain those countries' declines after 1990, and in the case of the United States, the declines in 2006 and 2011 may have also reflected growing public disapproval of the wars against Afghanistan and Iraq. The case of Germany requires more complex analysis due to reunification, but here Germany serves as a point of comparison with Japan, as explained below.

Japan and Germany both experienced extensive destruction and defeat in total war and transitioned from authoritarianism to democracy. Their relatively low willingness to fight may be partially attributable to that legacy, but that alone cannot explain the attitudinal difference between the two. For that matter, South Korea also experienced significant devastation during the Korean War, yet its willingness to fight has consistently been higher than that of the other three countries. The Korean War has never officially ended, and there are relatively few limitations on the South Korean state's war making potential. The country's policy of universal male conscription could also be a factor in its high affirmative response rate. Likewise, the United States has frequently engaged in war making since World War II and the Korean War and maintains the world's highest military budget.

One important difference between Japan and the others is that many Japanese citizens, as explained in the "Emergent Culture of Peace" subsection, valorize the peace constitution, and it continues to function, therefore, as a potential check on the state's challenges to restrictions on war making. In comparison to the Federal Republic of Germany, which re-established an army in 1955 and fully integrated into NATO, the JSDF continues to exist in the context of Article 9 and is not legally an army. It seems possible, then, that one of the functions of the peace constitution has been to deny the Japanese state a popular base of attitudinal support for war making. The peace constitution provides no authority for the state to order citizens to engage in killing and dying in war, and only a minority of citizens seems willing to do so. Insofar as this prevents state-level actors from war making, the right to peace is at least partially secured.

Emergent Culture of Peace

As a point of comparison to Costa Rica and other countries, Japan ranked eighth out of 162 countries on the 2015 Global Peace Index (Vision for Humanity 2015). According to the Index, Japan has low rates of weapons imports and exports, armed services personnel, violent crime, access to weapons, homicide, and death from external and internal conflicts. It has higher rates, however, of poor relations with neighboring countries and possession of heavy weapons. As for the first factor, the Japanese government has yet to resolve territorial disputes and historical issues dating from its imperial past. To improve its rating on the second indicator, the government must reduce its procurement and maintenance of heavy weapons. At present, this seems unlikely due to US–Japan Defense Guidelines that require increased integration of the JSDF with the US military.

While Japan ranks highly on the Global Peace Index, the attitudes and behaviors of people in Japanese civil society, especially their valorization and defense of the peace constitution, also demonstrate its emergent culture of peace. Whereas promotion of the culture of peace has had state support in Costa Rica, members of civil society have supported an emergent culture of peace in Japan in the postwar period by trying to defend the peace constitution against revisionist state-level actors. The postwar example that best demonstrated this was the social movement that opposed the US–Japan Security Treaty in 1960, as described above. A more recent example is the citizen network that arose to oppose constitutional revision and war making after 2001.

The 2001 terror attacks on the United States provided the Japanese government with a new opening to seek constitutional revision. Speaking to the Diet in October 2001, Prime Minister Junichiro Koizumi said that the JSDF did, in fact, have military capabilities, a statement that was the opposite of official constitutional interpretation. In 2004, Koizumi's political party recommended constitutional recognition of the JSDF as a national army and the right of collective self-defense. It also claimed that citizens should have a duty to defend the country. The following year, the party recommended changing the words "renunciation of war" to "security" in the title of Chapter II of the constitution and recommended renaming the JSDF the "Self-Defense Military" (Hughes 2006). At the same time, Koizumi was transforming Japan's defense-only policy. In 2001, Koizumi dispatched the Marine Self-Defense Force to the Arabian Sea in support of the war against Afghanistan. In 2003, Koizumi relaxed Japan's strict arms export restrictions. Finally, Koizumi sent the JSDF to Iraq in 2004. Although the JSDF was prohibited from fighting, it was the first time it had ever gone into a combat area. A majority of Japanese people opposed the policy.

In June 2004, nine prominent citizens announced the formation of the Article 9 Association (A9A), an organization designed solely to prevent constitutional revision. In response to their appeal to form a national network, citizens began forming local A9As. Within two years there were 4700 associations, and by 2008, the number of A9As in the network totaled 7294 (Article 9 Association 2008b). The A9As were often formed as national affinity groups (Scientists' A9A, Architects' A9A, Disabled Persons' A9A, etc.) or local units (Fukuoka Women's A9A, Municipal Heads' Group of A9A in Miyagi Prefecture, etc.).

As both a social movement and an embodiment of Japan's emergent peace culture, the A9As undertook several activities. For educational purposes, they conducted public lectures, convened constitution study groups, and held public demonstrations. They also held social and cultural events such as festivals and art shows with the peace constitution as their theme. In addition, they produced publicity in the form of websites, newsletters, and newspapers. Finally, they engaged in political lobbying to convince politicians to defend the peace constitution (Peters 2010).

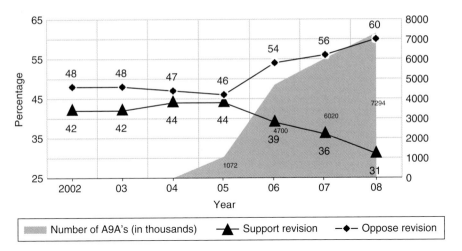

Figure 11.2 Attitudes toward revising Article 9 versus growth of A9As.

As Figure 11.2 shows, the number of A9As increased from 2004 until 2008 as the percentage of the public that opposed constitutional revision also increased (A9A and *Yomiuri Shinbun*). The highest level of support for revision of Article 9 occurred in 2004 and 2005, and then that support decreased as the number of A9As steadily increased. While it is impossible to prove a causal effect between the increasing number of A9As and the increase in opposition to revision, the A9A attributed the change in public opinion to its growing strength (Article 9 Association 2008a). Furthermore, the A9A's widespread support was a factor in the ruling party's election defeats in 2007 and 2009, which resulted, in turn, in suspension of parliamentary debate over constitutional revision. Absent a sense of urgency about defending the peace constitution, many A9As became defunct, and the national network receded from the public eye.

In 2012, the revisionist LDP returned to power, and in 2014, Prime Minister Shinzo Abe announced a reinterpretation of the constitution in order to claim the right of collective self-defense, despite the party's 40-year policy of rejecting such a claim as unconstitutional. In that same year, Japanese defenders of the peace constitution were nominated for the Nobel Peace Prize. They were nominated again, along with the people of Costa Rica, in 2015 by the Legislative Assembly of Costa Rica. Meanwhile, the Abe government readied security legislation to authorize the right of collective self-defense, which the parliament passed the same year despite the largest public demonstrations in recent decades. Faced with continuing public demonstrations and opposition not seen since the 1960s, it is not clear whether the Japanese state will finally succeed in removing constitutional limits on war making or if Japanese civil society will continue to preserve its emergent culture of peace and the constitution that makes it possible.

Conclusion

Any constitution that institutionalizes arrangements between state-level actors and members of society by patterning their interactions toward peaceful behaviors is a peace system. Peace constitutions are specific constitutional peace systems that establish peace as a foundational value, provide the basis for the right to peace, and ban a state's right to maintain an army and/or make war. The peace constitutions of Costa Rica and Japan exemplify these three standards. In Costa Rica, state-level actors have largely followed and reaffirmed these standards, so the peace constitution is functional. In Japan, by comparison, state-level actors have challenged the limits of the peace constitution through judicial interpretation, institution building, and policy making. Despite the majority of Japanese citizens' valorization of the peace constitution, state-level actors have rendered it semifunctional.

One way both the Costa Rican and Japanese constitutions have functioned is by strictly limiting war making. Neither state has a *de jure* military establishment, although the JSDF lacks only the legal status of a military, institutions of military justice, and the power of conscription. Because of the existence and scale of the JSDF, the US–Japan alliance, and revision through reinterpretation of the constitution, Japan is left with a semifunctional peace constitution. As for the Costa Rican peace constitution, it has proven highly functional even under circumstances as severe as invasion. The declaration of neutrality places a further constraint on the Costa Rican state and greatly limits its potential to carry out war making or violent state making against Costa Rican society. In other words, Costa Rica's peace system patterns interactions between state-level actors and members of society toward peaceful behaviors.

In terms of the right to peace, this right is realized in different ways in Costa Rica and Japan. Despite the absence of an explicit right to peace in the constitution, the Costa Rican judiciary has recognized that right, and such rulings have proven effective in reversing and limiting state making and war making. In Japan, on the other hand, citizens have not fully secured the right to peace stated in the preamble of the peace constitution. Despite lower court rulings, the Supreme Court of Japan has so far denied the right to peace as an actionable right. Still, as comparative WVS data indicate, the right to peace takes form in Japan through the individual attitude of refusal to fight in war. Such a widespread unwillingness to fight would be regarded as a serious constraint on war making by any state, and it indicates the degree to which most Japanese citizens believe they have a right to peace.

The third point of comparison is the degree to which each state has an emergent culture of peace. Costa Rica ranks 34th on the Global Peace Index, and it surpasses neighboring countries on measures of positive peace related to poverty reduction, income distribution, and social expenditures per capita. In addition, there is obvious state support for institutionalizing peace processes. Japan ranks eighth on the Global Peace Index, and civil society members who valorize the peace constitution are mainly responsible for Japan's emergent culture of peace. That has been particularly evident during historical moments when the state acted in ways that undermined the constitution, most recently in the first decade of the 21st century. The establishment of over 7000 autonomous Article 9 Associations in just four years proved that Japanese citizens would organize to protect their peace constitution in the face of revision, and so far the constitution remains unamended.

The constitutional peace systems of Costa Rica and Japan are only two examples of peace constitutions, although perhaps the best known. Constitutions are understudied as systems of peace, and there is little research on how they function to establish arrangements between state-level actors and members of society to pattern their interactions toward peaceful behavior. Considering that neither the Costa Rican nor the

Japanese state has ordered its citizens to kill in war for nearly 70 years, and considering that people in both countries enjoy relatively high levels of peace as indicated on the Global Peace Index, much more study is needed of the functions and benefits of these and other constitutional peace systems.

References

Article 9 Association. (2008a). *Article 9 Association Bulletin and News, 106*.

Article 9 Association. (2008b). *Article 9 Association Bulletin and News, 118*.

Barbey, C. (2015). *Non-militarisation: Countries without armies*. Mariehamn, Finland: The Åland Islands Peace Institute.

Benjamin, M. (2014). Suing for peace in Costa Rica. *Foreign Policy in Focus*. Retrieved from http://fpif.org/suing-peace-costa-rica/

Bird, L. (1984). *Costa Rica: The unarmed democracy*. London: Sheppard Press.

Brysk, A. (2009). *Global good Samaritans*. Oxford: Oxford University Press.

Buck, J.H. (1967). The Japanese Self-Defense Forces. *Asian Survey, 17*(9), 597–613.

Duke, B.C. (1973). *Japan's militant teachers: A history of the left-wing teachers movement*. Honolulu: University of Hawaii Press.

Eide, A. (1980). Towards a declaration on the right to peace. *Security Dialogue, 11*.

Fry, D.P. (2007). *Beyond war: The human potential for peace*. Oxford: Oxford University Press.

Fry, D.P. (2013). War, peace, and human nature: The challenge of achieving scientific objectivity. In D.P. Fry (Ed.), *War, peace, and human nature: The convergence of evolutionary and cultural views*. New York: Oxford University Press.

Grossman, D. (1996). *On killing; The psychological cost of learning to kill in war and society*. New York: Back Bay Books.

Hellegers, D.M. (2001). *We the Japanese people: World War II and the origins of the Japanese constitution* (Vols. 1 and 2). Stanford, CA: Stanford University Press.

Hook, G.D. (1996). *Militarization and demilitarization in contemporary Japan*. New York: Routledge.

Hughes, C.W. (2006). Why Japan could revise its constitution and what it would mean for Japanese security policy, *Orbis, 50*(4), 725–744.

Hughes, C.W. (2009). *Japan's remilitarisation*. New York: The International Institute for Strategic Studies and Routledge.

Katz, E., & Lackey, M. (2010). *Costa Rica as a peaceful state*. Washington, DC: Center on Hemispheric Affairs. Retrieved from http://www.coha.org/costa-rica-as-a-peaceful-state-one-costa-rican-lawyer's- odyssey-v-his-nation's-establishment/

Kruk, M.R., & Kruk-de Bruin, M. (2010). *Discussions on context, causes and consequences of conflict*. Leiden: The Lorentz Center, Leiden University.

Lehoucq, F. (2012). *The politics of modern Central America: Civil war, democratization, and underdevelopment*. New York: Cambridge University Press.

LeMoyne, J. (1987, March 22). Costa Rica's return to neutrality strains its ties to Washington. *New York Times*. Retrieved from http://www.nytimes.com/1987/03/22/weekinreview/costa-rica-s-return-to-neutrality-strains-its-ties-with-washington.html

Maeda, A. (2008). *Guntai no nai kokka [States without armies]*. Tokyo: Nihonhyōronsha.

Marujo, C. (2011). Threats to peace: The militarization of Costa Rica. *Peace and Freedom*, *71*(1), 14–15.

Matsui, S. (2011). *The constitution of Japan: A contextual analysis*. Oxford: Hart.

McCormack, G. (2007). *Client state: Japan in the American embrace*. London: Verso.

McNelly, T. (2000). *The origins of Japan's democratic constitution*. Lanham, MD: University Press of America.

New Economics Foundation (NEF). (2015). Happy planet index. Retrieved from http://www.happyplanetindex.org/data/

Packard, G.R. (1966). *Protest in Tokyo: The security treaty crisis of 1960*. Princeton, NJ: Princeton University Press.

Peters, B.A. (2010). *Democratic antimilitarism in postwar Japan: Institutions and the culture of peace*. PhD dissertation, Rutgers, The State University of New Jersey.

Peters, B.A. (2013). Security without deadly violence: Costa Rica's potential as a nonkilling state. In J. Evans Pim (Ed.), *Nonkilling security and the state*. Omaha, NE: Creighton University.

Roche, D. (2003). *The human right to peace*. Toronto: Novalis.

Rummel, R.J. (1994). *Death by government*. New Brunswick, NJ: Transaction Publishers.

Sasaki-Uemura, W. (2001). *Organizing the spontaneous*. Honolulu: University of Hawaii Press.

Sasamoto, J. (2010). *Seikai no heiwakempō aratana chōsen* [The new challenge of the world's peace constitutions]. Tokyo: Otsukishoten.

Schlichtmann, K. (2009). *Japan in the world: Shidehara Kijuro, pacifism, and the abolition of war* (Vols. 1 and 2). Lanham, MD: Lexington Books.

Shane, P.M. (2006). Analyzing constitutions. In R.A.W. Rhodes, S.A. Binder, & B.A. Rockman (Eds.), *The Oxford handbook of political institutions*. Oxford: Oxford University Press.

Shklar, J.N. (1998). Political theory and the rule of law. In S. Hoffmann (Ed.), *Judith Shklar: Political thought and political thinkers*. Chicago: The University of Chicago Press.

Shutts, S. (2009). Costa Rica creates Department of Peace. *YES! Magazine*. Retrieved from http://www.yesmagazine.org/peace-justice/costa-rica-creates-department-of-peace

Sodei, R. (2001). *Dear General MacArthur: Letters from the Japanese during the American Occupation* (M. Shizue, Trans.). New York: Rowman & Littlefield.

Spruyt, H. (2002). The origins, development, and possible decline of the modern state. *Annual Review of Political Science*, *5*, 127–149.

Tilly, C. (1985). War making and state making as organized crime. In P. Evans, D. Rueschemeyer, & T. Skocpol (Eds.), *Bringing the state back in*. Cambridge: Cambridge University Press.

United Nations Office of the High Commissioner on Human Rights. (2016). Open ended intergovernmental working group on a draft United Nations declaration on the right to peace. Retrieved from http://www.ohchr.org/EN/HRBodies/HRC/RightPeace/Pages/WGDraftUNDeclarationontheRighttoPeace.aspx

Verbeek, P. (2008). Peace ethology. *Behaviour*, *145*, 1497–1524.

Verbeek, P. (2013). An ethological perspective on war and peace. In D.P. Fry (Ed.), *War, peace, and human nature: The convergence of evolutionary and cultural views*. New York: Oxford University Press.

Vision for Humanity. (2015). Global peace index. Retrieved from http://www.visionofhumanity.org/#/page/indexes/global-peace-index

World Values Survey (WVS). (2015). Online data analysis portal. Retrieved from http://www.worldvaluessurvey.org

Zamora, L.R.B. (2010). *The lowest form of military aggression*. Washington, DC: Foreign Policy in Focus.

Zamora, L.R.B. (N.d.). *Costa Rica court invalidates presidential decree militarizing policy*. Available online at: http://www.peaceasahumanright.org/Peace/invalidates_Presidential_Decree_Militarizing_Police.html

12

Exploring the Village Republic: Gandhi's Oceanic Circles as Decentralized Peace Systems

Joám Evans Pim
Dedicated to Carlos Calvo Varela.

Introduction

In his efforts to translate the principles of ahimsa (nonviolence) into the realm of politics, Gandhi developed the vision of Swaraj, understood as full community self-government. In many of his writings, the practical manifestation of Swaraj is presented as a *village republic* following the traditional Indian panchayat and gram sabha assembly government. While Gandhi formulated the specifics for such village republics in some detail, the overall vision on how these self-governed units should relate to each other in a stateless context remained somewhat vague.

The Gandhian *oceanic circles* vision, a structure of innumerable villages with "ever-widening, never-ascending circles," not a "pyramid with the apex sustained by the bottom" but a "circle whose centre will be the individual" (1946/1998, p. 326), remains as one of the least explored aspects of Gandhi's political thought. Summy (2013, pp. 55–56) interpreted Gandhi's vision of the most "outer oceanic circle" as a world federation of interdependent units based on the small, self-sufficient village republics. Such a vision shares attributes with the concept of *peace systems*, as defined by Fry (2012).

Fry's (2012) concept presents peace systems as "groups of neighboring societies that do not make war on each other," providing a variety of examples such as the Upper Xingu River basin tribes of Brazil, the Iroquois Confederacy of North America, and the European Union. This is a significant contribution as it establishes the possibility and some factual characteristics of nonwarring societies (i.e., societies that do not make war either in absolute terms or at least against other societies within the system itself).

Considered together, the notion of peace systems sheds light on historical and current examples of structures that approach the idea of oceanic circles through expanded social identities, interconnectedness, interdependence, peaceful values and symbolism, and superordinate conflict management institutions that prevent escalation of potentially lethal group conflict. Likewise, Gandhi's Village Swaraj provides insights on how to move nonwarring societies beyond external peace to become internally nonviolent, or at least nonkilling, starting from the bottom – every single village – and moving up

Peace Ethology: Behavioral Processes and Systems of Peace, First Edition.
Edited by Peter Verbeek and Benjamin A. Peters.
© 2018 John Wiley & Sons Ltd. Published 2018 by John Wiley & Sons Ltd.

to a *global oceanic circle*. This synthesis of the ideas of Gandhi and Fry will be referred to as *decentralized peace systems*, encompassing nonwarring external arrangements and nonkilling/nonviolent internal organization.

This chapter will look into our past as a species to try to understand the evolutionary precursors of such decentralized peace systems and to provide recent political experiences that reflect their present validity. The findings of this chapter offer an array of heuristic models for the development of nonkilling social and political forms of organization locally, regionally, and globally.

The Political Prehistory of the Village Republic

Within anthropology, a general agreement exists that for over 95% of our existence as *Homo sapiens* (200,000 years of anatomically modern humans vs. a variable fraction of the last 10,000), we have lived and organized ourselves socially as small-band hunter-gatherers, also referred to as nomadic foragers (Bicchieri 1972; Giorgi 2010; Sponsel 2010; Fry 2013). Even though extrapolations always need to take into account the influence of surrounding state societies, contemporary small-band hunter-gatherers provide an extraordinary window to understand the species-typical social arrangements of humans during the Late Pleistocene (126,000 to 12,000 years ago). These societies have been characterized by an ethos of egalitarianism, cooperation, generalized reciprocity ("gift economy") or simply sharing, extended alloparenting, nonviolence, and embeddedness within nature. Economic self-sufficiency, small group size, and nonhierarchical and unsegmented social organization or lack of fraternal interest groups favored cooperative and egalitarian practices intended to safeguard harmonious nonkilling social relations (Sponsel 2010). In fact, Younger (2008) concluded on the basis of demographic and geographical analysis that small (under 1000 individuals) egalitarian or unsegmented societies, characterized by the relevance of face-to-face contact and close ties, have more chances for survival if they actively prevent violence, thus providing evolutionary grounds for forms of social organization that represent the basis for decentralized peace systems as described in this chapter.

In evolutionary terms, the most recent 10,000 years of human existence have seen the emergence of species-atypical behaviors, first in the Near East and then in other world regions, such as coordinated intergroup violence (i.e., warfare) and political structures that support and expand inequalities (i.e., the state). Structural violence is coincident with social stratification and socio-political organization at the state level (Sponsel 2010, p. 21). The distribution of this anomaly in geographical and historical terms has been uneven, from the first known transitions into the Neolithic occurring roughly 13,500–10,000 BP years ago during the "pre-agricultural revolution" (Knauft 1991) to the current existence of a small set of small-band hunter-gatherer societies, such as the Batek and Paliyan of Asia or the Mbuti and San of Africa, from which we derive much of our ethnographic knowledge of these forms of social organization. Taking these facts into account, some authors have argued that, from an evolutionary perspective, we are not well adapted, at least neurologically, to cope with our current existence in large, hierarchical, competitive, and violent communities where "the vast majority of human beings have become unhappy, ill and with limited material resources" (Giorgi 2009, p. 117; also see Narvaez 2013, this volume, Chapter 6).

In political terms, nomadic forager societies are characterized by egalitarianism, high levels of personal autonomy, and an absence of formal leaders (Boehm 1999, p. 67; Fry 2006, p. 181). No single person has political power over anyone else in the group, even if nominal headmen, respected elders of either sex, could exercise some degree of influence that was by no means binding, as coercion, violence, and aggression are considered unacceptable (Leacock 1978, p. 249; Woodburn 1988; Endicott 2013, pp. 245–246). This usually happens informally on a day-to-day basis as part of constant social interaction (playing, singing, joking, or sitting together). Otherwise, any incipient leadership is "very elaborately constrained to prevent them from exercising authority or using their influence to acquire wealth or prestige" (Woodburn 1988, p. 444).

Examples can be drawn from dozens of societies (for a comprehensive survey, see Fry 2006). One contemporary group is the South Indian Paliyan, a society that has endured – together with other groups – among the surrounding stratified Tamil society (Gardner 2010, 2013). While the Paliyan and other South Indian foragers clearly practice individual autonomy and self-reliance in their decision making (including self-restraint in the face of conflict), the coexistence of "assemblies of community members" (kuttan) among some groups shows a distinct form of collective deliberation "by which the principal parties can weigh public opinion and make personal decisions on whether or not to back down" (Gardner 2013, p. 306). Although these kinds of band or camp meetings do not actually have the power or authority for binding decision making, they seem to be a tool for responding to conflicts and fostering noncoercive collective deliberation while preserving each individual's responsibility for forming their own options.

Whereas Gardner considers that the kuttan could be a new institution among the Paliyan (i.e., a recent egalitarian adaptation of the Indian panchayat), it could also be argued that the panchayati raj, a traditional system of self-government present throughout the Indian subcontinent, could be the continuation of preexisting social arrangements such as these egalitarian assemblies with parallels in other nomadic hunter-gatherer social institutions such as the Australian Aboriginal "Big Meeting" (Tonkinson 2013, p. 268). Recalling Lee (1992, p. 40), egalitarian societies "have social and political resources of their own and are not just sitting ducks waiting to adopt the first hierarchical model that comes along." In fact, the panchayat, based on an assembly (ayat) of five (panch) respected elders chosen by the community in a gram sabha general assembly, had conflict resolution at the core of its original orientation until it was forced to serve as part of the state's tax-extracting apparatus. The early Rigvedic (1700–1100 BCE) vidatha assemblies, in which women also participated on equal terms (Sharma 1996; Rohman 2005, p. 23), suggest the distant and perhaps egalitarian origins of such institutions.

In both instances, community assemblies are probably more about "moralistic social control" (that can tackle antisocial deviance with gossip, mockery, ostracism, or shaming) in the form common to nomadic foragers (Boehm 2013, p. 318) than coercive authority to impose decisions. This is also the case for so-called "good heads," one to three people singled out in most Paliyan bands "who are able to step forward voluntarily to help when there is tension over social or ritual matters," using "word play, clowning, or soothing speech to distract and calm their fellows" but lacking any formal authority (Gardner 2010, p. 187). Clastres (1989, p. 30) argued that normal civil power in stateless societies is based on the consensus omnium, with the formal headman, council, or community gathering having the role of maintaining peace and harmony in the group with neither the authority nor the capacity to use coercive force. Such forms of noncoercive problem solving and conflict resolution were probably the common setting during the vast majority of our species' evolutionary past.

The slow decay of nomadic hunter-gatherer societies was ignited by the appearance of agriculture, which allowed for growing population densities, geographic concentration of resources, social and political hierarchies, monopolizable long-distance trade of valuable prestige goods, and food storage and management beyond the domestic units (Ferguson 2013, p. 192). Such state agricultural societies also played (and continue to play) an important part in the quick and violent marginalization or annihilation of neighboring hunter-gatherers. All these are preconditions for the development of the first forms of states and, with them, organized interpersonal violence, but they are not determining factors. In fact, during the Neolithic and moving into the Bronze and Iron Ages, there seem to have been extended periods where the practice of agriculture did not necessarily translate into structural violence and widespread warfare, perhaps overcome by mutual interdependence and cooperative efforts, social ties among groups, and peaceful attitudes and beliefs (Ferguson 2013, p. 193), settlement size being a critical factor. Strictly egalitarian societies have also been identified in groups with delayed-return horticultural or agricultural economies (vs. the immediate-return characteristic of hunter-gatherer societies) such as the Majangir of Ethiopia and among nomadic sea-dwellers such as the Sama or Bajau Laut (Macdonald 2011, p. 72). A good representation of the diversity of such manifestations can be found in *Anarchic Solidarity* (Gibson and Sillander 2011).

Surprisingly, most of the attributes that characterize early agro-pastoral societies in the literature are quite the opposite of those present in some modern agricultural societies up to the early 20th century. An example from Western Europe such as Galiza's rural society is, in general terms, representative of highly dispersed populations with relatively low densities, social egalitarianism, self-reliant families and villages, consistent mutual aid, consensual group decision making and problem solving, and extensive communal stewardship of the land and its resources (Peixoto 1908/1990; García Ramos 1912; Tenorio 1914/1982; Dias 1981, 1984; Rodrigo Mora 2013). So-called "feudal" systems overlapped with the existence of such horticultural and agro-pastoral societies in parts of Europe and elsewhere for the last millennia, but actual control was extremely limited beyond the extraction of certain rents and taxes. This is not to say such forms of social organization can be likened to those of nomadic hunter-gatherers, but they illustrate the endurance of many key aspects of our ancestral ethos even within rural communities enclosed by contemporary industrial states.

Dentan (2010, p. 131) argues that small, local, egalitarian, mutual-aid groups "occur spontaneously at every level of human biological and social evolution." These "primary groups" emerge even in the interstices of the state as temporal communities such as the "Rainbow Family" gatherings, disaster relief groups, or more stable endeavors. Prolonged interstices can appear in areas with weak state control due to lack of political and economic relevance, as it could be argued for rural areas such as Galiza throughout modern history, or where the state has actually lost control, such as the Zapatista Autonomous Municipalities in the context of indigenous insurgence, or Kurdish Democratic Confederalism in northern Syria. Contemporary intentional communities, in the form of ecovillages, communes, cooperatives, or semipermanent protest camps, replicate in many ways the basic characteristics of such "primary groups." In the case of disasters, the unplanned appearance of such localized groups suggests that "just as many machines reset themselves after a power outage, so human beings reset themselves to something altruistic, communitarian, resourceful and imaginative after a disaster, that we revert to something we already know how to do" (Solnit 2009, p. 18).

Anthropologist Marvin Harris (1990, p. 438) placed the emergence of what he described as "nonkilling religions," namely Jainism, Buddhism, Hinduism, and Christianity (at least with regard to fragments of the New Testament), in a common background of state failure to deliver "worldly benefits," the confluence of brutal and costly wars, environmental depletion, population growth, the rise of cities, food shortages, widespread poverty, and rigidified social stratification. The ethics of ahimsa (nonkilling/nonviolence), a basic tenet of the first three spiritual traditions mentioned by Harris, continued to reappear around the world over the following millennia as an important component of many social, political, and spiritual movements through a diversity of deeply rooted forms.

Harris' (1990) explanation of the emergence of the first known conceptualizations of nonkilling/noviolence can be related to Ferguson's archaeological account of how early episodes of intense warfare were followed by prolonged periods in which no material evidence of organized intergroup violence can be found. It could be hypothesized that relatively peaceful societies may have re-emerged around the egalitarian mutual-aid ethos proposed by Dentan from the ashes of some of the darkest periods in the past 10,000 years. Such societies did not need to abandon acquired agricultural techniques and other technologies but perhaps resumed coexisting values in order to practice the small-scale, self-sufficient, subsistence economy that was the most common form of production right up to the 20th century.

We currently confront some of the most complex problems that we have faced as a species. With the confluence of peak oil (also applicable to coal, gas, phosphorus, and other crucial resources for industrial society), climate change, economic instability, and a global population of over 7 billion, the question remains: will the current forms of state organization, and indeed capitalism, be able to solve (and survive) such challenges? Civil society efforts, such as the Transition Towns, Degrowth, Permaculture, or Integral Revolution movements, have called for the need to radically shift the way we relate to the environment and fellow humans (Trainer 2010; Rodrigo Mora 2012). Dentan (2010, p. 170) argued, "It seems likely that nation states will disintegrate into progressively smaller and smaller local social formations, as people revert to their usual response to disaster." Current emerging forms of intentional rural communities can be illustrative of future arrangements. Giorgi (2010, p. 93) is also convinced that, after this 7000-year violent interlude, "nonkilling cultures will soon develop in some regions of the earth and their superior life-style and level of humanity will become a model to imitate." There is reason to believe this is possible and perhaps less catastrophic than what some have previously suggested if the functional potential of some proposals along the lines of decentralized peace systems are considered.

As Ferguson points out, "Even at relatively advanced levels of sociocultural evolution, there is no reason, theoretically, to deny the possibility of peaceful societies" (2013, p. 192). The relatively recent patterns of warfare, social inequality, and centralized authority "are not rooted deep in our evolutionary past but rather are capacities facilitated by the changing demographic, technological, and structural realities of human populations" (Fuentes 2013, p. 90). The socially and culturally adaptive plasticity of our species provides room for change: removing the social mechanisms that enable direct and structural violence and regaining the values and practices of solidarity, cooperation, and mutual aid, while retaining the appropriate knowledge and technologies that have been developed in the past 10,000 years (Giorgi 2010).

Our evolutionary baseline is not about romanticizing the past but about understanding if our current behaviors "are normative for human beings ... or maladaptations that emerge from a mismatch between evolved needs and current environments" (Narvaez 2013, p. 353). Gandhi's call for a nonviolent society based on self-governed village republics is a practical example of a political attempt to transfer the cooperative, egalitarian, relatively peaceful, and, in many instances, nonviolent ethos of our hunter-gatherer past, where the whole planet perhaps supported about 6 million people, to our contemporary world of over 7 billion. But rather than building such an alternative on a neo-Malthusian argument (i.e., reducing world population back to 6 million), Gandhi identified extreme aggregation of people in small territories (large cities) as the main problem, mainly because such a pattern is not a response to human needs but a financial convenience to exploit intensive production and consumption at the expense of environmental and human degradation. In contrast, the self-sufficient but interdependent village republics are based on small, scattered populations bioregionally integrated in decentralized peace systems.

Swaraj and Swadeshi as Cornerstones of Gandhi's Village Republic

Gandhi's vision of building nonviolent societies relied upon two basic principles: swaraj (non-hierarchical community self-governance) and swadeshi (self-sufficiency), which were presented as mutually interdependent.

In imagining a political system based on nonviolence, Gandhi's ideas were very close to those of Tolstoy and Thoreau, two figures he admired, though providing a vision that was richer in detail. In an interview prepared by Harold Williams and published in *The Manchester Guardian* on February 9, 1905 ("A Visit to Count Tolstoy," pp. 7–8), Tolstoy insisted, "All governments are maintained by violence or the threat of violence and violence is opposed to freedom." Gandhi would repeatedly insist on this Weberian definition of the state, thus considering this form of political organization as incompatible with his vision of nonviolence: "The State represents violence in a concentrated and organized form. The individual has a soul, but as the State is a soulless machine, it can never be weaned from violence to which it owes its very existence" (1934/1998, vol. 65, p. 318). Tolstoy's nonviolent society would be based on small-scale agrarian self-sufficiency, without division of labor, without cities, without factories, without laws enforced by coercion, and without governmental rule or courts (McKeogh 2009, pp. 165–166), thus setting the basis for Gandhi's village swaraj.

From Thoreau's *Civil Disobedience*, Gandhi borrowed the motto "That State will be the best governed which is governed the least," adding, "That is why I have said that the ideally non-violent State will be an ordered anarchy" (1940/1998, vol. 79, p. 122). Interestingly, the term *ordered anarchy* also appeared the same year in Evans-Pritchard's classic anthropological study of the Nuer of Southern Sudan (1940). Yet Gandhi's idea of self-government, understood as both individual self-government and community self-government, is also one of Thoreau's most significant contributions expressed in *Walden* (Thoreau 1854), where self-governance is presented as a deeply political everyday experience emerging out of freedom from, or indifference to, the state, thus implying the absolute decentralization of political commitments (see Lane 2005; Jenco 2009). Gandhi agreed: "Centralization as a system is inconsistent with non-violent structure of society" (1942/1998, vol. 81, p. 424).

Gandhi labeled the socio-political structure that would support a nonviolent society as "village republic" or "village swaraj," following the traditional panchayat local government (see Gandhi 1962). Gandhi's definition of swaraj, or self-government, involves a "continuous effort to be independent of government control, whether it is foreign government or whether it is national" as no government should take care of the regulation of everyday life (1925/1998, vol. 32, p. 258). Swaraj, characterized as "true democracy" and "individual freedom," will be achieved "only when all of us are firmly persuaded that our Swaraj has got to be won, worked and maintained through truth and Ahimsa alone" (1939/1998, vol. 75, p. 176), "outward freedom" being obtained only to the extent that "inward freedom" has been self-grown.

Every individual and community should autonomously practice swaraj. Gandhi argued in 1946: "Independence must begin at the bottom. Thus, every village will be a republic or panchayat having full powers. It follows, therefore, that every village has to be self-sustained and capable of managing its affairs even to the extent of defending itself against the whole world" (1946/1998, vol. 91, p. 325). The village republic, as a societal unit, would be naturally based not on social status or property titles but on truth, nonviolence, and equal labor. An outline of the village swaraj is presented as that of "a complete republic, independent of its neighbors for its own vital wants, and yet interdependent for many others in which dependence is a necessity" (1942/1998, vol. 81, p. 113). This model was evidently inspired by the traditional panchayat:

> [E]very village's first concern will be to grow its own food crops and cotton for its cloth. It should have a reserve for its cattle, recreation and playground for adults and children.... As far as possible every activity will be conducted on the co-operative basis. There will be no castes such as we have today with their graded untouchability. Non-violence with its technique of satyagraha and non-co-operation will be the sanction of the village community.... The government of the village will be conducted by a Panchayat of five persons annually elected by the adult villagers, male and female, possessing minimum prescribed qualifications. These will have all the authority and jurisdiction required. Since there will be no system of punishments in the accepted sense, this Panchayat will be the legislature, judiciary and executive combined to operate for its year of office.... Here there is perfect democracy based upon individual freedom. The individual is the architect of his own government. The law of non-violence rules him and his government. He and his village are able to defy the might of a world. (1942/1998, vol. 81, p. 113)

In practical terms, Gandhi argues that the establishment of such a form of independent village swaraj does not require external authorization and need not wait for any major political revolution to happen in the surrounding state. Therefore, it sets a clear precedent for contemporary intentional communities, such as ecovillages, that are able to flourish in the interstices of the state. Initiating a village swaraj is an individual obligation that should expand to involve and commit the whole community:

> Any village can become such a republic today without much interference even from the present Government whose sole effective connection with the villages is the exaction of the village revenue.... To model such a village may be the work of a lifetime. Any lover of true democracy and village life can take up a village, treat it as his world and sole work, and he will find good results. He begins by being the village scavenger, spinner, watchman, medicine man and schoolmaster all at once. If nobody comes near him, he will be satisfied with scavenging and spinning. (1942/1998, vol. 81, pp. 113–114)

As early as 1910, Gandhi warned that if India replicated the British political, economic, administrative, legal, educational, and military institutions, it would be ruined, as it was these institutions, regardless of who controlled them, that posed the greatest barrier to the development of nonviolent swaraj and swadeshi (1910/1998, vol. 10, p. 258). The freedom of India's peoples could not be reduced to transferring the administration of the state apparatus but should, above all, mean the complete removal of such structures. Unfortunately, this was not the case, as Gandhi clearly stated in "His Last Will and Testament" (January 29, 1948):

> India having attained political independence through means devised by the Indian National Congress, the Congress in its present shape and form, i.e., as a propaganda vehicle and parliamentary machine, has outlived its use. India has still to attain social, moral and economic independence in terms of its seven hundred thousand villages as distinguished from its cities and towns. (1948/1998, vol. 98, pp. 333–334)

In fact, even if the Indian National Congress placed village swaraj, as envisioned by Gandhi, at the core of its political platform from the 1920s to the country's independence, the concept of swaraj was almost completely neglected thereafter. The panchayats were included in the Constitution (Art. 40) but in non-justiciable terms ("The State shall take steps to organise village panchayats"), only to later be completely distorted in Part IX (Bates 2005, pp. 177–178; Swain 2008, p. 8). It could be argued that Gandhi's vision of village swaraj is incompatible not only with the Western configuration of the Indian state but also with the industrial and urban ethos that currently rules it: "You cannot build non-violence on a factory civilization, but it can be built on self-contained villages.... You have therefore to be rural-minded before you can be non-violent, and to be rural-minded you have to have faith in the spinning-wheel" (1939/1998, vol. 77, p. 43). From a neurobiological perspective, this relates to Narvaez's (2014) argument that immersion in the natural world (vs. urban isolation) is crucial to develop "receptive intelligence" from early childhood onward, as an environment that embeds children with natural agents and companions (and not just objects) is important in the process of creating a common human nature of cooperation, empathy, self-regulation, and small I-ego.

Gandhi argued that two divergent schools of thought challenged each other to move the world in opposing directions: that of the rural village, based on handicrafts, and that of cities, dependent on machinery, industrialization, and war (1944/1998, vol. 85, p. 233). Modern cities are presented as an "excrescence" with the only purpose of "draining the life-blood of the villages," being "a constant menace to the life and liberty of the villagers" (1927/1998, vol. 38, p. 210). As Thoreau and Tolstoy marked Gandhi's vision of politics, his correspondence with Edward Carpenter, author of *Civilisation, Its Cause and Cure* (1921), influenced the opposition established by Gandhi between satyagraha and industrial civilization, understood as a "malady which needed a cure." Industrialism was based on the "capacity to exploit," and the "cure" for urban populations would be to "become truly village-minded" (1946/1998, vol. 91, p. 390). Gandhi sharply stated: "The blood of the villages is the cement with which the edifice of the cities is built" (1946/1998, vol. 91, pp. 56–57). There was no place for exploitation or coercion in the context of village self-sufficiency and self-government.

Much of the malaise that Gandhi attributed to industrialism did, in fact, affect India in the hands of the new independent state in spite of his continuous warnings. The consequences are evident in Vandana Shiva's book *The Violence of the Green Revolution* (1991) that exposes the tragic results of India's governmental agricultural development programs launched with the technical and economic support of international agencies under "quick-fix" promises. Such measures left a deadly trail of violence, with approximately 15,000 killed between 1986 and 1991 in associated conflicts, destruction of soil fertility, suppression of genetic and ecological diversity, and indebted farmers. While Gandhi stated without doubt that "tractors and chemical fertilizers will spell our ruin" (1947–1948/1998, vol. 98, p. 88), he publicly supported contemporary efforts to develop organic agriculture. In fact, the principles of organic agriculture developed by Balfour (1944) and Howard (1940) during the 1940s and still current today were based mainly on the observation of traditional agricultural methods in India, an experience also facilitated by Gandhi and his associates.

This places Gandhian thinking on integral, simple living as a clear precedent for many proposals that were advanced in the last quarter of the 20th century in the fields of economy (Schumacher 1973; Ostrom 1990), technology (Mumford 1967–1970; see also Glendinning's Neo-Luddite Manifesto), energy (Trainer 2010), and politics (Bookchin 2003). The practical application of such principles in intentional communities such as the ashrams also spread into numerous experiments around the world. Lanza del Vasto's "Community of the Ark" is one early example of what would later develop into a global ecovillage movement (see Drago & Trianni 2008). Vinoba Bhave, another disciple of Gandhi, also continued the vision of decentralized administration in India in spite of the lack of political support for such a program (see Bhave 2007). In the following section, some of these movements and their own diverse historical roots are explored.

The Village Republic beyond Gandhi

While nonviolence has sometimes been portrayed as a seemingly passive attitude, Gandhi's political development of the term through satyagraha and the view expressed in this chapter are the opposite: nonviolence entails proactively resisting injustice, violence, and oppression through individual and community self-government and self-sufficiency. Even though it is clear that the traditional panchayati raj was an immediate inspiration for Gandhi's vision of village swaraj, it has sometimes been argued that such forms of local political organization would be relevant only to (and possible in) the Indian subcontinent or similar cultural settings, and certainly not urbanized Western societies. Such a view fails to see the connection between Gandhi's proposal and other social and political movements around the world that have included similar conceptualizations as a key component of their aspirations. Gandhi's writings on this topic were probably unknown to many agrarian, libertarian, ecological, spiritual, or indigenous movements but, nevertheless, the same thread binding self-government, self-sufficiency, and nonviolent conflict resolution (even if expressed in different terms) appears abundantly.

Early agrarian movements such as the "Diggers" in mid-17th-century England defended the egalitarian, simple lifestyle of rural communities embedded in nature by resisting enclosures of communal land (that placed in the hands of individuals or the state what previously was open territory under community stewardship) and other impositions of the expanding state. Other Christian-based rural sects, such as the Amish, Hutterites, or Mennonites (with obvious individual nuances), developed similar patriarchal interpretations of egalitarianism and rural community government, while more recent Christian pacifist communities such as the Tzotzil Maya "Las Abejas" from Chiapas, Mexico, practice a more integrative approach along similar lines to the Zapatista efforts toward community autonomy (Tavanti 2003). Another relevant example of a spiritually based community self-government initiative is Sri Lanka's Sarvodaya Shramadana Movement, founded by A.T. Ariyaratne, which clearly combines Buddhist and Gandhian practices for village swaraj based on ahimsa (see Ariyaratne 1999).

In other contexts, it was national liberation or indigenous movements that placed traditional and renewed forms of community self-government at the core of their political aspirations, which were naturally and frequently confronted by centralizing nation-state ideology. The inclusion of rural assembly democracy, including the participation of men and women, was a feature of Galizan nationalism in the first half of the 20th century, seeking the reestablishment of village and parish "open councils" (concelho aberto) and mutual aid practices (ajudas and rogas). Similar practices (such as the Basque batzarre and auzolan) have also been considered as tenets for the reorganization of these societies in opposition to the decaying nation-state framework (Santos Vera and Madina Elguezabal 2012; Sastre 2013; Escalante Ruiz 2014). This trend is recurrent among contemporary movements that seek to support their calls for grassroots democracy on historical or traditional popular institutions, ranging from the veche assemblies reclaimed by the Slavic Rodnoverie (Aitamurto 2008) to the New England town meetings upheld by some political movements in the United States (Bryan & McClaughry 1990).

An especially interesting case is that of Kurdistan, where Kurdistan Workers' Party leader Abdullah Öcalan proposed to abandon violence, embracing a new model of "democratic confederalism," striving for community self-sufficiency and self-government as a "democratic system of a people without a State" (Öcalan 2011). The implementation of such ideas since 2005 in the "low-intensity war" context of northern Kurdistan and since 2012 in the context of outright warfare in Syrian Rojava by the Koma Civakên Kurdistan (Kurdistan Communities Confederation) is a practical example of an effort to establish an extensive system of village and neighborhood councils incorporating the principles of ecology, gender liberation, and direct democracy (TATORT Kurdistan 2013, 2014). Such efforts share commonalities with the new forms of rural community governance in the Zapatista Autonomous Municipalities and Caracoles of Chiapas, which also represent an ongoing large-scale model of alternative social and political organization. The fact that such structures are able to emerge and thrive in the war-torn contexts of Chiapas and Kurdistan reaffirms the hypotheses of a surfacing egalitarian mutual-aid ethos, as discussed in this chapter and following Harris (1990) and Dentan (2010).

Arguably, libertarian-inspired village republics also share much of Gandhi's vision of swaraj. Early anarchist theorists such as Proudhon, a pacifist, conceived a stateless society organized through a federation of "free communes" (see *The Principle of Federation*; Proudhon 1863). In his *Revolutionary Catechism*, Bakunin (1866) also argues that "the basic unit of all political organization in each country must be the completely autonomous commune," in similar terms to those of Kropotkin in his 1880 *The Paris Commune* (Kropotkin 1880/1895). Some creative contemporary proposals include P.M. (1985), Fotopoulos (1997, p. 224), Bookchin (2003), Herod (2007), and Rodrigo Mora (2012, 2013). The autonomy of Swiss communes (municipalities) with their well-established assembly governments was an inspiration for such proposals and continues to serve as a relevant example of community self-government today (see Ladner 2002). A fair number of contemporary intentional communities set out in rural areas (ecovillages) represent attempts to implement such libertarian principles. The Federation of Egalitarian Communities in several US states is one example.

Ecovillages with diverse sources of inspiration, yet including a common set of principles that usually integrate voluntary simple living, permaculture, consensus decision making, and ecological sustainability, have grown exponentially since the late 1960s and early 1970s. Trainer (2010, p. 285) went so far as to suggest that the ecovillage movement was the most significant event in the 20th century as the "first significant attempt to build settlements that are ecologically, socially and spiritually satisfactory." Of course, many communities with these attributes were already in existence if we look beyond the Western industrialized world, but ecovillages certainly represent an experimental demonstration of how individual and collective transformations of city dwellers back to a nonviolent rural ethos are possible, as Gandhi asserted. The larger Transition Towns Movement is a recent attempt to take the values of such small intentional communities to larger town and city settings, with a fairly positive start. Peasant organizations in the Via Campesina movement and agrarian authors such as Berry (2002) have also shifted to advocate for family-farm-based sustainable agriculture and food sovereignty in opposition to the system of the so-called Green Revolution. All in all, such efforts reinforce Gandhi's vision of (re)building societies based on nonviolence from the small efforts of individuals and communities throughout the world.

Decentralized Peace Systems as Oceanic Circles

Gandhi envisioned oceanic circles as a global federation of small, self-sufficient, but interdependent village republics, a "structure of innumerable villages ... [where] there will be ever-widening, never-ascending circles" (1946/1998, p. 326). To try to grasp such a vision, some of the examples of peace systems presented by Fry (2012) will be discussed, highlighting the contrasts between their centralized or decentralized nature and their combination of nonwarring external arrangements and nonkilling/nonviolent internal organization. The review of actual institutional settings and political proposals, even if not along the lines of Gandhian thinking, illustrates the potentiality of this model. As Fry concludes,

> Constructing a peace system for the entire planet would involve many synergistic elements, including the transformative vision that a new peace-based global system is in fact possible, the understanding that interdependence and common challenges require cooperation, an added level of social identity that includes all human beings and encompasses more than mere national patriotism, the creation of effective democratic and judicial procedures at a supranational level, and the development of values and symbols that not only sustain peace and justice for all but also relegate the institution of war, like slavery before it, to the pages of history. (2012, p. 882)

Fry (2012, p. 881) provides three diverse examples of peace systems that were able to suppress warfare within the systems themselves: the Upper Xingu River basin tribes in Brazil, the Iroquois Confederacy, and the European Union, which are considered as *active peace systems* as they have been actively created and maintained. Warfare outside the system and violence within, including homicide, occur nevertheless. By comparison, "passive systems" are those in which nonwarring is a behavioral default, more often presenting not only internal and external restraint from war but also a nonviolent societal ethos, as is the case for the Central Peninsular Orang Asli societies, such as the Batek, Chewong, or Semai.

In spite of the huge differences between the three active peace systems described by Fry, ranging from the originally nomadic hunter-gatherers and now mostly sedentary Upper Xingu tribes or the agriculturalist Iroquois to the 28 member states of the European Union, they provide examples of how decentralized peace systems could operate.

Fremion (2002, p. 34) described the Iroquois Confederacy, also referred to as the League of Peace and Power (although not exactly the same), as a decentralized federation. In the absence of any permanent centralized hierarchy, and considering that each of the 50 Hoyenah or Sachems (chiefs) appointed to the Grand Council of the League was open to being deposed by the women of each clan who elected them in the first place, the Confederacy resembles the idea of widening circles from village, to clan, to each of the Six Nations and the wider Confederacy. As the Confederacy operated by consensus, the Hoyenah delegates of the clans of each nation should find agreement among themselves, to be followed by a unanimous decision at the council. Keeping the peace for over three centuries among its members, the Iroquois League exemplifies Kant's system of perpetual peace through the use of consultation and negotiation (Crawford 1994, p. 346).

But the Iroquois Confederacy is not only relevant as an experience of the past, as "it survives to this day and guides the political life of some of the most radical and self-reliant indigenous communities of North America" (Day 2005, p. 193). Current Kanien'kehá:ka or Mohawk, one of the Confederacy's nations, "conceptualized a path of self-determination that involves neither a recovery of a partial remnant of sovereignty lost in the past, nor a future project of a totalizing nation-state," considering as Gandhi did that "while redistribution of sovereignty may indeed challenge a particular colonial oppressor, it will not necessarily challenge the tools of his oppression" (Day 2005, p. 194). Such reflection has moved communities to devise and implement forms of self-government that do not depend on devolution of authority from the existing states and that are actually able to elude their control, such as various indigenous initiatives in North America, the Zapatista Autonomous Municipalities in Chiapas, or the Kurdistan Communities Confederation, ultimately envisioned to act as decentralized peace systems.

A rather different example of a peace system proposed by Fry (2012) is the European Union (EU). Although the EU has been successful in preventing war among its members, its structure remains as a compromise between a bureaucratic superstate and a federal arrangement between nation-states. Stemming from the European Coal and Steel Community (1951), the European Economic Community (1957), and the European Atomic Energy Community (1958), the EU is to a great extent the opposite of a "decentralized peace system" if compared to the Iroquois, Xingu, or Zapatista arrangements. In spite of the "principle of subsidiarity" that calls for decision making being made as closely as possible to the citizens (Art. 5 of the Treaty on European Union), which would theoretically enhance the role of municipal and regional governments, in practice all core decision making is dependent on state governments through the EU institutions and an expanding class of senior EU bureaucrats. While problem solving at the smaller communities should take precedence over any other level, EU policy recommendations, including anticrisis measures imposed on Southern European states such as Spain or Portugal, demanded the suppression of submunicipal administrations (parish or community councils), the only level in which direct assembly democracy in the form of village, parish, or town assemblies was legally possible. In political terms, the greatest contrast between the EU (or, for that case, any other federal nation-state that maintains peace within its borders) and decentralized peace systems is the horizontality of public decision making derived from social egalitarianism in the latter arrangements, versus the concentration of political power in small elites that characterizes nation-states in general.

The fact that this is actually the case does not mean that alternative decentralized models could not be considered instead. Indeed, they have been considered in the past. Beyond the economistic and industrialist views that inspired Jean Monnet's European integration proposals, the crucial years after 1945 saw a variety of different views on how to build a European Peace System. Swiss liberal theorist Adolf Gasser had written and published during the Second World War his book *Communal Freedom as Salvation of Europe* (1943), arguing that non-authoritarian states are only viable if they are grounded on strong communal (municipal) self-government, establishing a direct link between stable peace and free municipal self-government. Starting from the Swiss tradition and distilling the bottom-up concept of federalism offered by Proudhon without giving up his own liberal view of the state, Gasser defined his vision of a peaceful interdependent Europe as a "voluntary contractual federation of communities" built up from the small, self-governing units (Roca 2010).

Such views on bottom-up federalism were shared not only by some liberals and most libertarians (at least partially), but also by peripheral nationalists, such as the French National Minorities Committee, formed in 1927 under a charter that asserted:

> Modern States, based on force, will become invalid due to the world's increasing economic interdependence, as the antagonisms in which their existence is based only lead to increasingly terrible wars. It would be best to substitute them by a federation of peoples … providing the two most essential needs: freedom and peace. (Quoted in Castelao 1944/2010, p. 59)

To illustrate this, in the 1930 novel *Arredor de si* by Galizan nationalist Outeiro Pedralho, the author's literary alter ego Adrião Solóvio imagines the future map of Europe, in which "instead of States, each land is a free grouping of municipalities where no one dominated anyone else," "a fraternity of small happy communes" opposed to the "monstrous unnatural growth of large cities" (1930/1985, p. 192).

Both Proudhon's *The Principle of Federation* and Gasser's principle of "communal autonomy" stimulated a variety of grassroots groups across Europe, including the French "integral federalists," such as Alexandre Marc and Michel Mouskhely, and the overarching Union of European Federalists. The 1949 Permanent Committee for European Municipalities and Regions led to the establishment in 1951 of the Council of European Communes, which was envisioned by Gasser as a first step toward a federal Europe based on self-governing municipalities (Gouzy 2004).

The success of the top-down, nation-state-based approach steered by Monet and Schuman led to the formation of the pivotal European Economic Community, eventually relegating the Council of Communes into a politically irrelevant institution. Nevertheless, the building of a peaceful Europe based on the internal freedom of thousands of democratically self-governed communities remains as a vision for a continental-wide decentralized peace system.

Beyond Europe, the tension between centralism and community decentralism is present in other polities that share some of the EU's federal features, such as the United States. Two decades prior to the Occupy Movement, interest in the town meeting as an institution of direct assembly democracy brought about political proposals such as those by Frank Bryan and John McClaughry (1990) in *The Vermont Papers: Recreating Democracy on a Human Scale*. The authors defend the transformation of the State of Vermont into a federation of small shires (with an average of 10,000 people), in turn made up of self-governing towns. Bryan (2004), in a seminal book on town meetings significantly titled *Real Democracy*, continued to defend decentralized self-government as the only meaningful form to empower people and surpass the crisis of nation-states. Recalling Dahl and Tufte's (1973) argument that transnational organizations require very small units where people can become politically involved, the author suggests that the United States should shift toward a commonwealth of small self-governing political units organized through bioregions. Others suggest outright independence through secession to realize such a platform (see Miller & Williams 2013).

Political experiences such as the Iroquois League, the Council of European Communes, the Zapatista Autonomous Rebel Municipalities, and the Kurdistan Communities Confederation represent different attempts to build peace systems based on the principle of decentralized community self-government. Additional insights are offered by cases such as the Icelandic Commonwealth (930–1262 CE), a stateless system for conflict resolution based on decentralized democratic consensus, and contemporary proposals. Templer's (2008) "No-state solution" for Palestine–Israel incorporates the vision of a decentralized peace system as a novel solution to a seemingly intractable conflict, consisting of a multicultural and multifaith "Cooperative Commonwealth" built on the basis of "new forms of decentralized direct democracy, people's participation and horizontalism, neighborhood autonomy" that would go beyond historical Palestine to encompass other territories of the Fertile Crescent region (Iraq, Kuwait, Syria, Lebanon, Jordan, and Egypt), following a bioregional perspective that considers the need for common management of increasingly scarce resources such as fresh water, gas, and oil. Andrej Grubačić (2010, p. 208) describes a Balkans Federation along similar lines: "a transethnic society with polyculturalist outlook that recognizes multiple and overlapping identities and affiliations based on voluntary cooperation, mutual aid, a direct democracy of nested councils and a self-managed economy." Grubačić calls for a balkanization of Europe from below, considering his Balkans example as a "basis for the regeneration and reconstruction of social and political life of Europe" (2010, p. 209) through a Commonwealth grounded on direct local decision making (see also Ali & Walters, this volume, Chapter 5).

These actual examples, together with multiple proposals on direct democracy currently stemming from social movements such as Occupy, Transition Towns, the 15 M Movement in Spain, the Catalan Integral Cooperative, and a variety of nonviolent grassroots initiatives around the world, indicate that Gandhi's oceanic circles are indeed plausible and that the time for their consideration is ripe.

Conclusion

Some authors seem to support the idea that while our nomadic hunter-gathering past was marked by chronic carnage, it is the "forces of modernity" that have played a key role in the reduction of violence, in contrast with nostalgic elements of a peaceable past such as "communitarian solidarity," "ecological sustainability," or "harmony with the rhythms of nature." Needless to say, this stance encompasses the idea that we are somehow on the right path and that the coercive state is a necessity if we are to control the innate violence of our human nature. While some claim that the state, as it currently stands, is somehow an inevitability in our path toward "civilization," Clastres (1989) defended the view that stateless societies remained so, retaining their egalitarian and largely nonviolent forms of organization, because of the mechanisms they had in place to prevent the accumulation of power and emergence of hierarchies. Both views would probably concur in considering the forces of modernity – the state and its most notorious physical manifestation, the city – as agents designed to simultaneously remove us from nature and remove any nature within us.

But this is not necessarily a one-way road. As Eisler (1987) reminds us, the life-sustaining and enhancing chalice of partnership cultures is an alternative to the lethal blade of domination systems. Top-down control, rigid male dominance, and cultural conditioning to accept violence and domination can be replaced by egalitarian family and political structures, equal partnership between women and men, and recognition of nonviolence, nonkilling, and nondomination as normative social behaviors. The forces of the domination model, "which in the case of states, is ultimately their capacity to inspire terror," as Graeber (2004, p. 63) argues, can be diverted, frozen, transformed, or deprived of their substance. In the case of the modern state, this would occur both from above, through the development of international organizations, and from below, through the revitalization of local self-governance.

The peace system that operates in Peninsular Malaysia among the neighboring Chewong, Semai, Jahai, Btsisi, Batek, and other Semang nomadic hunter-gatherer groups, not only keeping at peace with each other but also upholding internal nonviolence and renouncing the use of violence toward encroaching outsiders, is a contrast with the violent clashes among, between, and within neighboring state societies in the same region (Satha-Anand and Urbain 2013). If we consider that such peace systems were common in Zomia, the great mountain uplands of mainland Southeast Asia, China, India, and Bangladesh (see Scott 2009), it is perhaps the "forces of modernity" and its dominant nation-state political system that should start to be placed under scrutiny.

As we move toward the peak of a historical process of centripetal concentration of political power, the idea of decentralized peace systems offers an alternative for the creation of new structures that enable both the participatory resolution of conflicts in large regional or global settings and reengaging communities with a sense of ownership over decision making and problem solving. Increased political participation and involvement in decision-making processes have been linked with community happiness, well-being, resilience, and cohesion, which in turn are correlated with reduced levels of violence. Evidence from the direct assembly democracy practices of Swiss local communes and participatory procedures at canton and federal levels (Frey & Stutzer 2002) and observations from the New England town meetings in the United States (Bryan 2004) point in this direction and invite further study on the psychology and neurophysiology of political participation.

The development of decentralized peace systems may happen from below, as Gandhi predicted, but the acute social transformations linked to peak oil and associated energy problems may well drive a certain degree of decentralization from above (Trainer 2010). In either case, the existence of an active, inclusive, and participatory citizenship, from the community level to the international arena, is the cornerstone for the emergence of such systems. Participation beyond purely electoral politics requires not just decentralization allowing communities to make decisions on their own, but also the reconsideration of the principle of subsidiarity in international organizations (such as the EU) as a means to enable real citizen participation in decision-making processes.

In practical terms, decentralized peace systems would require community constituencies of 1000 or fewer members where face-to-face political participation is possible, rebuilding a culture of deliberative, consensus-driven problem-solving practices. Oceanic circles in the form of federal, confederal, and commonwealth arrangements should have such self-governing village republics as their foundations. The establishment of a pilot Chamber of Communes in Europe facilitating consultations at the local level through electronic or assembly direct democracy tools could be a test for such a build-up, moving the EU toward a decentralized peace system. Alternatively, lack of state responsiveness to calls for democratization could bring about parallel institution-building based on grassroots movements, as some of the examples in this chapter illustrate. Revolution or reform, collective will or systemic collapse, or any combination thereof may produce an approximation of Gandhi's vision of a decentralized polity capable of sustaining a truly nonviolent society in peaceful relations with the rest of the world.

Acknowledgments

I thank Kirk M. Endicott, Piero Giorgi, Charles Macdonald, Darcia Narvaez, Glenn Paige, R. Brian Ferguson, Douglas P. Fry, Benjamin A. Peters, and Robert Tonkinson for their comments on a draft of this chapter and their valuable ideas to improve it. I am also thankful to the editors for the opportunity to generate and develop these ideas.

References

Aitamurto, K. (2008). Egalitarian utopias and conservative politics. Veche as a societal ideal within Rodnoverie Movementmore. *Axis Mundi, 2*. Retrieved from http://www.religionistika.sk/files/file/Axis%20Mundi/Kaarina_Aitamurto-Axis_Mundi_2-2008.pdf

Ariyaratne, A.T. (1999). *Collected works*. Ratmanala: Sarvodaya Lekha Publishers.

Bakunin, M. (1866). *Revolutionary catechism*. St. Petersburg: Author.

Balfour, E.B. (1944). *The living soil*. London: Faber & Faber.

Bates, C. (2005). The development of Panchayati Raj in India. In C. Bates & S. Basu (Eds.), *Rethinking Indian political institutions* (pp. 169–184). London: Anthem Press.

Berry, W. (2002). *The art of the commonplace: The agrarian essays of Wendell Berry*. Berkeley, CA: Counterpoint.

Bhave, V. (2007). *I valori democratici. La politica spirituale di Gandhi attraverso le parole di Vinoba Bhave*. Verona: Il Segno di Gabrielli.

Bicchieri, M.G. (Ed.). (1972). *Hunters and gatherers today*. New York: Holt, Rinehart and Winston.

Boehm, C. (1999). *Hierarchy in the forest. The evolution of egalitarian behavior*. Cambridge, MA: Harvard University Press.

Boehm, C. (2013). The biocultural evolution of conflict resolution between groups. In D.P. Fry (Ed.), *War, peace, and human nature* (pp. 315–340). New York: Oxford University Press.

Bookchin, M. (2003). The communalist project. *Harbinger, 3*(1). Retrieved from http://www.social-ecology.org/2002/09/harbinger-vol-3-no-1-the-communalist-project/

Bryan, F. (2004). *Real democracy: The New England town meeting and how it works*. Chicago: University of Chicago Press.

Bryan, F., & McClaughry, J. (1990). *The Vermont papers: Recreating democracy on a human scale*. White River Junction, VT: Chelsea Green Publishing.

Carpenter, E. (1921). *Civilisation: Its cause and cure*. London: Allen & Unwin.

Castelao, A.D.R. (2010). *Sempre em Galiza*. Compostela: Através. (Original work published in 1944)

Clastres, P. (1989). *Society against the state: Essays in political anthropology*. New York: Zone Books.

Crawford, N.C. (1994). A security regime among democracies: Cooperation among Iroquois nations. *International Organization, 48*(3), 345–385.

Dahl, R.A., & Tufte, E. (1973). *Size and democracy*. Stanford, CA: Stanford University Press.

Day, R.J.F. (2005). *Gramsci is dead: Anarchist currents in the newest social movements*. London: Pluto Press.

Dentan, R.K. (2010). Nonkilling social arrangements. In J. Evans Pim (Ed.), *Nonkilling societies* (pp. 131–182). Honolulu: Center for Global Nonkilling.

Dias, J. (1981). *Vilarinho da Furna. Uma aldeia comunitária*. Lisboa: Imprensa Nacional.

Dias, J. (1984). *Rio de Onor. Comunitarismo Agro-pastoril*. Lisboa: Presença.

Drago, A., & Trianni, P. (Eds.). (2008). *La filosofia di Lanza del Vasto: un ponte tra Occidente ed Oriente*. Milan: Jaka Book.

Eisler, R. (1987). *The chalice and the blade: Our history, our future*. San Francisco: Harper Collins.

Endicott, K. (2013). Peaceful foragers. In D.P. Fry (Ed.), *War, peace, and human nature* (pp. 243–261). New York: Oxford University Press.

Escalante Ruiz, A. [Egin Ayllu]. (2014). *Las vecindades vitorianas*. Barcelona: Ned ediciones.

Evans-Pritchard, E.E. (1940). *The Nuer: A description of the modes of livelihood and political institutions of a Nilotic people*. Oxford: Clarendon.

Ferguson, R.B. (2013). The prehistory of war and peace in Europe and the Near East. In D.P. Fry (Ed.), *War, peace, and human nature* (pp. 191–240). New York: Oxford University Press.

Fotopoulos, T. (1997). *Towards an inclusive democracy: The crisis of the growth economy and the need for a new liberatory project*. London: Cassell.

Fremion, Y. (2002). *Organs of history*. Edinbirgh: AK Press.

Frey, B.S., & Stutzer, A. (2002). *Happiness and economics: How the economy and institutions affect well-being*. Princeton, NJ: Princeton University Press.

Fry, D.P. (2006). *The human potential for peace: An anthropological challenge to assumptions about war and violence*. New York: Oxford University Press.

Fry, D.P. (2012). Life without war. *Science, 336*, 879–884.

Fry, D.P. (2013). War, peace and human nature: The challenge of achieving scientific objectivity. In D.P. Fry (Ed.), *War, peace, and human nature* (pp. 1–21). New York: Oxford University Press.

Fuentes, A. (2013). Cooperation, conflict, and niche construction in the Genus *Homo*. In D.P. Fry (Ed.), *War, peace, and human nature* (pp. 78–94). New York: Oxford University Press.

Gandhi, M. (1962). *Village Swaraj*. Ahmedabad: Navajivan Publishing House. Retrieved from http://www.mkgandhi.org/ebks/village_swaraj.pdf

Gandhi, M. (1998). *The collected works of Mahatma Gandhi* [Online]. New Delhi: Publications Division Government of India. Retrieved from http://www.gandhiserve.org/e/cwmg/cwmg.htm (Original works published 1910–1948)

García Ramos, A. (1912). *Arqueología jurídico-consuetudinaria-económica de la región gallega*. Madrid: Jaime Ratés.

Gardner, P.M. (2010). How can a society eliminate killing? In J. Evans Pim (Ed.), *Nonkilling societies* (pp. 185–194). Honolulu: Center for Global Nonkilling.

Gardner, P.M. (2013). South Indian foragers' conflict management in comparative perspective. In D.P. Fry (Ed.), *War, peace, and human nature* (pp. 297–314). New York: Oxford University Press.

Gasser, A. (1943). *Gemeindefreiheit und Zukunft Europas* [Communal freedom as salvation of Europe; not translated into English]. Basel: Bücherfreunde.

Gibson, T., & Sillander, K. (Eds.). (2011). *Anarchic solidarity: Autonomy, equality and fellowship in Island Southeast Asia*. New Haven, CT. Yale University Southeast Asia Studies Monographs.

Giorgi, P.P. (2009). Nonkilling human biology. In J. Evans Pim (Ed.), *Toward a nonkilling paradigm* (pp. 95–122). Honolulu: Center for Global Nonkilling.

Giorgi, P.P. (2010). Not killing other people: The origin and other future of *Homo sapiens*. In J. Evans Pim (Ed.), *Nonkilling societies* (pp. 83–98). Honolulu: Center for Global Nonkilling.

Gouzy, J.P. (2004). The saga of the European Federalists during and after the Second World War. *The Federalist, 46*(1). Retrieved from http://www.thefederalist.eu/site/index.php?option=com_content&view=article&id=521&lang=en

Graeber, D. (2004). *Fragments of an anarchist anthropology*. Chicago: Prickly Paradigm Press.

Grubačić, A. (2010). *Don't mourn, Balkanize! Essays after Yugoslavia*. Oakland, CA: PM Press.

Harris, M. (1990). *Our kind*. New York: Harper Perennial.

Herod, J. (2007). *Getting free: Creating an association of democratic autonomous neighborhoods*. Boston: Lucy Parsons Center.

Howard, A. (1940). *An agricultural testament*. Oxford: Oxford University Press.

Jenco, L.K. (2009). Thoreau's critique of democracy. In J. Turner (Ed.), *A political companion to Henry David Thoreau* (pp. 68–96). Lexington: University Press of Kentucky.

Knauft, B.B. (1991). Violence and society in human evolution. *Current Anthropology, 32*, 391–428.

Kropotkin, P. (1895). *The Paris Commune* (Freedom Pamphlets, No. 2). London: W. Reeves. (Original work published in 1880)

Ladner, A. (2002). Size and direct democracy at the local level: The case of Switzerland. *Environment and Planning, 20*, 813–828.

Lane, R. (2005). Standing "aloof" from the state: Thoreau on self-government. *The Review of Politics, 67*(2), 283–310.

Leacock, E. (1978). Women's status in egalitarian society: Implications for social evolution. *Current Anthropology, 19*, 247–55.

Lee, R.B. (1992). Art, science, or politics? The crisis in hunter-gatherer studies. *American Anthropologist, 94*(1), 31–54.

Macdonald, C.J.H. (2011). Primitive anarchs: Anarchism and the anthropological imagination. *Social Evolution & History, 10*(2), 67–86.

McKeogh, C. (2009). *Tolstoy's pacifism*. Amherst, NY: Cambria Press.

Miller, R., & Williams, R. (Eds.). (2013). *Most likely to secede: What the Vermont Independence Movement can teach us about reclaiming community and creating a human scale vision for the 21st Century*. Waitsfield: Vermont Independence Press.

Mumford, L. (1967–1970). *The myth of the machine* (Vols. 1 and 2). New York: Harcourt Brace Jovanovich.

Narvaez, D. (2013). The 99 percent: Development and socialization within an evolutionary context. In D.P. Fry (Ed.), *War, peace, and human nature* (pp. 341–357). New York: Oxford University Press.

Narvaez, D. (2014). *Neurobiology and the development of human morality: Evolution, culture, and wisdom*. New York: W.W. Norton & Co.

Öcalan, A. (2011). *Democratic confederalism*. London: Transmedia Publishing.

Ostrom, E. (1990). *Governing the commons: The evolution of institutions for collective action*. Cambridge: Cambridge University Press.

Outeiro Pedralho, R. (1985). *Arredor de si*. Vigo: Galaxia. (Original work published in 1930)

Peixoto, R. (1990). Survivances du regime communautaire en Portugal. In *Etnografia Portuguesa*. Lisboa: Dom Quixote. (Original work published in 1908)

P.M. (1985). *Bolo'bolo*. New York: Autonomedia.

Proudhon, P.-J. (1863). *The principle of federation*. Paris: E. Dentu.

Roca, R. (2010). Community freedom and democracy. *Current Concerns, 17*. Retrieved from http://www.currentconcerns.ch/index.php?id=1109

Rodrigo Mora, F. (2012). *¿Revolución integral o decrecimiento?* Barcelona: El grillo libertario.

Rodrigo Mora, F. (2013). *O atraso político do nacionalismo autonomista galego.* Galiza: Edições da Terra.

Rohman, W. (2005). *Historical development of legal literature on customary laws in Assam.* Delhi: Kalpaz Publications.

Santos Vera, S., & Madina Elguezabal, I. (2012). *Comunidades sin Estado en la Montaña Vasca.* Antsoain: Hagin.

Sastre, P. (2013). *Batzarra, gure gobernua.* Donostia: Elkar.

Satha-Anand, C., & Urbain, O. (Eds.). (2013). *Protecting the sacred, creating peace in Asia Pacific.* New Brunswick, NJ: Transaction Publishers.

Schumacher, E.F. (1973). *Small is beautiful: A study of economics as if people mattered.* London: Blond & Briggs.

Scott, J.C. (2009). *The art of not being governed: An anarchist history of upland of Southeast Asia.* New Haven, CT: Yale University Press.

Sharma, R.S. (1996). *Aspects of political ideas and institutions in ancient India.* Delhi: Motilal Banarsidass Publishers.

Shiva, V. (1991). *The violence of the green revolution.* London: Zed Books.

Solnit, R. (2009). *A paradise built in hell: The extraordinary communities that arise in disaster.* New York: Viking.

Sponsel, L.E. (2010). Reflections of the possibilities of a nonkilling society and a nonkilling anthropology. In J. Evans Pim (Ed.), *Nonkilling societies* (pp. 17–52). Honolulu: Center for Global Nonkilling.

Summy, R. (2013). Changing the power paradigm. From mainstream to nonkilling politics. In J. Evans Pim (Ed.), *Nonkilling security and the state* (pp. 35–65). Honolulu: Center for Global Nonkilling.

Swain, P.C. (2008). *Panchayati Raj.* New Delhi: APH Publishing.

TATORT Kurdistan. (2013). *Democratic autonomy in North Kurdistan: The council movement, gender liberation, and ecology – In practice.* Porsgrunn, Norway: New Compass Press.

TATORT Kurdistan. (2014, October 10). Democratic autonomy in Rojava. *New Compass.* Retrieved from http://new-compass.net/articles/revolution-rojava

Tavanti, M. (2003). *Las Abejas: Pacifist Resistance and Syncretic Identities in a Globalizing Chiapas.* London: Routledge.

Templer, B. (2008). Reclaiming the commons in Palestine/Israel: ¡Ya Basta!/Khalas! *Monthly Review Zine.* Retrieved from http://mrzine.monthlyreview.org/2008/templer230708.html

Tenorio, N. (1982). *La aldea gallega.* Vigo. Xerais. (Original work published in 1914)

Thoreau, H.D. (1854). *Walden: Or, life in the woods.* Boston: Ticknor and Fields.

Tonkinson, R. (2013). Dream-spirits and innovation in Aboriginal Australia's Western Desert. *International Journal of Transpersonal Studies, 32*(1), 127–139.

Trainer, T. (2010). *The transition to a sustainable and just world.* Canterbury: Envirobook.

Woodburn, J. (1988). African hunter-gatherer social organization: Is it best understood as a product of encapsulation? In T. Ingold, D. Riches, & J. Woodburn (Eds.), *Hunter-gatherers vol. 1. History, evolution and social change* (pp. 31–64). Oxford: Berg.

Younger, S. (2008). Conditions and mechanisms for peace in precontact Polynesia. *Current Anthropology, 49*(5), 927–934.

13

Building Peace Benefits
Daniel Hyslop and Thomas Morgan

Conceptualizing a Peace Process

A peace process nominally refers to "a political process in which conflicts are resolved by peaceful means" (Saunders 2001). Understood simply, many see it as the process that ends a conflict, the search for what Johan Galtung first defined simply as negative peace, the absence of violence. This chapter aims to characterize a peace process as a long-term dynamic process characterized by violence minimization and elimination and by the progressive cultivation of institutions that deliver positive peace – a more equitable social order that meets the basic needs and fulfills the rights of all people.

While Peace and Conflict Studies is a well-established discipline, some (e.g., Coleman 2012) have criticized it for being too focused on the conditions that promote war, violence, aggression, and conflict. For decades, research has been problem oriented and focused on better understanding why wars break out, why terrorism occurs, or the prevention of nuclear war. The research agendas of the major peace research institutes reflect this analytical bias toward conflict and war. For instance, the Peace Research Institute of Oslo (PRIO) has three key research questions in its agenda: why do wars break out? How are wars sustained? What does it take to build durable peace? Hence, many of the current research agendas in the field have been engaged in understanding responses to armed conflict, its impacts, and how people cope with it.

Systematic understanding of the drivers and consequences of conflict and war is undoubtedly a critical starting point for analysis. The damage wrought by violence, conflict, and war on societies is immense. We know from empirical research that conflict takes a tremendous and crippling toll on the social, structural, and psychological well-being of nations and individuals. Recent studies such as the World Bank's *World Development Report 2011* brought together this research to show the significant impacts of conflict and war on long-term development prospects (World Bank 2011). Very few conflict-affected states, for example, have met Millennium Development Goals (MDGs)

Peace Ethology: Behavioral Processes and Systems of Peace, First Edition.
Edited by Peter Verbeek and Benjamin A. Peters.
© 2018 John Wiley & Sons Ltd. Published 2018 by John Wiley & Sons Ltd.

with progress in target indicators like MDG 1.A (halving the proportion of people whose income is less than $1.25 a day between 1990 and 2015). Those nations affected by major violence have seen very little progress in this regard. Aside from the immediate costs in terms of lost life, this underlines the longer term loss of attainment in human capabilities and well-being.

However, peace processes for conflict cessation are only the first part of the broader peacebuilding process. It was Galtung (1985) who defined this distinction by going further than the elimination of violence underpinning negative peace to further elucidating a definition of positive peace, a state of affairs in which people address injustice, oppression, and inequality in order to achieve "a more equitable social order that meets the basic needs and rights of all people." More recently, the field has moved to use the term *sustainable peace* as an attempt to capture the dynamic and interdependent nature of both negative and positive peace. Solomon (1997) defined sustainable peace as "requiring that long-time antagonists not merely lay down their arms but that they achieve profound reconciliation that will endure because it is sustained by a society-wide network of relationships and mechanisms that promote justice and address the root cause of enmity before they can regenerate destabilisation tensions." The network of relationships and mechanisms required to maintain profound reconciliation seemingly depends on more than just the cooperation of the conflict actors themselves but explicitly infers the needs for positive society-wide support (see Wessells & Kostelny [this volume, Chapter 9] for a discussion of the role of communities in supporting sustainable peace after conflicts).

More recent interdisciplinary contributions such as those from Coleman (2012) have aimed to define the psychological characteristics of sustainable peace with six key requirements: (1) a strong sense of positive interdependence among members of society; (2) a strong sense of global, as well as local, patriotism and loyalty; (3) the sharing of basic common values as recognitions that human beings have to be treated with respect, dignity, and justice; (4) mutual understanding, and freedom to communicate and be informed; (5) a sense of fair recourse; and (6) social taboos against the use of violence to solve problems. Understanding the linkages and mechanisms underlying these psychological components of sustainable peace is undoubtedly complex and takes the definition of sustainable peace well beyond the hypothetical starting point of negative peace or the absence of violence and fear of violence. While these psychological components of sustainable peace intuitively go hand in hand, the significant practical challenges of measuring and observing these factors in operation limit our ability to further empirically verify the theory.

This chapter aims to conceptualize a peace process as a broader dynamic process in which institutions aim to both prevent violence and sustain a virtuous cycle of societal interactions and behaviors that develop a more equitable social order that meets the basic needs and fulfills the rights of all people. While the foremost aim of any long-term peacebuilding process is to repair the damage of conflict, this more fundamentally means the cultivation of the institutions, attitudes, and structures that create a more equitable social order to result in less violence and more just, equitable, and nonviolent social outcomes. In order to demonstrate the positively reinforcing nature of progress in positive peace, this chapter illustrates the economic dimensions of the impact of violence, conflict, and war and also underlines the systemic nature of peace processes as well as their economic benefits.

The Economic Benefit of Peacebuilding

This analysis should not be considered an attempt to extract, isolate, or separate the economic cost of violence from its social and psychological costs. It is not merely a matter of financial quantification of the suffering brought about by war or a way of rendering intangible losses as tangible – and palatable – numbers and figures. The economic cost of violence does not sit above or apart from its other costs; rather, an understanding of the economic cost of violence is crucial to understanding the extent of the damage caused by war and how the resultant economic devastation can amplify or extend that damage. Violence itself is harmful to the economy at both the macro and micro levels. Large-scale armed conflict leads to capital destruction and the disruption of normal economic activity. This in turn means that countries in a postconflict environment cannot easily return to pre-war levels of both wealth and economic growth. In the wake of especially devastating conflicts, per capita income can often take decades to return to pre-war levels. While microlevel violence is not usually as immediately destructive to income, high levels of interpersonal violence (e.g., homicide and violent crime) can severely depress Gross Domestic Product (GDP) growth by destruction of human capital and depression of consumptive behavior. Economic uncertainty and inequality can give rise to political and social conflict, and a failure to address economic recovery in the wake of war can lead to renewed conflict. Thus, any theory of peace is incomplete without an understanding of the economics of peace.

Peace economics begins with the understanding that direct spending on violence is generally both unproductive and destructive. Military weapons (missiles, tanks, etc.) are outputs that largely remain unused; their value depreciates over time during which they generate little to no additional productivity, like capital stock; and, worse, when they are deployed, they destroy existing capital stocks. While there is economic value in spending on violence containment (police, judicial system, private security companies, etc.), such spending reduces the resource allocation for other public services, such as education and health, which have far larger public benefits and which directly and indirectly have greater capacity for leading to future increases in productivity. Expenditure on violence containment is economically efficient when it effectively prevents violence for the least amount of outlay. However, public spending on surplus violence containment or inefficient programs has the potential to constrain a nation's economic growth. The theoretical basis of this approach sees the purpose of economic study as premised on the need to understand the design of political, economic, and cultural institutions; understand their interrelations; and then uncover policies to prevent, mitigate, or resolve any type of latent or actual destructive conflict within and between societies (Brauer & Caruso 2011).

Even when military spending stimulates economic growth in the short run, its long-run consequences are often indistinguishable from previous growth trends or deleterious to long-run performance. During World War II, military spending in the United States increased exponentially. However, GDP per capita declined as this spending was scaled back, and the growth in GDP per capita returned to its pre-war trend. Empirical research by the Institute for Economics and Peace (IEP) has shown military action taken by the US government has corresponded with reduced consumption and investment growth rates, increased inflation, and increased government debt (IEP 2012a).

Undoubtedly, heightened military spending during armed conflict does create employment and additional economic activity, and it contributes to the development of new technologies, which can then filter through into other industries. These are some of the often-discussed positive benefits of heightened government spending on military outlays. However, programs specifically targeted at accelerating research and development (R&D) or creating employment directly would potentially have the same effect at a lower cost. Pollin and Garrett-Peltier (2009) recently reviewed the multiplier effect of US investments in military spending to show the potential job creation effect of transferring spending from the military to other areas that simply generate more jobs. Their modeling shows that investments in health and education, and even in equivalent tax cuts for personal consumption, generate more jobs than spending on the military. Hence, even prior to the use of weapons and the impact of their destructive capacity, there is the effect of a general undermining of economic productivity.

One illustration of the long-run economic impacts of direct violence is an example of a nation that has recently experienced civil war. The Sierra Leone Civil War lasted for 11 years, beginning in 1991 and ending in 2002. Even though the end of the war brought back economic growth, by 2010 the level of GDP per capita was still approximately 31% lower than what would have been expected in a linear growth model in the absence of conflict. The negative economic impacts from conflict in Sierra Leone have also been mirrored by the trends in human development as measured by the United Nations Development Programme (UNDP) Human Development Index (HDI), with Sierra Leone's levels of human development lagging behind regional averages and only improving after the cessation of conflict.

Although such examples provide a powerful illustration of the economic and development impacts of violence, the benefits of peace extend beyond the absence of violence. That is, peace must be conceptualized beyond its negative definition of the absence of violence to include its positive characteristics involving the creation of those institutions and structures that develop a more equitable social order that meets the basic needs and fulfills the rights of all people.

Thus, peace processes that are successful in reducing or ending conflict in the short run and also restoring the factors most closely associated with resilient, peaceful societies in the long run will have the greatest economic impact on postconflict societies. This in turn helps mitigate the likelihood of the peace process breaking down. Economic prosperity goes hand in hand with development and social resilience. Peace is the key factor that underlies both.

Measuring Negative Peace

In order to understand peace and the economic benefits that stem from successful peace processes, it is first necessary to provide an accurate understanding and measurement of peace. Peace is usually thought of as the absence of war or large-scale conflicts within states, usually between governments and militias or rebel groups. However, IEP uses a broader definition of peace that encompasses both the absence of violence and the absence of the fear of violence. A single composite measure combining 23 indicators provides a common denominator for different types of violence, a tool for comparing levels of peace in 163 countries, and a baseline for tracking increases and decreases in peace over time. While there is some conceptual variation between the different types of violence captured by the Global Peace Index (GPI), there is a strong correlation between the different GPI indicators. Countries with high levels of internal conflict, distrust, and interpersonal violence are also more likely to have higher levels of political instability, terrorist activity, and internal and external conflict. This in turn is associated with higher levels of militarization. This suggests that the combination of these different measures of violence into a single composite index is conceptually appropriate.

The 23 indicators in the GPI fall into three quantifiable domains: ongoing domestic and international conflict, societal safety and security, and militarization (for a full list of the GPI indicators, see www.economicsandpeace.org). Each of these domains captures an aspect of violent conflict that any peace process must address to reduce conflict in both the short and long run. This provides an important baseline for analysis and statistical research to understand the key factors that are common to nations that have less direct violence.

Measuring Positive Peace: The Pillars of Peace

Similar to the measurement of negative peace, there are few guiding theories and frameworks for measuring Galtung's conceptualization of positive peace. Like negative peace, positive peace is also multidimensional in nature but describes a more complex interdependent network of institutions, attitudes, behaviors, and norms that sustain nonviolence and also develop a more equitable social order.

In order to derive a framework for conceptualizing such a complex network of institutions, IEP developed a model that correlated thousands of indicators of development and institutional strength against the GPI. It selected these indicators based on a comprehensive review of Peace and Conflict and Developmental Studies literature regarding the drivers of conflict. The Pillars of Peace are the outcome of this analysis (IEP 2013).

The Pillars of Peace provide a framework for assessing the "positive peace" factors that create peaceful societies. These positive peace factors can also indicate how supportive the underlying environment is toward development as well as equitable and just societal outcomes. The framework provides a benchmark against which to measure the performance of the broader aspects of social development and a country's overall resilience when confronted with social upheaval.

The Pillars of Peace are an eight-part taxonomy of key factors that sustain positive peace:

- *Well-functioning government*: A stable, functioning government is necessary for the enforcement of contracts and the creation of an appropriate regulatory environment. Governance can be further divided into three subcategories. Effectiveness is a measure of a government's ability to implement policy, the Rule of Law is a measure of the formal and informal constraints on government and the functioning of justice, and *voice* and *accountability* refer to how responsive the government is to its citizens and how well it protects their basic liberties.
- *Sound business environment*: A sound business environment maximizes economic growth and a diverse, growing economy. A sound business environment will have the

appropriate supporting infrastructure (transportation, etc.), large and well-developed markets, and business with sophisticated supply chain processes, innovation, and R&D. A sound business environment amplifies the economic benefits that stem from successful peacebuilding.

- *Low levels of corruption*: Corruption has a very strong association with peacefulness. Corruption among elites, in the judiciary, and in police forces is strongly correlated with the outbreak of large-scale violence. Corrupt institutions are a particular problem in the postconflict environment. Corruption can be both venal (where low-level bribery is an implicit or explicit part of daily transactions) and systematic.
- *High levels of human capital*: The human capital pillar includes both education and healthcare. Education is essential for long-run growth, stability, and competitiveness. Efficient health services are vital for human flourishing in both the short and long term.
- *Free flow of information*: Free and independent media help increase transparency and reduce the likelihood of corruption. A sophisticated and responsive press with a variety of outlets and sources can also help offset problems associated with low, incorrect, or asymmetrical information contributing to social unrest and the outbreak of violence.
- *Equitable distribution of resources*: Unstable and unequal economic growth can exacerbate income inequality, which increases the likelihood of societal unrest and violent conflict. This pillar also encompasses the distribution of education and health.
- *Acceptance of the rights of others*: If laws, attitudes, and societal norms acknowledge the rights of others, tensions between groups are reduced and higher levels of output, productivity, and human capital can be reached when the rights of women, for example, are recognized and respected.
- *Good relations with neighbors*: Reducing tensions between groups and between different nations is key to sustaining long-term peacefulness. This is particularly important in the postconflict environment.

These pillars are best thought of as "characteristics" of more peaceful societies. They all have associated inputs, processes, and outcomes and have a collinear relationship with measures of violence. As such, there are no pure or set policy prescriptions that arise from the existence of these pillars: targeting one pillar in isolation is not possible, especially over the long run. Rather, context- and country-specific factors will shape the interaction between the pillars. Increased economic performance resulting from successful peace processes can help "kickstart" a virtuous cycle, whereby resources are freed up to strengthen institutions and processes, which in turn leads to interaction effects between the pillars over the long run, as shown in Figure 13.1.

Investing in the pillars of peace can create a long-term virtuous cycle of lower unproductive violence containment spending and greater investment in the underlying institutions that create a more peaceful, just, and equitable society.

More highly developed contexts demonstrate the longer term virtuous cycle. Countries that reach a certain level of peacefulness are far more resilient to shocks (both conflict related and otherwise) and much more likely to recover quickly from short-term destabilization. This "stickiness" is apparent when comparing GPI and Pillars of Peace measures and their movements over time: while more peaceful countries do occasionally experience shocks (for example, the global financial crisis, riots, and

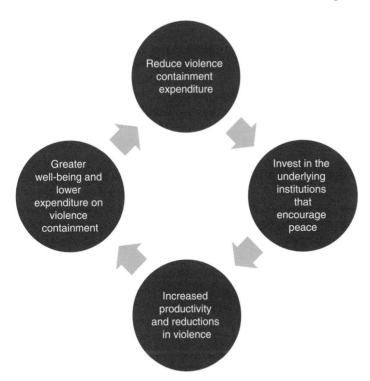

Figure 13.1 The virtuous cycle of peace

natural disasters) that decrease their level of peacefulness in the short run, they quickly return to previous levels of peacefulness. Similarly, violence is also "sticky," as most of the least peaceful countries have remained mired in conflict over the long run or shown a propensity to relapse into conflict.

When aggregated, key indicators of these eight pillars are associated with peaceful environments and are both interdependent and mutually reinforcing, such that improvements in one factor would tend to strengthen others and vice versa. Therefore, the relative strength of any one pillar has the potential to influence the others, thereby influencing the overall condition of peace. To demonstrate how factors impact each other, consider the example of increases in corruption. An increase in corrupt behavior by elites would likely have an effect on well-functioning government and the free flow of information. Alternatively, consider restrictions on the free flow of information: they may affect societal understanding of key issues, lower the acceptance of the rights of others, and potentially increase corrupt behavior.

Due to the interdependent nature of these factors, the weakening or strengthening of any one pillar will also weaken or strengthen the other pillars. A peaceful environment therefore depends on the strength of all pillars. This is analogous to a brick wall: taking out one brick diminishes the strength of the entire wall. Examination of states' progress on key indicators over time via growth analysis bears this out. This is where the average change in GPI scores is compared for nations according to the initial strength of each of the pillars. That is, we separated states into groups according to their relative strength in a pillar, such as the free flow of information, and then we observed their average change in GPI scores. From this, we found that the past strength of pillars tended to indicate whether a state would experience a more virtuous cycle of peace.

The purpose of this framework is not to isolate causality; rather, it aims to describe the optimal environment for peace to flourish. When considering the environment that underlies a peaceful society, it is useful to apply systems thinking to the way government, the economy, and culture might interact. For the Pillars of Peace, this means that any one pillar cannot be considered alone.

This also means that defining causality is difficult, as it may not be possible to isolate simple lines of causation that interact with one another to make a country more peaceful. Similarly, rather than linear progress, this implies systemic thinking in terms of virtuous or vicious cycles. The Pillars of Peace are mutually interdependent, meaning that the best improvements in peace result from improvements in the entire system.

Measuring the Economic Impact of Conflict and Violence

To measure the economic benefits that come from peace properly, it is necessary to have both a baseline measurement of peacefulness (which the GPI provides) and a methodology that allows for the quantification of all the costs of violence. While a total and completely accurate account of all of the economic costs associated with violence is not possible, the GPI provides a framework for a comprehensive assessment of the economic benefits of peace.

Accurately measuring all the costs associated with violence is difficult. Past conflicts may have costs that are still being paid off in the present. Private sector spending on violence containment is difficult to disaggregate. Estimates of future productivity increases from reductions in violence vary significantly. This section provides a review of the various prior approaches to capturing the economic impact of violence.

Much of the literature on the economic cost of violence comes from academic and government sources on the cost of crime. Most of the studies identify different types of costs associated with crime, emphasizing the existence of tangible and intangible costs and their measurement methods. Cohen (2000) reviewed some of the methodologies used to measure a society's responses to crime and its costs, identified a number of different approaches to measure a society's response to the costs of crime, and classified costs as either tangible or intangible and measurement methods as either direct or indirect. He defined tangible costs as those that involve monetary payments such as medical costs, stolen or damaged property, wage losses, prison cells, and police expenditures. On the other hand, he classified as intangible or nonmonetary those costs not normally exchanged in private or public markets, such as fear, pain, suffering, and lost quality of life.

The intangible costs of violence are much more difficult to capture with accuracy, which has led to a number of different approaches to try to quantify fear, pain, lost productivity, and so on. A recent trend in estimating the costs of crime has been toward a "willingness to pay" methodology (Webber 2010). This involves asking the public what they would be willing to pay to reduce the likelihood of becoming a victim of a specific crime and then combining this with information about the risk of victimization to calculate the implied cost of one crime from the results. However, one limitation of willingness-to-pay measures is that the methodology assumes that people are knowledgeable about the risks of crime. If there are misperceptions regarding crime in the community, then willingness-to-pay estimates may not be completely accurate (Mayhew 2003). Another method for estimating the intangible costs of violence is the "jury compensation method," which takes into account jury award data to estimate the monetary value of pain, suffering, and lost quality of life for nonfatal injuries (Cohen 2000).

Other studies have tried to combine the tangible and intangible costs of violence. One study (McCollister *et al.* 2010) uses a two-pronged approach that employs both cost-of-illness and jury compensation methods. The cost-of-illness approach estimates the tangible costs of crime, including lost productivity for the perpetrator and victim as well as short-term medical expenses, lost earnings, and property damage or loss for the victim. The tangible costs also include a "crime career cost," defined as the opportunity costs associated with the criminal's choice to engage in illegal rather than legal and productive activities. The difference between the jury's total award and the direct economic loss to the victim, including medical expenses and lost earnings incurred by the victim that are determined during the trial, are the intangible costs. Research conducted in countries other than the United States and United Kingdom has followed a similar approach. One Australian study took into account the underreporting of crimes (including violent crimes) by using estimates based on extrapolations from surveys (Mayhew & Adkins 2003).

In addition to all these tangible and intangible costs identified in most of the literature, crime and violence have significant "multiplier" effects on the economy by depressing savings, investments, earnings, productivity, labor market participation, tourism, and ultimately growth. Morrison *et al.* (2003) presented a typology of many of the costs that may be associated with violence, which reflect not only direct monetary and nonmonetary costs but also other so-called economic multiplier effects, including macroeconomic, labor market, and intergenerational productivity effects, as well as social multiplier effects that refer to the impact on interpersonal relations and quality of life. As an example, they mentioned a case study of Colombia, which suggested that for every additional 10 homicides per 100,000 residents, the level of investment falls by approximately 4%, or, alternatively, if homicide rates in Colombia had remained unchanged since the 1960s, total annual investment in Colombia today would be around 20% higher.

The second major component of the cost of violence is the cost of conflict in the form of war, civil war, or terrorism. Terrorist activity not only has the tangible and intangible costs mentioned previously but also a substantial multiplier effect on the rest of the economy. Terrorist activity can have a significant macroeconomic impact, leading to reduced consumption and investment, although this impact is much lower than the impact of warfare (Blomberg *et al.* 2004). One study found that the economic consequences of terrorism are visible only in the short term and dissipate quickly, even after one year, while the effects of wars take up to three years and internal conflict takes up to six years to dissipate (Collier 1999).

Many studies have tried to capture the economic costs of internal and external conflict. However, most of these studies tend to express the economic consequences of conflict as a proportion of GDP, including effects that are directly attributable to the conflict but omitting the indirect costs (De Groot *et al.* 2009). There are two main approaches to assessing the cost of conflict. One is an accounting technique that attempts to calculate the total value of capital destroyed during a conflict, while the other uses counterfactual analysis to try to estimate GDP if no conflict had occurred, then compares actual GDP to the counterfactual estimate. A counterfactual case study on the Basque Country suggested that conflict between the government and Basque separatist groups reduced actual GDP by 10 percentage points. Another study that looked at internal conflict in Sri Lanka disaggregated lost GDP. It used data on military expenditure to calculate the amount of forgone investment and calculated the influence of forgone investment on the growth rate of GDP. While these and other country case studies are valuable, there is little consistency of approach among them, making it difficult to generalize about the economic benefits of peace processes.

Other approaches have tried to address this issue by looking at how the length of a conflict affects its economic impact. One of the most influential studies in the literature to survey the economic consequences of conflict is from Collier (1999), who focuses on civil war (see also Collier 1998). He argued that civil wars affect growth through the destruction of resources, the disruption of infrastructure and social order, budgetary substitution, dissaving, and portfolio substitution of foreign investors highlighting. The first four of these channels influence an economy only during conflict, whereas the final one is likely to continue having an effect after the restoration of peace. In particular, he argued that long-running conflicts are more likely to be followed by an increase in growth, whereas short-lasting conflicts will suffer reduced growth rates over a longer period of time. During civil conflict, the annual growth rate decreases by 2.2%. After a one-year conflict, the five postconflict years will have a growth rate 2.1% below the growth path in the absence of conflict. On the other hand, after a 15-year conflict, the postwar growth rate is 5.9% higher.

Some more novel approaches have tried to mirror the willingness-to-pay approach by looking at how much income people would be willing to give up to live in a peaceful world. Hess (2003) presented an interesting methodology to measure the economic welfare cost of conflict, very different from the standard Collier-style regressions. He estimates how much income people would be willing to give up to live in a peaceful world. He employs a technique developed by Lucas (1987) and compares the actual consumption path of the world's citizens with a hypothetical consumption path in a world in which there is no conflict at all. He found that individuals who live in a country that has experienced some conflict during the 1960–1992 period would permanently give up to approximately 8% of their current level of consumption to live in a purely peaceful world.

The Economic Cost of Violence Containment to the United States

Assessing the economic benefit of specific peace processes is difficult, for the reasons outlined in this chapter. The resumption of steady economic growth following the cessation of violence may take years or even decades. The methodology outlined in this section aims to give a generalized account of the total cost of violence, which can if necessary be disaggregated into its component parts. This allows for specialized estimates for specific peace processes.

While the existing literature on the costs of violent crime, conflict, and warfare highlights the many dimensions of violence and their economic impact, it does not provide a comprehensive methodology that allows for the generalization of costs of violence across all countries. IEP has developed an approach that allows for such a comparison. Using the existing literature, we calculate a "unit cost" for as many of the GPI indicators as possible. This unit cost can be multiplied by the raw data for each indicator and then summed to produce an estimate of the total economic impact of violence for any given country.

One of the problems with such an approach is that the existing literature is not evenly distributed around the world. Some countries have multiple studies that estimate the cost of violence, while others have none or only have studies for a subset of the GPI indicators. Thus, to illustrate the methodology for calculating the cost of violence, this section focuses on a single country (the United States), as most of the studies examining the cost of violence were from the United States, and it also has the broadest available range of violence data.

IEP refers to all the spending on issues related to violence in the United States as the "violence containment industry" (IEP 2012b). This includes all expenditures related to violence, including but not limited to medical expenses, incarceration, police, the military, insurance, and the private security industry. It consists of local, state, and federal government expenditures as well as private spending by corporations, households, and individuals. Violence containment spending includes both the public and private sectors and is represented in terms of net value added. It shows that the federal government spends over $1.3 trillion, or approximately 9% of GDP, on violence containment. This is more than the amount spent on pensions and more than double infrastructure spending in 2010. National defense spending includes the Department of Defense, Homeland Security, Veterans Affairs, and the debt servicing on these expenditures, which is based on the proportion of military-related government expenditure. Private sector spending on violence containment is conservatively estimated to be $605 billion, or 4.2% of 2010 GDP. State and local governments spend the remaining amount on police, justice, the prison system, and other security-related measures.

Public sector spending includes government expenditures at the federal, state, and local (county or municipal) levels. At the federal level, the most obvious component is the US armed forces or military expenditure in general. This includes the cost of the Afghan and Iraq wars, US bases and troop deployments, ongoing military and nuclear activity, the military use of outer space, the intelligence agencies, arms procurement, military aid to foreign powers, virtually all of homeland security, veterans' affairs, and more. Because the US federal government budget is usually in deficit, the shortfall must be debt financed.

Thus, in proportion to the violence containment industry's contribution to the federal debt, a portion of the interest paid is a legacy cost of previous violence containment spending coming due in the present budget year. Law enforcement includes police protection, judicial and public sector legal expenses, and correctional facilities at the federal, state, and local levels. *Corrections* refers to the cost of operating prisons and jails, even if contracted out to the private sector prison industry, and the cost of dealing with parole, probation, and the court system. In addition are security-related expenditures for labor, services, and equipment associated with infrastructure and events such as municipal airports, public schools and universities, and publicly sponsored or supported events. The private sector consists of activities in private households and businesses, such as deadbolt locks; building and car alarms; security guards; insurance premiums paid to insure against loss of life, limb, or property; and so on.

According to estimates made by the Small Arms Survey (2011), the private security sector employs some 20–25 million people worldwide and 2 million people in the United States. This compares to approximately 11 million police personnel worldwide, including 883,600 in the United States in 2007. In addition, there are the security costs at passenger and commercial transportation hubs, which, in the United States, are mostly privately owned and operated. There are also private sector legal costs associated with violence against property or persons. While mostly captured in the legal services industry, these costs are also partly captured by large corporations' internal legal departments, making them difficult to count and therefore excluded from this study. Additionally, there are the costs imposed on the health, medical, and rehabilitation sectors that are associated with violence.

The private sector also covers arms producers, of both major conventional weapons and weapons of mass destruction. The federal budget includes the cost of weaponry manufactured for and sold to the US Department of Defense (DoD). The cost accounting also includes small arms such as bladed weapons, handguns, and long guns, and associated accessories and supplies such as gun sights, night scopes, ammunition, accoutrements, shooting ranges, and the gun magazine/publishing industry, but this portion makes up a very minor component of the final figure. Neither public nor private accounting systems currently separate security from nonsecurity items, and it would take a major change in government and corporate accounting standards to capture the correct numbers.

All of the costs listed above pertain to violence containment spending in a single year. While transferring this spending to other areas (education, infrastructure building, etc.) would have no immediate impact on the size of GDP in a given year, it would have a long-term impact on economic growth, as spending on violence containment is fundamentally unproductive. In addition, there are multiple costs associated with long-term productivity loss from violence that are not included when looking only at violence containment. Estimates come from McCollister *et al.* (2010), who used a range of methods to estimate both the tangible and intangible costs attributable to violence and homicides. Specifically, their analysis used the "cost of illness" and extent of "jury compensation" to estimate the costs of crime in the United States. The jury compensation method is a more comprehensive measure because it attempts to take into account both the direct costs of violence and its associated pain and suffering. This method does not include punitive damages that US courts award in civil cases.

These estimates were therefore used as the underlying assumption for the cost of a homicide, violent assault, death from external conflict, and terrorism-related fatality or injury. Specifically, the estimated cost of a homicide was $8,888,692, while each violent assault and terrorism-related injury was $120,622. Because it was assumed that many of the costs related to deaths from conflict would be accounted for in military expenditure, only direct costs were included; that is, the cost was assumed as $1,370,449 (McCollister *et al.* 2010). These estimates may be relatively conservative but are located near the middle of estimates by similar studies (Cohen 1988, 2004; Miller *et al.* 1993, 1996;

Rajkumar & French 1997; Aos *et al.* 2001). Other costs associated with productivity loss included in the methodology include the cost of internally displaced people, the loss of life from terrorism, the lost economic activity associated with the fear of violence, and capital destruction from conflict.

The Total Cost of Violence to the Global Economy

Unfortunately, reliable data and comparable studies are available for only a handful of countries, making a detailed assessment of the total global cost of violence almost impossible. Thus, the global total is a product of a subset of the violence-related factors listed above, specifically those factors that correspond to GPI indicators. Where country-specific estimates were not available, a scaling system was used, such that the US-to–other country GDP ratio was used as the basis for assessing the unit cost of specific types of violence (for example, as Thailand's GDP is approximately 18% of US GDP, the cost of a homicide in Thailand was assumed to be 18% of the cost of a homicide in the United States). When this same scaling methodology described here is applied to the world as a whole, the estimated benefit to the global economy of perfect peacefulness is at least $9.8 trillion (over the long run, in 2013 US dollars). To put this in perspective, this is equivalent to around US$1350 per person and is twice the size of Africa's economy. Compared to estimates for 2012, this represents an increase of US$179 billion or a 3.8% rise in violence containment costs globally.

While this methodology takes into account multiple types of violence, including many that would not usually be the direct subject of a peace process, the high correlation between the different GPI indicators suggests that successful peace processes will lead to reductions in all or most forms of violence over the long run.

Virtuous Cycle of Economic Growth and Peace in the Long Run

There is a strong association between GDP per capita and peacefulness. Countries with GDP per capita below a certain threshold are much more likely to suffer from high levels of internal violence and are more susceptible to the outbreak of conflict. Therefore, a successful peace process should result in sustained, higher levels of peacefulness in the long run. However, this is often not the case in practice. Peace processes might stall, take years or even decades, or seem to succeed only for a country to lapse back into conflict. Therefore, while economic growth is a likely outcome of a sufficient peace process, it is not, in and of itself, enough to guarantee future peace and prosperity.

Existing approaches to peacebuilding and the risks of conflict relapse tend to focus on short-term shocks or economic complications that might endanger a peace process, such as lingering youth unemployment, income inequality, aid flows, and so on. Such approaches place a premium on macroeconomic policy aimed at promoting stability: sustained, equitable growth; low inflation; avoiding austerity in the short term; and so on. While any peace process should target such short-term factors, in the long term, macroeconomic stability by itself is not enough to sustain peacefulness.

The economic benefits of increases in peacefulness have already been outlined in this chapter. Some of these benefits are almost immediate as production shifts from spending on violence and violence containment and capital destruction is halted and reversed. However, some economic gains from peace processes only accrue in the long run as the economic climate improves and GDP growth is restored. Economic renewal is not only a benefit of a successful peace process; it is essential for maintaining peaceful environments in the long run.

The economic benefit associated with peacefulness makes the case for investing in peacebuilding in the long term. In an era when public debt is increasingly limiting government spending, a sharper focus on the costs and benefits of government programs is necessary. Under these conditions, programs that alleviate the need to contain violence become more economically viable over the medium term, making the case for peace even stronger, as any violence-alleviating program also has many positive spin-off effects, such as encouraging education, better health, and a more competitive business environment, which in turn help improve social cohesion and human capital. This then helps in reducing the need for policing, judiciary, and incarceration, as well as increasing labor market productivity and increasing taxation receipts. Peace seen in this light is the prerequisite for sustained economic development as well as the general well-being and flourishing of human potential.

By understanding the social and economic drivers of violence, policymakers and business leaders can better understand the costs and benefits of particular social and economic policies. Furthermore, by directing resources toward addressing the root causes of violence, society can begin to make long-term investments in the creation of a virtuous cycle of peace and economic prosperity.

References

Aos, S., Phipps, P., Barnoski, R., & Lieb, R. (2001). The comparative costs and benefits of programs to reduce crime: A review of research findings with implications for Washington State. In B. Welsh, D. Farrington, & L. Sherman (Eds.), *Costs and benefits of preventing crime*. Boulder, CO: Westview Press.

Blomberg, S., Hess, G., & Orphanides, A. (2004). The macroeconomic consequences of terrorism. *Journal of Monetary Economics, 51*(5), 1007–1032.

Brauer, J., & Caruso, R. (2011). *Peace economists and peace economics* (MPRA Paper 34927). Munich: University Library of Munich.

Cohen, M. (1988). Pain, suffering, and jury awards: A study of the cost of crime to victims. *Law and Society Review, 22*, 537–555.

Cohen, M. (2000). Measuring the costs and benefits of crime and justice. *Criminal Justice, 4*, 263–315.

Cohen, M.A., Rust, R.T., Steen, S., & Tidd, S.T. (2004). Willingness-to-pay for crime control programs. *Criminology, 42*, 89–110. doi:10.1111/j.1745-9125.2004.tb00514.x

Coleman, P. (2012). The missing piece in sustainable peace (weblog). New York: The Earth Institute, Columbia University. Retrieved from http://blogs.ei.columbia.edu/2012/11/06/the-missing-piece-in-sustainable-peace/

Collier, P., & Hoeffler, A. (1998). On economic causes of civil war. *Oxford Economic Papers, 50*(4), 563–573.

Collier, P., & Hoeffler, A. (1999). On the economic consequences of civil war. *Oxford Economic Papers, 51*(1), 168–183.

De Groot, O., Brück, T., & Bozzoli, C. (2009). *How many bucks in a bang: On the estimation of the economic costs of conflict* (Economics of Security Working Paper 21). Berlin: Economics of Security.

Galtung, J. (1985). Twenty-five years of peace research: Ten challenges and some responses. *Journal of Peace Research, 22*(2), 141–158.

Hess, G. (2003). *The economic welfare cost of conflict: An empirical assessment* (CESifo Working Paper 852). Munich: CESifo Group.

Institute for Economics and Peace (IEP). (2012a). *Economic consequences of war on the U.S. economy.* Sydney: Institute for Economics and Peace.

Institute for Economics and Peace (IEP). (2012b). *Violence containment spending in the United States.* Sydney: Institute for Economics and Peace.

Institute for Economics and Peace (IEP). (2013). *Pillars of peace.* Sydney: Institute for Economics and Peace.

Lucas, R. (1987). *Models of business cycles.* Oxford: Blackwell.

Mayhew, P. (2003). *Counting the costs of crime in Australia: Technical report (AIC Technical and Background Paper Series 4).* Canberra: Australian Institute of Criminology.

Mayhew, P., & Adkins, G. (2003). *Counting the costs of crime in Australia (Trends and Issues 247).* Canberra: Australian Institute of Criminology.

McCollister, K., French, M., & Fang, H. (2010). The cost of crime to society: New crime specific estimates for policy and program evaluation. *Drug and Alcohol Dependence, 108*(1–2), 98–109.

Miller, T.R., Cohen, M.A., & Rossman, S.B. (1993). Victim costs of violent crime and resulting injuries. *Health Affairs, 12*, 186–197.

Miller, T., Cohen, M., & Weirsema, B. (1996). *Victim costs and consequences: A new look.* Washington, DC: National Institute of Justice.

Pollin, R., & Garrett-Peltier, H. (2009). *The U.S. employment effects of military and domestic spending priorities: An updated analysis.* Amherst, MA: Political Economy Research Institute (PERI).

Rajkumar, A., & French, M. (1997). Drug abuse, crime costs, and the economic benefit of treatment. *Journal of Quantitative Criminology, 13*(3), 291–323.

Saunders, H. (2001). *Prenegotiation and circum-negotiation: Arenas of the multilevel peace process.* Washington, DC: US Institute of Peace.

Small Arms Survey. (2011). *Small Arms Survey 2011: States of security.* Cambridge: Cambridge University Press.

Solomon, R. (1997). *Building peace: Sustainable reconciliation in divided societies.* Washington, DC: US Institute of Peace Press.

Webber, A. (2010). *Literature review: Cost of crime.* Sydney: Attorney General and Justice, NSW Government. Retrieved from http://www.crimeprevention.nsw.gov.au/agdbasev7wr/_assets/cpd/m66000112/cost%20of%20crime%20literature%20review.pdf

World Bank. (2011). *World Development Report 2011: Conflict, security, and development.* Washington, DC: World Bank.

Part Four

Evolution

14

The Evolutionary Logic of Human Peaceful Behavior

Douglas P. Fry

This chapter situates war and peace within the mammalian world. One premise is that when it comes to understanding human aggression and peace, a careful consideration of the mammalian context is critically important. This chapter, therefore, explores what our mammalian and primate heritage suggests about the nature of human aggression and peacefulness. I will argue that patterns of human aggression – which include noncontact threats and bluffs as well as physical altercations – correspond in most ways with the recurring patterns observed across mammalian species, wherein risks are minimized, restraint is primary, within-species deadly conflict is the exception rather than the rule, and evolutionary costs and benefits to fitness reflect a long evolutionary legacy of adaptations for keeping aggression within bounds, thus contributing to a relatively non-injurious, even peaceful, social life. Such recurring features of limited, within-species aggressive competition can be called the *mammalian model*.

This comparative and evolutionary vantage point leads to the conclusion that whereas the aggressive behaviors of mammals – including humans – reflect an evolutionary legacy of adaptations, human war as a social institution and group-level phenomenon does not. To the contrary, evidence from different quarters suggests that the multiple origins of organized violence in humans were made possible due to a bundle of demographic, ecological, and social changes in the direction of increasing organizational complexity, beginning just prior to the agricultural revolution 10,000 years ago. We will consider the evidence that shows that war is a very recent development in the evolutionary chronology of humanity – a species-atypical behavior – and consequently that any intergroup killing would have constituted the rare exception, not the rule, when compared to the absolutely crucial peaceful exchanges of cooperation, sharing, and caring ubiquitously manifested in the earliest human societies.

The evolutionary comparative mammalian model also provides a very different perspective from that of the often-cited *chimpanzee model* with its central tenet that natural selection, exceptionally, has favored intergroup killing in chimpanzees (*Pan troglodytes*) and humans (Wrangham 1999; Wrangham & Glowacki 2012; Wilson *et al.* 2014). First, the chimpanzee model's proposal that killing is an adaptation is discordant with a large corpus of mammalian data. Second, the attempted application of the

Peace Ethology: Behavioral Processes and Systems of Peace, First Edition.
Edited by Peter Verbeek and Benjamin A. Peters.
© 2018 John Wiley & Sons Ltd. Published 2018 by John Wiley & Sons Ltd.

chimpanzee model to humans is contradicted by data on the oldest form of society, nomadic forager bands. And, third, the arguments used to apply the chimpanzee model to humans are insufficient and inadequate to demonstrate that intercommunity killing could be an evolved adaptation. Merely observing that chimpanzees and humans are capable of killing is not a sufficient basis for proposing that intergroup killing has evolved to fulfill specific evolutionary functions in these species. In this chapter, we will consider how the chimpanzee model's fixation on war and violence actually diverts attention from the predominant pattern of peaceful behavior and controlled aggression within mammalian species overall, including humans.

The New Paradigm of the 5Rs

There are indications that a new paradigm is emerging that acknowledges the importance and predominance of cooperation, restrained aggression, and peaceful behavior (Aureli & de Waal 2000; Fuentes 2004, 2013; Verbeek 2008, 2013; de Waal 2009; Sussman & Cloninger 2011; Fry 2012, 2013). The new perspective challenges a traditional violence-centric view of humanity of which the chimpanzee model, constructed around coalitional killing, is but one of many manifestations. By contrast, the new paradigm recognizes the central role of the 5Rs in human and nonhuman primate sociality: Restraint, Ritualization, Relationship, Resolution, and Reconciliation. The five Rs, with the exception of Relationship, differ from the four Rs of indigenous values proposed by Harris and Wasilewski (2004)– Relationship, Reciprocity, Responsibility, and Redistribution– although these other Rs also can be seen as important to maintaining peaceful social life. Restraint and ritualization are recurring themes in mammalian within-species competition, whereas "no-holds-barred" fighting and killing are not (Jones 2002; Fry 2006; Roscoe 2007; Fry *et al.* 2010; Fry & Szala 2013). Pan-mammalian patterns of within-species aggression – in correspondence with game-theoretic simulations (Maynard Smith & Price 1973) – reveal that escalated fighting among same-species rivals is very rare compared to the widespread practice of noncontact displays and restrained and ritualized contests.

The members of social species rely on others within the group, and consequently the development and maintenance of positive, cooperative relationships can be a matter of life or death (van Schaik & Aureli 2000; Fry 2006; Watts 2006; Bekoff & Pierce 2009; Silk *et al.* 2009; Verbeek 2013). The landmark work of de Waal (e.g., 1989, 2009) and his colleagues (e.g., Aureli & de Waal 2000) on conflict resolution shows that the members of many primate species readily reconcile following aggression. Thus, evolved capacities to resolve conflicts, reconcile following aggression, and maintain valuable relationships are not unique to humans. The new paradigm rests on the observation that the 5Rs predominate in the behavioral repertoires of social species of mammals, including human and nonhuman primates.

Turning to a peace ethology of humans (Verbeek 2008, 2013), the first points to emphasize are that *Homo sapiens* as a species is greatly reliant upon learning, has a relatively long juvenile stage of development, and thus displays marked behavioral variation across cultural contexts. Spoken language and symbolic communication are hallmarks of the human species, and consequently it is not surprising that peacefulness and aggressiveness exhibit considerable cross-cultural variability.

Ross (1993) developed measures for internal and external conflict that include physical and nonphysical elements. Ross' (1993) data can be used to illustrate the existence of a cross-cultural peacefulness-to-aggressiveness continuum. For instance, the Lepcha of the Himalayas are near the peaceful end of the continuum. "Theft is virtually unknown and the last authenticated murder took place two centuries before Gorer's fieldwork in the 1930's" (Ross 1993, p. 90). The peacefulness–aggressiveness continuum shows the wide range of societal possibilities, including the realization of very peaceful social life (Fry 2006).

Within the worldwide ethnographic range of societies, internally peaceful and nonwarring societies exist (Howell & Willis 1989; Bonta 1996, 2013; Fry 2006, 2012; Souillac & Fry 2014). This raises the practical question: how can the high levels of peace that some cultures already manage to achieve also be developed in other social contexts? Studies of internally peaceful societies and nonwarring societies suggest that the answer is multifaceted. There is no single path to peace, in other words, but there do seem to be some recurring elements. In terms of maintaining internal peacefulness, a social belief system is central that actively promotes nonviolence, strongly condemns aggression, or both (Bonta 1996; Fry 2006). The socialization of children so that they internalize nonviolent beliefs and practices as the valued and culturally accepted way to live one's life can be very effective in the perpetuation of peaceful social life (Montagu 1978; Howell & Willis 1989; Fry 2006; Endicott & Endicott 2014). Of the peaceful Semai of Malaysia – who do not war or feud, rarely murder, and avoid striking anyone, even when angry – Robarchek (1980, p. 114) observes that due to the paucity of aggressive adult models, for any child raised in such nonviolent societies "the learning of aggressive behavior by observation and imitation is almost entirely precluded."

The societies at the peaceful end of the peacefulness–aggressiveness continuum tend to have effective conflict management practices. Elders may mediate or arbitrate disputes; beliefs in eminent justice may prevent physical responses, since people believe that wrongdoers will be punished by supernatural beings; or disputes may be channeled through contests that allow for the venting of emotional steam, the restoration of honor, or the determination of a winner and loser in such a way that serious injuries are precluded (Kemp & Fry 2004; Fry 2006; Fry & Szala 2013).

It is sometimes assumed in the West that war is simply part and parcel of human nature, as reflected for instance in Barack Obama's Nobel Peace Prize acceptance speech, where he said, "War, in one form or another, appeared with the first man." The existence of nonwarring societies and peace systems shows that this assumption is not valid (Fry 2006, 2012). The nonwarring Semai, for example, are part of a larger nonwarring peace system in aboriginal Peninsular Malaysia that also includes neighboring societies such as the Batek and Chewong (Robarchek 1980; Endicott & Endicott 2014). The Iroquois also represent an intriguing case of a peace system that was formed when five previously warring neighbors – the Mohawk, Oneida, Onondaga, Cayuga, and Seneca – created a nonwarring union among themselves, although they continued to engage in warfare beyond their peace system (Dennis 1993; Fry 2012). Features that have been hypothesized to be important in the development and maintenance of peace systems are the creation of an overall social identity, connectivity among subgroups, intergroup interdependence, nonwarring value orientations, cultural symbols that promote peace, overarching institutional authority, and mechanisms for intergroup conflict resolution (Fry 2012).

In the remainder of this chapter on the evolutionary logic behind a peaceable view of human nature, first the evolutionary costs and benefits of aggression in a mammalian context will be considered. Second, the restrained nature of primate aggression, including the paucity of within-species killing within the taxonomic order to which humans belong, will be highlighted. Third, archaeological and nomadic forager data will be presented that contradict bellicose assumptions about humanity. Fourth, the chimpanzee model of intergroup killing will receive a focused methodological and data-grounded critique, for although it is based on the shallowest evidence, the chimpanzee model has received an inordinate amount of attention (e.g., Johnson 2014; NBC News 2014). The closing thought of the chapter will be to emphasize the necessity of forsaking the institution of war and drawing upon the human capacity to create conflict resolution mechanisms, legal systems, and governing institutions to deal with conflicts justly and without violence.

Contextualizing Human Behavior: Evolutionary Costs and Benefits of Aggression

Primatologist Irwin Bernstein (2008, p. 60) writes, "The potential costs of fighting are such that natural selection has favored individuals that avoid taking risks when the cost to themselves is likely to exceed the benefits of anything obtained by engaging in that interaction." Fry and Szala (2013) point out that costs include mortality; physical injury; injuring one's own relatives if they are opponents; losing friends and allies by damaging relationships; taking time and energy away from important activities such as finding food, watching for predators, and seeking mates; and being expelled from the social group as a troublemaker (Archer & Huntingford 1994, p. 10; Riechert 1998, p. 82; van Schaik & Aureli 2000; Jones 2002; Bernstein 2007, 2008). On the other side of the cost–benefit scale, benefits of aggression vary from one species to the next and from one context to another, but they include the attaining of sustenance, territory, and mates; protecting one's offspring and oneself from aggression; and maintaining social dominance, which in turn correlates with obtaining other resources (Alcock 2005). The take-home message is that although aggression can be risky, it also can be beneficial to individual fitness in certain circumstances.

Consideration of within-species competitive and aggressive behavior across the mammalian class reveals a variety of mechanisms through which individuals minimize risks. First, noncontact displays are used in place of actual fighting. Second, when fighting does occur, it tends to consist of ritualized aggression (e.g., the sparring contests of many ungulate species), wherein serious injuries and death are rare. Third, in territorial species, once boundaries have been established, threats and fights markedly decrease among neighbors (Kokko 2008). Fourth, dominance hierarchies within social groups greatly reduce fighting on a daily basis, as individuals know their place relative to the other group members (Preuschoft & van Schaik 2000). Fifth, animals practice avoidance and hence eliminate the possibility of confrontations.

Fry and Szala (2013) put forth a sequence of increasingly escalated aggressive behavior consisting of (1) avoidance of other individuals; (2) noncontact displays such as vocal calls, facial threats, and, in humans, verbal comments, harangues, threats, and so forth; (3) restrained contests that are sometimes ritualized; and (4) unrestrained, serious, or lethal fighting. The first three in the sequence dramatically reduce risks to opponents and greatly outnumber the fourth in the sequence, that is, severe fights or injurious and lethal altercations. This empirically supported model suggests that it is in the interests of both competitors to "follow the rules" of restrained fighting so as not to expend unnecessary energy or increase the risk of injury and death through severe fighting. At the point where one animal signals submission or attempts to flee, continuation of the fight merely increases the risks to both contestants and wastes energy without any gain to either individual (Bernstein 2007; Roscoe 2007). Restrained forms of within-species aggression make theoretical sense and dramatically predominate in mammalian species.

For the most part, individuals do not escalate to more serious forms of fighting. Male mule deer (*Odocoileus hemionus*) "fight furiously but harmlessly by crashing or pushing antlers against antlers, while they refrain from attacking when an opponent turns away, exposing the unprotected side of its body" (Maynard Smith & Price 1973, p. 15; see also Verbeek 2013). Researchers recorded 1308 ritualized sparring matches between pairs of male caribou (*Rangifer tarandus*) who followed the restrained rules of engagement, compared to only six escalated fights (Alcock 2005). That is a ratio of one escalated fight to every 218 ritualized contests. In a study of white-ear kobs (*Kobus kob leucotis*), serious injuries occurred only four times in 305 ritualized contests (Enquist & Leimar 1990). Similarly, among red deer stags (*Cervus elaphus*), of 107 observed ritualized contests, only two resulted in injury (Enquist & Leimar 1990).

Primate Patterns: The Prevalence of Peace

Across the order Primates, physical aggression constitutes only a minute portion of all behavior (Sussman & Garber 2007), and within-species killing is very rare, aside from the special case of infanticide in some species, for which there appears to be clear evolutionary benefits and minimal risks to the killers (Angst & Thommen 1977; Hrdy 1977; Huntingford & Turner 1987). Lethal aggression in chimpanzees (*Pan troglodytes*) has received the most attention (Wrangham & Glowacki 2012; Wilson *et al.* 2014), and isolated cases of within-species killing in capuchin (*Cebus capucinus*) and spider monkeys (*Ateles geoffroyi yucatanensis*) have also been reported (Gros-Louis *et al.* 2003; Valero *et al.* 2006).

Bekoff and Pierce (2009) point out that the aggressiveness of chimpanzees has been exaggerated by all the attention being paid to rare incidences of lethal aggression, and they quote Jane Goodall to this effect:

> It is easy to get the impression that chimpanzees are more aggressive than they really are. In actuality, peaceful interactions are far more frequent than aggressive ones; mild threatening gestures are more common than vigorous ones; threats per se occur much more often than fights; and serious wounding fights are very rare compared to brief, relatively mild ones. (Goodall 1986 quoted in Bekoff & Pierce 2009, p. 4)

Even when *aggression* is defined broadly as "mild spats, displacements, threats, stares, and fighting," a definition that catches both noncontact and contact behaviors, a solid majority of primate social interactions are affiliative and peaceful (for example, food sharing, huddling together, and grooming), with aggression accounting for only a tiny fraction of social behavior (Sussman & Garber 2007, p. 640). Across nearly 60 species of prosimians, monkeys, and apes, less than 1% of all interactions involve within-species aggression, as broadly defined here by Sussman and Garber (2007, p. 640).

Only rarely does nonhuman primate aggression result in death. Nagel and Kummer (1974) conclude, "Inflicting damage (wounds) or even killing is so rare in Old World monkeys under natural conditions that it may be regarded as an accident." During fighting, the most vulnerable parts of the body are injured less frequently, again suggesting the overall pattern of restrained aggression wherein attacking animals avoid biting vulnerable areas (Bernstein 2007). And across species of primates, restrained aggression outnumbers unrestrained aggression many times over (Fry & Szala 2013). For example, Deag (1977, table II) reported that 1740 observed aggressive interactions among wild Barbary macaques (*Macaca sylvanus*) involved only avoidance and noncontact threats; therefore, Deag concluded that contact aggression is relatively rare in this species. Of more than 15,000 aggressive events (noncontact threats and chases plus unidirectional attacks and bidirectional fights) recorded for rhesus monkeys (*Macaca mulatta*), a trifling 0.4% consisted of mutual fighting (Symons 1978, p. 166, reporting data collected by Southwick). In another study, Cooper and Bernstein (2008) found that of 1001 aggressive interactions among rhesus macaques, only six bites were inflicted, which amounts to 0.6%. The same researchers report for Assamese macaques (*Macaca assamensis*) an even lower percentage of biting, 0.4% for a pool of 262 observed aggressive encounters (Cooper & Bernstein 2008). For a variety of Old World monkey species, Nagel and Kummer (1974) summarized reported findings on noncontact versus contact aggression. The noncontact-to-contact aggression ratios paint a fairly consistent picture across primate species. The ratios differ from one species to the next, but as a generalization the data show the occurrence of much more noncontact than contact aggression and a paucity of actual wounding. Such data from nonhuman primates illustrate the overarching pattern of restrained aggression that recurs with regularity during within-species aggressive interactions in mammals.

Humans are primates. Contra the impression that might be gained from the news and entertainment media, cooperative and helping behaviors characterize most social interactions from one society to the next. As in other social primates, restraint and noncontact behavior predominate during conflict situations (Fry 2006; Fry *et al.* 2010; Fry & Szala 2013). The overwhelming majority of human behavior is nonviolent; across the cross-cultural landscape, conflicts are managed without physical aggression through withdrawal of one or both opponents; toleration of a situation; discussion of the issues; reprimands; withdrawal of support; negotiation of agreements; payment of compensation; application of mediation, arbitration, or adjudication; sulking; apologizing; forgiving; reconciling; and so forth (Fry 2006). As in other primates, aggression is restrained.

In correspondence with the within-species behavioral pattern among other mammals, human disputes rarely entail violence. This is not a denial that humans fight and kill. There is bountiful evidence that humans are capable of within-species killing. Rather, the point is that serious, injurious fighting is exceptional within mammalian species, and the same restraint in aggressive competition that is a hallmark of mammalian behavior generally also predominates in humans. Human killing is exceptional behavior. Homicides, after all, are tallied per 100,000 per annum, not per 100 per annum. Even during World War II, all the European war deaths per annum between 1939 and 1945, coupled with domestic homicides, constituted less than 1% of the population (Malby 2010). It is important to keep in mind that the overwhelming majority of human beings never kill anybody in the course of their lives. Killing, while possible, is atypical human behavior (Verbeek 2013). The relative infrequency of killing behavior across mammalian species, including in humans, argues against a claim that within-species killing is an evolved adaptation.

The Recent Origins of War

The worldwide archaeological record offers data centrally relevant to drawing inferences about the origins of war and peace. It shows that the recent origins of war in different locations at different times correspond with transitions away from the nomadic forager lifestyle. Sedentism, resource accumulation, population growth, social hierarchy, and leadership constitute significant changes away from the age-old human condition of low population density, nomadism, day-to-day foraging, egalitarianism, and non-surplus-accumulating social life (Haas 1996). Dramatic changes in human social organization and demographics began only about 12,500 years ago.

With the possible exception of one case, a site in Africa called Jebel Sahaba that dates between 12,000 and 14,000 years ago, the earliest archaeological evidence of warfare worldwide is within the last 10,000 years. Within this archaeologically recent time frame, archaeology shows transitions from warlessness to warfare occurring at different places at different times (Fry 2006). For example, on the northwest coast of North America, the Valley of Oaxaca in Mexico, the Anasazi region of the southwestern United States, and the Southern Levant and other parts of the Near East, prehistoric time sequences show shifts from peace to war over time – in other words, multiple origins of war (Haas 1996; Maschner 1997; Flannery & Marcus 2003; Fry 2006; Ferguson 2013). The worldwide archaeological evidence also shows how war became more common and destructive with the rise of the state as recently as 4000 and 6000 years ago.

For most of humanity's existence, the ancestral nomadic foraging societies simply lacked large population concentrations, permanently settled communities, stored plunderable resources, authoritative leadership, distinct social rankings, and, significantly, warfare (Fry 2006; Ferguson 2013; Fry & Söderberg 2013a, 2013b). With the development of greater social complexity, beginning about 12,500 years ago, humanity embarked on a set of monumental changes, which made the development of war possible (Haas 1996; Fry 2006; Ferguson 2013). These changes toward social complexity and the subsequent agricultural revolution provided the foundations for new forms of social institutions such as rigid social classes, slavery, and war.

Nomadic Forager Data on War and Peace

Extant nomadic foragers also provide insights about war and peace. For hundreds of societies of different social types from around the world, Murdock (1967) coded ethnographic data pertaining to subsistence, settlement, social equality and inequality, and other variables. In a study of war and peace among foragers of different types, Fry

(2006) used Murdock's (1967) codes to define a sample of foraging societies that fulfilled the operational definition of having no more than 5% subsistence dependence on agriculture and animal husbandry. Within a preexisting sample of 186 societies called the Standard Cross-Cultural Sample (SCCS), which was derived by Murdock and White (1969) to represent worldwide cultural provinces, 35 societies meet the operational definition of foragers generally. By examining other variables coded by Murdock (1967), the 35 forager societies were sorted into three subtypes.

The first are nomadic or seminomadic forager societies that lack social class distinctions and domestic animals (including horses). They are egalitarian in ethos and behavior (Fry 2006). Nomadic foragers approximate the oldest type of human social organization; until just a couple of millennia before the agricultural revolution, humankind engaged in the nomadic forager lifeway (Fry 2006, 2013). Second, complex or non-egalitarian forager societies live in settled communities and have developed social distinctions based on class or wealth. This social type is actually recent, generally appearing within the last 12,500 years as some human groups began to settle in one place and undergo transitions away from a nomadic foraging way of life. Finally, societies that rely on the horse for hunting constitute a third type of forager society of very recent origin, arising after the Spanish introduced horses into the Americas a few hundred years ago.

The 35 forager societies in the SCCS consist of 21 nomadic forager societies, nine complex forager societies, and five equestrian hunting societies. All of the complex, non-egalitarian societies and all of the equestrian foragers engage in war, whereas a majority (13/21) of the nomadic foragers in the sample do not (Fry 2006). This finding suggests that changes associated with the development of social complexity – settling down, social hierarchies, population increase, and accumulation of stored food and other items to plunder – greatly increase the likelihood of warfare over that encountered in nomadic forager social organization.

In short, a systematic sampling methodology suggests that most nomadic forager band societies do not make war. Three additional relevant points deserve mention. First, the kind of war described for nomadic forager societies tends to be less severe in comparison to accounts of war for complex, non-egalitarian foragers (Fry 2006). Second, at least some cases of war seem attributable to the arrival of Europeans, the displacement of indigenous societies, and a plethora of resulting recent disruptions (Guenther 2014). And third, among nomadic foragers, conflicts tend to be interpersonal in nature, stemming from adultery, jealousy, insults, and revenge. This observation raises the question as to whether all cases of reported "war" for nomadic foragers really do involve group-versus-group conflict as opposed to manifestations of interpersonal aggression. In fact, there are reasons to suspect that the label *war* has been too loosely and too readily applied in the nomadic forager context to disputes that are actually interpersonal, not intergroup, in nature (Fry 2006). All these factors combine to suggest that nomadic forager social organization is not particularly conducive to war.

Recently, Fry and Söderberg (2013a, 2013b) examined the details of all incidences of lethal aggression ethnographically reported in the SCCS-derived sample of 21 forager societies. Again, the SCCS and Murdock codes were used to avoid the sampling bias that can easily occur if ethnographic cases are simply self-selected from the literature (Murdock 1967; Murdock & White 1969; White 1989). Fry and Söderberg (2013a, 2013b) used only the recommended primary ethnographic sources (White 1989), which provide time/place orientation information and the oldest high-quality data about each society in the sample. Fry and Söderberg (2013a, 2013b) did not classify *a priori* lethal events such as manslaughter, homicide, feud, or war but rather examined the fundamental characteristics of the events, such as the relationships of killers to victims, reasons for the killings, whether lethal events occurred within or between groups, and so on.

The core results are as follows (Fry and Söderberg 2013a, 2013b). Three societies had no reported lethal events. For the remaining 18 societies in the sample, there were 148 lethal aggression events. One society, the Tiwi of Australia, was exceptionally violent, accounting for 47% (69/148) of all lethal events. A majority of the 148 events, 55%, involved only one perpetrator killing only one victim. Women were rarely perpetrators or co-perpetrators of lethal violence. Nearly half of the societies (10/21) had no lethal events enacted by more than one killer. Most reasons for killing were personal, involving sexual rivalries, jealousies, quarrels, and vengeance as well as in-group executions, accidents, wife killings, and revenge-based feuding. Many of the lethal events took place within the same community between brothers, father and son, mother and child, in-laws, wives and husbands, companions, friends, clan "brothers" (Tiwi only), neighbors, and so forth. Six percent of the lethal events involved husbands killing their wives. Fry and Söderberg (2013a, p. 270) conclude that among nomadic forager societies, most incidents resulting in lethal aggression "may be classified as homicides, a few others as feuds, and a minority as war."

It must be emphasized that the overwhelming majority of conflicts, disputes, and grievances in nomadic forager contexts do not lead to killings or even to physical altercations (Fry 2006). The manifestation of the 5Rs is ubiquitous in nomadic forager societies: relationships are relevant, restraint and ritualization recur with regularity, and rivals routinely resolve, reconcile, and resume regular relations.

A very common response to conflict is for one or both parties to simply walk away, sometimes called "voting with one's feet" in the ethnographic literature, an easy option given the nomadic lifestyle and fission–fusion nature of nomadic forager groups (Fry 2006). Another typical and nonviolent approach to conflict in nomadic forager society is to "talk it out." Among the Ju/'hoansi, for example, "Differences are resolved by talk. Relatives and friends may intervene in quarrels between the husbands and wives and help them to stop the quarrel" (Marshall 1976, p. 177). In nomadic forager society, third parties and sometimes an entire camp act as mediators between disputants (Fry 2006). Alternatively, disgruntled persons may vent their feelings through harangues as the camp clusters around campfires at night, as reported for the Mardu of Australia and the G/wi of Africa, for instance (Fry *et al.* 2010).

For the 21 nomadic forager societies in the SCCS sample, avoidance in response to conflict was reported for 76% of the sample (Fry & Söderberg 2013b). Third parties were noted in the primary authority ethnographies to intervene to separate or distract disputants in 57% of the societies. More active mediation efforts, wherein third parties attempted to facilitate a resolution of a conflict, were noted in 48% of the sample. Finally, contests (ritualization) were reported for 43% of the sample (Fry & Söderberg 2013b).

Wrestling was the most typical type of contest. Although contests involve physical aggression, they provide a socially acceptable way for disputants to "fight it out" with little danger of serious injury due to the self-restraint of the rivals and the presence of referees. For instance, the Ona of South America and the Ingalik of North America wrestled to resolve disputes; Canadian Slave and Dogrib bands practiced the custom of allowing rivals to wrestle over a wife (Fry & Szala 2013). As we have seen, the restraint and ritualization of aggression in many mammalian species check serious injuries among contestants; corresponding examples of restraint and injury prevention are apparent in some nomadic forager contexts as well (Fry 2006; Fry *et al.* 2010; Fry & Szala 2013).

There are several peace-related "take home" messages from nomadic forager research (Fry & Söderberg 2014). People living in this nomadic form of social organization are not inclined to practice war. Occasionally, lethal aggression may take place within and between groups, but default patterns are to get along with others and deal with conflicts through avoidance, restraint, and various nonviolent approaches such as delivering harangues or letting mediators step in. Peaceful behaviors are evident in the daily sharing of food, cooperative childrearing, and the handling of most conflicts without violence. The data across nomadic forager ethnographies show that the 5Rs predominate in nomadic forager social life. Furthermore, when conflicts occur, they tend to be between individuals over personal matters. Violence, while possible, does not predominate, and intergroup killing is rare. One implication, which we will next consider, is that there is an immense incongruity between the nomadic forager data and attempts to overlay the chimpanzee model onto nomadic forager societies.

The Chimpanzee Model

Prompted by field observations that the members of some chimpanzee groups attack and kill members of neighboring communities, primatologist Richard Wrangham put forth a series of propositions about human aggression that has become known as the *chimpanzee model* (Wrangham 1999; Wrangham & Peterson 1996; Wrangham & Glowacki 2012). Wrangham (1999, p. 22) proposed the existence of psychological mechanisms – a lethal raiding psychology – that underpin coalitionary lethal aggression in chimpanzees and humans, such as "the experience of a victory thrill, an enjoyment of the chase, a tendency for dehumanization … ready coalition formation, and sophisticated assessment of power differentials." The chimpanzee model proposes that "both chimpanzees and humans evolved a tendency to kill members of other groups in safe contexts" (Wrangham & Glowacki 2012, p. 7). The use of the word *tendency* in the foregoing quotation is very important, for it asserts the existence of evolved, adaptive propensities toward within-species killing of members of a rival political group when the risks are low. "Selection has accordingly favored male tendencies to search for and take advantage of safe circumstances to cooperate in killing members of neighboring rival groups" (Wrangham & Glowacki 2012, p. 6).

A recent article in the journal *Nature* (Wilson *et al.* 2014) proposes again that killing within chimpanzees reflects an evolved adaptation derived from natural selection. Although the authors refrain from mentioning humans in the article itself, McCoy (2014), reporting in the *Washington Post*, summarizes and quotes the lead author Michael Wilson: "The research feeds into a lengthy debate over the nature of chimp violence, and what it means for humanity's own propensity for murder. 'We're trying to make inferences about human evolution,' [said] lead researcher Michael L. Wilson." The article contains no data on human killing, yet in the media, Wrangham as well as Wilson suggested that observations of chimpanzee lethality shed light on the evolution of human warfare and killing (see also Johnson 2014; McCoy 2014).

However, the nomadic forager findings on war and peace, which are based on a systematically derived sample of 21 nomadic forager societies (Fry 2006; Fry & Söderberg 2013a, 2013b), contradict the essential features of the chimpanzee model. First, the fact that over half of all lethal events involved only one killer and one victim shows that those events had nothing to do with hypothesized coalitions of cooperating males (Fry & Söderberg 2013a). The findings show that the majority of lethal events in nomadic forager society are one-on-one killings, not coalitionary deeds. This point is also reflected in the distribution of the single-author and single-victim events across the sample; 10 out of 21 societies in the sample lacked even a single event that included more than one killer. A one-on-one killing is not a war. Neither does a one-on-one killing meet the basic coalitionary tenet of the chimpanzee model.

Second, many killings take place within the same society, and this fact is discordant with the chimpanzee model's focus on killings being directed at members of neighboring communities as "independent political units" (Wrangham & Glowacki 2012, p. 8). The fission–fusion band dynamics and shifting membership of local bands are well established in the ethnography of nomadic foragers, meaning that bands are interlinked within the same society (Marlowe 2010). Even among the comparatively violent Tiwi, the clans are interlinked through kinship ties, and hence they are not independent political groups (Hart & Pilling 1979).

Third, the simultaneous occurrence of two basic tenets of the chimpanzee model is rare. Fry and Söderberg (2014) tallied the exact number of lethal attacks that simultaneously were directed at different political units (as defined by Wrangham & Glowacki 2012) and involved multiple attackers (in other words, possible coalitions). Only 8.8% of the 148 lethal attacks simultaneously involved more than one perpetrator and were directed at members of a rival political unit (Fry & Söderberg, 2013b, table S3; 2014).

Fourth, the reported reasons for killing do not jive with the chimpanzee model (Fry & Söderberg 2013a, 2013b). By definition, a *tendency* is an inclination, a predisposal, a propensity. If the psychological tendency existed to attack rival groups whenever the risks are low, then evidence of such practices should be reflected with some regularly in lethal aggression scenarios. However, the material in the primary ethnographic sources describes jealousy, competition, adultery, insults, revenge, punishment, accidents, resource defense, family feuds, cattle rustling, and other reasons that are far afield from the chimpanzee model's assertion that, due to evolved tendencies, rival communities simply will be attacked whenever risks are low.

Finally, Wrangham and Glowacki (2012) make reference to "21 nomadic hunter-gatherer societies listed by Fry (2007) as being peaceful." But, as odd as it might seem, the Wrangham and Glowacki (2012) list of 21 societies and the SCCS list of 21 nomadic foragers used by Fry (2006, 2007; Fry & Söderberg 2013a, 2013b, 2014) are not the same. Only six societies are on both lists. Furthermore, some of the societies on the Wrangham and Glowacki (2012) list do not meet the operational definition of a nomadic forager society (Fry 2006; Fry & Söderberg 2013b). Wrangham and Glowacki (2012) maintain that nomadic foragers engage in a lot of warfare, but by developing their own separate list of 21 societies, they simply sidestep and ignore the findings that the majority of the nomadic foragers in the SCCS sample do not engage in war (Fry 2006, 2007).

Obviously, Wrangham and Glowacki (2012) and Fry and Söderberg (2013a, 2013b, 2014) reach very different conclusions about the peacefulness or aggressiveness of nomadic foragers. How can this be explained? First, the possibility of sampling bias merits consideration. Wrangham and Glowacki (2012) self-select six forager examples, whereas Fry and Söderberg (2013a, 2013b) do not self-select cases. Fry and Söderberg (2013a, 2013b) use preexisting codes to define the sample of 21 nomadic foragers as a subsample of the SCCS. Self-selecting ethnographic examples is no methodological match for the systematic sampling of cases.

A second point involves the quality of sources used. In some cases, Wrangham and Glowacki (2012) use secondary sources in place of primary ethnographies. Sampling bias regarding the self-selection of sources presents another concern. For three of their six cultural examples (western Alaska, the Andaman Islands, and Tierra del Fuego), Wrangham and Glowacki (2012) rely on only one self-chosen source per area, even when bountiful ethnographic material exists. To avoid secondary literature and sampling bias issues, Fry and Söderberg (2013a, 2013b) use only the SCCS-recommended best quality primary literature (primary authority sources) published in the SCCS bibliography by White (1989).

A third and specific sampling concern pertains to one of the six cases that Wrangham and Glowacki (2012) simply label "Canada/Great Lakes" (although they mention that parts of the United States are included as well). It is unclear which societies, from which locations, and at which time periods are under consideration for this immense geographical area. Wrangham and Glowacki (2012, p. 12) provide a set of decontextualized quotations on hostility and violence that mention the Chippewa, Sioux, Eskimos, Cree, Iroquois, and unspecified societies. Although Wrangham and Glowacki (2012) propose focusing only on nomadic forager societies, the equestrian Sioux and horticultural Iroquois are included.

A fourth point has interwoven epistemological and methodological elements. This concern focuses on Wrangham and Glowacki's (2012) implicit assumption that merely listing examples of hostility and violence between nomadic forager communities could be used to support the functional interpretation that natural selection has favored "a tendency to kill members of neighboring groups when killing could be carried out safely" (Wrangham & Glowacki 2012, p. 5). To have credence, such a proposal about evolved human tendencies requires more than self-chosen illustrations of hostility and violence. As Williams (1966, p. 4) cautions, "Adaptation is a special and onerous concept that should be used only where it is really necessary." Adaptations must be demonstrated on the basis of evolutionary theory and logic as well as design and function, not by merely cherry-picking examples that seem to fit one's speculations.

Because within-species lethal contests are exceedingly rare among mammals, an important implication is that any claim that escalated, unrestrained fighting is species-typical in humans – resulting from an evolved adaptation – must be strongly justified, rather than simply assumed or asserted, because such a claim flies in the face of a well-documented mammalian pattern of restrained aggression wherein killing within the species has been selected against. The logical default proposition would be that human aggression fits within the typical mammalian framework of limited and controlled aggression, rather than constitutes a reversal of selection pressures to favor war. Humans, after all, are mammals, so in an evolutionary context it makes sense to begin with the presumption that our aggression is typically mammalian – until convincingly demonstrated otherwise. For many reasons, the chimpanzee model fails to demonstrate otherwise.

The Prevalence of Peace

There is bountiful evidence from numerous disciplines that the preponderance of human behavior is oriented toward getting along with others without violence, showing restraint against lethal aggression, cooperating toward shared goals, feeling empathy toward others, and resolving disputes peaceably, often with reconciliation and forgiveness as part of the process. In proposing that in behavioral science we are seeing a paradigm shift toward a greater appreciation of cooperation, peacemaking, empathy, and restraint, the point is not to deny the obvious human capacity to engage in war and acts of violence, but rather to balance the traditional overemphasis on competition and violence with a brighter view of human nature that is consistent with the evidence from anthropology to zoology (Fry 2006, 2012; de Waal 2009; Ferguson 2011; Hughbank & Grossman 2013; Verbeek 2013).

This evidence-based perspective that acknowledges cooperation as well as competition, sharing as well as selfishness, empathy as well as envy, and beneficence as well as bellicosity has implications for security in today's geopolitical world. The new perspective highlights the evolved capacities for humans and other animals to get along with each other, work together peacefully, and resolve their disputes and differences without bloodshed. Such behavioral capacities have been important over the course of human evolution and are necessary in the twenty-first century. Competition does play a central role in Darwinian evolutionary thinking, and rightly so, but the new perspective shows that sometimes the best competitive strategy in a fitness-enhancing sense is actually to be cooperative in a behavioral sense. Similarly, aggression is dangerous, and so evolution clearly has selected for the judicial use of aggression rather than unbridled fury.

One's views of humanity affect one's views about how best to seek security. Perceptions that humans are naturally competitive and aggressive can facilitate fear of others, distrust, and reluctance to cooperate. Such perceptions also can lead to pessimism about ending the institution of war or preventing particular wars. If human nature were nasty and aggressive, it would follow that there might be only a slim chance of achieving a more peaceful and secure world. By contrast, perceptions of humans as cooperative as well as competitive, and peaceful as well as warlike, open the human mind to conceptualizing and implementing a different type of security strategy. When the peaceable tendencies of humans and other social species are recognized, then a vision of abolishing war and the handling of disputes justly and without violence becomes possible. The new view offers hope that human nature does not stand in the way of peace. This is an important real-world implication stemming from the growing corpus of evidence on mammalian restraint, primate peacefulness, the capacity to reconcile, peaceful societies, nonwarring peace systems, and the cross-cultural diversity of nonviolent conflict resolution mechanisms and social institutions; findings from psychiatry and military sciences that killing in combat does not come naturally and often leaves lifelong psychic scars; and, last but not least, the philosophy of science observation that the traditional paradigm, although still espoused in some quarters, is being increasingly challenged and critiqued as culturally biased, lacking in empirical support, and contradicted by a wealth of evidence from numerous disciplines.

References

Alcock, J. (2005). *Animal behavior*. Sunderland, MA: Sinauer.

Angst, W., & Thommen, D. (1977). New data and discussion of infant killing in Old World monkeys and apes. *Folia Primatologia, 27*, 198–129.

Archer, J., & Huntingford, F. (1994). Game theory models and escalation of animal fighting. In M. Potegal & J. Knutson (Eds.), *The dynamics of aggression* (pp. 3–31). Hillsdale, NJ: Lawrence Erlbaum.

Aureli, F., & de Waal, F. (Eds.). (2000). *Natural conflict resolution*. Berkeley, CA: University of California Press.

Bekoff, M., & Pierce, J. (2009). *Wild justice: The moral lives of animals*. Chicago: University of Chicago Press.

Bernstein, I. (2007). Social mechanisms in the control of primate aggression. In C. Campbell, A. Fuentes, K. MacKinnon, *et al.* (Eds.), *Primates in perspective* (pp. 562–571). New York: Oxford University Press.

Bernstein, I. (2008). Animal behavioral studies: Primates. In L. Kurtz (Ed.), *Encyclopedia of violence, peace, and conflict* (Vol. 1, pp. 56–63). New York: Elsevier/Academic Press.

Bonta, B. (1996). Conflict resolution among peaceful societies: The culture of peacefulness. *Journal of Peace Research, 33*, 403–420.

Bonta, B. (2013). Peaceful societies prohibit violence. *Journal of Aggression, Conflict, and Peace Research, 5*, 117–129.

Cooper, M., & Bernstein, I. (2008). Evaluating dominance styles in Assamese and rhesus macaques. *International Journal of Primatology, 29*, 225–243.

Deag, J. (1977). Aggression and submission in monkey societies. *Animal Behaviour, 25*, 465–474.

Dennis, M. (1993). *Cultivating a landscape of peace*. Ithaca, NY: Cornell University Press.

de Waal, F. (1989). *Peacemaking among primates*. Cambridge, MA: Harvard University Press.

de Waal, F. (2009). *The age of empathy*. New York: Harmony.

Endicott, K., & Endicott, K. (2014). Batek childrearing and morality. In D. Narvaez, K. Valentino, A. Fuentes, *et al.* (Eds.), *Ancestral landscapes in human evolution* (pp. 108–125). New York: Oxford University Press.

Enquist, M., & Leimar, O. (1990). The evolution of fatal fighting. *Animal Behaviour, 39*, 1–9.

Ferguson, R.B. (2011). Born to live: Challenging killer myths. In R. Sussman & C. Cloninger (Eds.), *Origins of altruism and cooperation* (pp. 249–270). New York: Springer

Ferguson, R.B. (2013). The prehistory of war and peace in Europe and the Near East. In D. Fry (Ed.), *War, peace, and human nature* (pp. 108–125). New York: Oxford University Press.

Flannery, K., & Marcus, J. (2003). The origin of war: New 14C dates from ancient Mexico. *Proceedings of the National Academy of Sciences, 100*(11), 801–805.

Fry, D. (2006). *The human potential for peace*. New York: Oxford University Press.

Fry, D. (2007). *Beyond war*. New York: Oxford University Press.

Fry, D. (2012). Life without war. *Science, 336*, 879–884.

Fry, D. (2013). Cooperation for survival: Creating a global peace system. In D. Fry (Ed.), *War, peace, and human nature* (pp. 543–558). New York: Oxford University Press.

Fry, D., Schober, G., & Björkqvist, K. (2010). Nonkilling as an evolutionary adaptation. In J. Evans Pim (Ed.), *Nonkilling societies* (pp. 101–128). Honolulu, HI: Center for Global Nonkilling.

Fry, D., & Söderberg, P. (2013a). Lethal aggression in mobile forager bands and implications for the origins of war. *Science, 341*, 270–273.

Fry, D., & Söderberg, P. (2013b). Supplementary material for lethal aggression in mobile forager bands and implications for the origins of war. Retrieved from http://www.sciencemag.org/content/341/6143/270/suppl/DC1

Fry, D., & Söderberg, P. (2014). Myths about hunter-gatherers redux: Nomadic forager war and peace. *Journal of Aggression, Conflict and Peace Research, 6*, 255–266.

Fry, D., & Szala, A. (2013). The evolution of agonism: The triumph of restraint in nonhuman and human primates. In D. Fry (Ed.), *War, peace, and human nature* (pp. 451–474). New York: Oxford University Press.

Gros-Louis, J., Perry, S., & Manson, J. (2003). Violent coalitionary attacks and intraspecies killing in wild white-faced capuchin monkeys (*Cebus Capucinus*). *Primates, 44*, 341–346.

Guenther, M. (2014). War and peace among Kalahari San. *Journal of Aggression, Conflict and Peace Research, 6*, 229–239.

Haas, J. (1996). War. In D. Levinson & M. Ember (Eds.), *Encyclopedia of cultural anthropology* (Vol. 4, pp. 1357–1361). New York: Henry Holt & Company.

Harris, L., & Wasilewski, J. (2004). Indigeneity, an alternative worldview: Four Rs (relationship, responsibility, reciprocity, redistribution) vs. two P's (power and profit): Sharing the journey towards conscious evolution. *Systems Research and Behavioral Science, 21*, 489–503.

Hart, C., & Pilling, A. (1979). *The Tiwi of North Australia*. New York, Holt, Rinehart, & Winston.

Howell, S., & Willis, R. (1989). *Societies at peace*. London: Routledge.

Hrdy, S. (1977). *The langurs of Abu*. Cambridge, MA: Harvard University Press.

Hughbank, R., & Grossman, D. (2013). The challenge of getting men to killing: A view from military science. In D. Fry (Ed.), *War, peace, and human nature* (pp. 495–513). New York: Oxford University Press.

Huntingford, F., & Turner, A. (1987). *Animal conflict*. London: Chapman and Hall.

Johnson, C. (2014, September 17). Is war innate or a modern invention? *Boston Globe*. Retrieved from http://www.bostonglobe.com/news/science/2014/09/17/war-innate-chimpanzees-lethal-attacks-may-provide-advantage-eliminating-rivals/V3lwx38s1RId3ZACQmnuiJ/story.html

Jones, C. (2002). Negative reinforcement in primate societies related to aggressive restraint. *Folia Primatologica, 73*, 140–143.

Kemp, G., & Fry, D. (2004). *Keeping the peace*. New York: Routledge.

Kokko, H. (2008). Animal behavioral studies: non-primates. In L. Kurtz (Ed.), *Encyclopedia of violence, peace, and conflict* (Vol. 1, pp. 47–56). New York: Elsevier/Academic Press.

Malby, S. (2010). Homicide. In S. Harrendorf, M. Heiskanen, & S. Malby (Eds.), *International statistics on crime and justice* (HEUNI Publication Series No. 64, pp. 7–19). Helsinki: European Institute for Crime Prevention and Control.

Marlowe, F. (2010). *The Hadza: Hunter-gatherers of Tanzania*. Berkeley, CA: University of California Press.

Marshall, L. (1976). *The !Kung of Nyae Nyae*. Cambridge, MA: Harvard University Press.

Maschner, H. (1997). The evolution of Northwest Coast warfare. In D. Martin & D. Frayer (Eds.), *Troubled times: Violence and warfare in the past* (pp. 267–302). Amsterdam: Gordon and Breach.

Maynard Smith, J., & Price, G. (1973). The logic of animal conflict. *Nature, 246,* 15–18.

McCoy, T. (2014, September 18). Chimpanzees are natural born killers, study says, and they prefer mob violence. *Washington Post.* Retrieved from http://www.washingtonpost.com/news/morning-mix/wp/2014/09/18/chimpanzees-are-natural-born-killers-study-says-and-they-prefer-mob-violence/

Montagu, A. (Ed.). (1978). *Learning non-aggression: The experience of non-literate societies.* Oxford: Oxford University Press.

Murdock, G. (1967). Ethnographic atlas: A summary. *Ethnology, 6,* 109–236.

Murdock, G., & White, D. (1969). Standard cross-cultural sample. *Ethnology, 8,* 329–369.

Nagel, U., & Kummer, H. (1974). Variation in Cercopithecoid aggressive behavior. In R. Holloway (Ed.), *Primate aggression, territoriality, and xenophobia: A comparative perspective* (pp. 159–184). New York, Academic Press.

NBC News. (2014). Chimps are naturally violent, study suggests. Retrieved from http://www.nbcnews.com/science/science-news/chimps-are-naturally-violent-study-suggests-n205651

Preuschoft, S., & van Schaik, C. (2000). Dominance and communication: Conflict management in various social settings. In F. Aureli & F. de Waal (Eds.), *Natural conflict resolution* (pp. 77–105). Berkeley, CA: University of California Press.

Riechert, S. (1998). Game theory and animal contests. In L. Dugatkin & H. Reeve (Eds.), *Game theory and animal behavior* (pp. 64–93). New York: Oxford University Press.

Robarchek, C. (1980). The image of nonviolence: World view of the Semai Senoi. *Federated Museums Journal (Malaysia), 25,* 103–117.

Roscoe, P. (2007). Intelligence, coalitional killing, and the antecedents of war. *American Anthropologist, 109,* 485–495.

Ross, M. (1993). *The culture of conflict.* New Haven, CT: Yale University Press.

Silk, J., Beehner, J., Bergman, T., Crockford, C., Engh, A., Moscovice, L., & Cheney, D. (2009). The benefits of social capital: Close social bonds among female baboons enhance offspring survival. *Proceedings of the Royal Society B, 276,* 3099–3104.

Souillac, G., & Fry, D. (2014). Indigenous lesions for conflict resolution. In P. Coleman, M. Deutsch, & E. Marcus (Eds.), *The handbook of conflict resolution* (pp. 604–622). San Francisco: Jossey Bass.

Sussman, R., & Cloninger, C. (2011). *Origins of altruism and cooperation.* New York: Springer.

Sussman, R., & Garber, P. (2007). Cooperation and competition in primate social interaction. In C. Campbell, A. Fuentes, K. MacKinnon, M. Panger, & S. Bearder (Eds.), *Primates in perspective* (pp. 636–651). New York: Oxford University Press.

Symons, D. (1978). *Play and aggression: A study of rhesus monkeys.* New York: Columbia University Press.

Valero, A., Schaffner, C., Vick, L., Aureli, F., & Ramos-Fernandez, G. (2006). Intragroup lethal aggression in wild spider monkeys. *American Journal of Primatology, 68,* 732–737.

van Schaik, C., & Aureli, F. (2000). The natural history of valuable relationships in primates. In F. Aureli & F. de Waal (Eds.), *Natural conflict resolution* (pp. 307–333). Berkeley, CA: University of California Press.

Verbeek, P. (2008). Peace ethology. *Behaviour, 145,* 1497–1524.

Verbeek, P. (2013). An ethological perspective on war and peace. In D. Fry (Ed.), *War, peace, and human nature* (pp. 54–77). New York: Oxford University Press.

Watts, D. (2006). Conflict resolution in chimpanzees and the valuable-relationships hypothesis. *International Journal of Primatology, 27*(5), 1337–1364.

White, D. (1989). Focused ethnographic bibliography: Standard cross-cultural sample. *Behavior Science Research, 23*, 1–145.

Williams, G. (1966). *Adaptation and natural selection: A critique of some current evolutionary thought.* Princeton, NJ: Princeton University Press.

Wilson, M., et al. (2014). Lethal aggression in *Pan* is better explained by adaptive strategies than human impacts. *Nature, 513*, 414–417.

Wrangham, R. (1999). Evolution of coalitionary killing. *Yearbook of Physical Anthropology, 42*, 1–30.

Wrangham, R., & Glowacki, L. (2012). Intergroup aggression in chimpanzees and war in nomadic hunter-gatherers. *Human Nature, 23*, 5–29.

Wrangham, R., & Peterson, D. (1996). *Demonic males: Apes and the origin of human violence.* Boston: Houghton Mifflin.

15

Trans-Species Peacemaking: Our Evolutionary Heritage

Harry Kunneman

This chapter focuses on the evolutionary heritage we could turn to in view of developing more peaceful relations with other species. In view of our indifference to, exploitation of, and destructiveness toward many other life forms, I argue that a deeper understanding of evolutionary resources for trans-species peacemaking could be of great importance for the future evolution of life, both human and nonhuman (Verbeek 2013). I designate these resources as our evolutionary trans-species peace heritage, or ETPH for short. In this chapter, I outline a transdisciplinary approach, aiming to clarify the content and the moral importance of our ETPH. I distinguish three evolved social patterns and analyze their significance for our trans-species peace heritage. In conclusion, I argue that we should concentrate on the elements of this heritage that we share with many other life forms in order to develop more peaceful trans-species social relations.

As a backdrop for my main argument, I start with a short reflection on the disciplinary prejudices that, in my experience, stand in the way of a transdisciplinary approach to our ETPH, followed by a short indication of the minimal conceptual and empirical requirements to be met by such an approach.

The Weight of Disciplinary Prejudices

The problem of peaceful cooperation studied by ethologists also manifests itself in communication and cooperation between academic disciplines. This goes especially for the relation between the sciences and humanities. For the development of peace studies, however, communication and cooperation on the basis of equality and mutual respect are necessary. In the domain of peace studies, we cannot separate behavioral data, ethological and evolutionary theories, and (explicit or implicit) moral values. The search for a deeper scientific understanding of peacemaking and its evolution is not just an interdisciplinary intellectual endeavor but also relevant to contemporary civic and professional efforts to develop more peaceful and sustainable relations with other species. Such practical efforts aim to change the *behavior* of present-day human beings in order to influence the *future* evolution of life on our planet. This practical influence

Peace Ethology: Behavioral Processes and Systems of Peace, First Edition.
Edited by Peter Verbeek and Benjamin A. Peters.

concerns the way the human species acts as an environmental agent that causally influences the adaptation, modification, and extinction of other species. But it also involves humans' self-perceptions and moral self-evaluations, for instance contemporary appeals to regulate or even ban specific forms of genetic modification on the basis of moral and ecological values. Whether such appeals gain widespread support or not may determine the future genetic properties of a whole range of "useful" species, in particular economically profitable plants and animals. In both cases, culturally transmitted self-perceptions and moral self-evaluations act as causal forces on the evolution of life. The same applies with regard to our scientifically and culturally mediated understanding of our evolutionary peace heritage. Will we consider it natural and unavoidable to compete against and exploit other species, or will we consider our natural peace heritage as grounded also in the evolution of life? With regard to this last point, debates in the sciences and humanities do not seem to offer much ground for optimism. Coming from the humanities and from critical social science, I have been wrestling during the last decade to acquaint myself, as a committed amateur, with central insights from general biology, ethology, and evolutionary theory. This intellectual journey has led me to the conviction that the humanities have to overcome anthropocentrism and antinaturalism in order to find more adequate answers to pressing moral and ecological questions. But the same goes for disciplinary prejudices on the other side of the fence. Coming from a humanistic background, popular metaphors such as "the selfish gene" or Wilson's contention that the whole of evolution can be understood in terms of *competition*, both between individuals and between groups (Dawkins 1976, 1986; Wilson 2000, 2014), really worry me. Although these views are under serious critique within mainstream biology, such authors lend an influential and authoritative voice to public debate concerning the primacy of competition and conflict in the evolution of life.

Wilson's *The Meaning of Human Existence* (2014) provides a case in point. In his view, the development of our social capacities under the pressure of group selection provides the ultimate explanation of the evolutionary success of the human species:

> The origin of the human condition is best explained by the natural selection for social interaction – the inherited propensities to communicate, recognize, evaluate, bond, cooperate, compete, and from all these the deep warm pleasure of belonging to your own special group. (Wilson 2014, p. 75)

In view of our tremendous success in "outcompeting" other life forms on our planet, this competitive framework poses a great problem. Deep misgivings about our singular destructiveness seem to be more warranted than the "deep, warm pleasure" of belonging to such a successful species. In fairness, Wilson has been – and continues to be – a leading advocate for the continuation of biodiversity. But his deep worries with regard to the destructive consequences of the competitive success of the human species can have no *conceptual* place in his scientific framework given his emphasis on between-group competition as a driving force in the evolution of life. The possibility that altruism, or more precisely peacemaking, *between* groups and between species could also be a constitutive force in the evolution of life is precluded by his Neo-Darwinian framework.

Comparable disciplinary prejudices originate from the humanities. Most philosophers and ethicists of our time disregard evolution and the manifold ways it continues to shape human behavior. This humanistic antinaturalism presupposes that humankind is "elevated" above the rest of nature, especially in its cultural achievements: language, art, philosophy, and, of course, ethics and morality. At the level of knowledge, this implies that the natural sciences (although indispensable as a source of control over natural forces) are far less important than the humanities, understood as the true guardians of our humanity, in particular our moral capabilities. At the level of moral values, this humanistic prejudice implies that human suffering and flourishing are of a completely different order from the suffering and flourishing of other life forms: the ideas of "humanity" and "humanization" specify a moral horizon for future development that only applies to animals with a human body (Rawls 1971; Ricoeur 1992).

The work of Jürgen Habermas, one of the most famous humanists of our times, provides a good example. Just like the majority of humanistic scholars, Habermas presupposes a deep split between human culture and the rest of nature. Our evolutionary heritage is considered as a kind of instinctual material that loses most of its relevance with the advent of human culture, in particular language. According to Habermas, "communicative action" aimed at bringing about shared understanding is a uniquely human capability. If need be, human actors can offer *good reasons* for their claims and actions. Thus, they can create consensual forms of social interaction, based on the distinction between externally imposed and voluntarily agreed-upon choices:

> [A] communicatively achieved agreement has a rational basis; it cannot be imposed by either party, whether instrumentally through intervention in the situation directly or strategically through influencing the decisions of opponents … what comes to pass manifestly through outside influence … cannot count subjectively as agreement. Agreement rests on common convictions. (Habermas 1987, p. 287)

What Habermas considers a *unique* human capability, anchored in the communicative deep-structure of human language, is seen by ethologists as a capacity we share with many other animals. Consider Bekoff on the behavior of apes wanting to play:

> When gorillas or chimpanzees want to play, they stroll up to each other with a loose, bouncy walk, called gamboling. Their shoulders and head sway from side to side. One may hit the other lightly on the shoulder and then run away as though saying, "Chase me." Their playing is like contact sports – they wrestle and roll on the ground. They "mouth" each other – gripping one another with their mouths, like dogs do. They even use a play face that resembles smiling. They might pant softly … to show that they want to play, not fight. (Bekoff 2008, pp. 14–15)

It is clear from this description that the apes reach a "communicatively created shared understanding," to use Habermas' words. It follows that reaching such an understanding is not a uniquely human capability.

Disciplinary prejudices such as these clearly stand in the way of a more adequate understanding of our evolutionary peace heritage. In this chapter, I will outline a transdisciplinary approach to our ETPH. I use the notion of transdisciplinarity here in the specific sense of an approach that, just like the Trans-Siberia Express, travels through and beyond different disciplines, taking aboard findings and insights from each "country" on the way to its destination. This approach does not strive for some overarching synthesis, but explores one *possible* route connecting the different countries, leaving open the possibility – and underlining the desirability – of alternative routes that *together* would provide a network of meaningful connections between the different disciplines.

The transdisciplinary approach to our ETPH outlined in this chapter thus aims to connect on *equal* footing:

- General biological theories,
- Behavioral data, and
- Cultural traditions and moral perspectives with regard to the present state and future evolution of life on our planet.

These aims also specify the structure of my argument. I start with a coarse-grained description of recent developments in biology that provide important building blocks for such a transdisciplinary approach. Putting these together, I then introduce three different social patterns in the evolution of life and analyze their significance for our trans-species peace heritage. Finally, I argue that we share this peacemaking heritage with many other life forms but do not sufficiently appreciate its importance for the future evolution of life on our planet.

Building Blocks for a Transdisciplinary Approach of Our ETPH

In this section, I focus on four shifts and the corresponding building blocks for a trans-disciplinary approach to our ETPH:

1) A shift from a mechanistic perspective to an agency-centered perspective
2) A shift from an aggression-centered perspective to a perspective combining aggression and peacemaking as equally constitutive for intra- and trans-species relations
3) A shift from a gene-centered evolutionary perspective to a perspective connecting development at the level of genotypes and phenotypes
4) A shift from an individualizing to a social and relational evolutionary perspective.

These four shifts create conceptual space within a *naturalistic* framework for the causal significance of agency, intelligence, morality, and emergence on the levels of both behavior and evolutionary development.

From a Mechanistic to an Agency-Centered Perspective

The first building block provides a foundation for the others, as it creates conceptual space for the experiences, initiatives, and intelligence of life forms as a causal force in the evolution of life. The core idea underlying this development has been articulated in different forms. Important authors include Maturana and Varela (1980, 1992), Rosen (1991, 2000), Ulanowicz (1997, 2009), and Kauffman (1995, 2000, 2008). For reasons of conceptual "economy," I use Maturana and Varela's notion of "autopoiesis" as an overarching concept of a shared, core idea: the central importance of agency for understanding the nature and evolution of life. I remark in passing that the notion of self-organization would have been another likely candidate as overarching concept. I prefer the notion of autopoiesis, however, because in the notion of self-organization the idea of organization tends to overshadow the idea of self, whereas the notion of autopoiesis foregrounds organization as the *result* of auto-poiesis, or self-production.

Maturana and Varela's main thesis hinges on the idea that living beings cannot be adequately understood with mechanistic models. A machine can be taken apart and re-assembled using a set of instructions, or rules of composition, that exist prior to the machine itself. Although machine-based models are applicable in part to living beings, they also miss out on a decisive point: the fact that living beings are not assembled into functioning wholes by an outside actor or force, but follow a specific form of "self-assembling." As Maturana and Varela say: "living beings are characterized in that, literally, they are continually self-producing" (Maturana & Varela 1992, p. 12). A certain measure of autonomy is thus the most important defining characteristic of life, from the first cells up to primates, dolphins, and elephants. They use the word autonomy "in its current sense":

> [A] system is autonomous if it can specify its own laws, what is proper to it…. Thus if a cell interacts with molecule X and incorporates it in its processes, what takes place as a result of this interaction is determined not by the properties of molecule X but by the way in which that molecule is 'seen' or taken by the cell as it incorporates the molecule in its autopoietic dynamics. (Maturana & Varela 1992, pp. 48, 51–52)

This has far-reaching implications. To start with, it connects life with agency. With the emergence of the first cells, understood as autopoietic entities, agency comes into play as a new natural force, generating new patterns of interaction. The physical and chemical properties of cells anchor these patterns but also differ from them in decisive ways: they result from the self-production of living beings, beings that produce themselves as *autos*. That is, each is an entity that has a certain measure of autonomy, that is capable of specifying up to a point "its own laws, what is proper to it," as Maturana and Varela famously say.

This view on "the autopoiesis of the living" also points to the emergence of a new, social, and relational problem that is co-extensive with the emergence of autopoietic entities. This problem is connected with the environmental *dependence* that forms the flipside of the autopoiesis of living beings, in particular their relations with other autopoietic life forms and the ways they "incorporate" them in their autopoietic dynamic. As Morin says, "Autonomy … involves a profound energetic, informational and organizational dependence with respect to the outer world" (Morin 2008, p. 70). This energetic dependency stems from the fact that autopoietic life forms need a constant influx of energy to produce and metabolically maintain the difference between their inner organizational complexity and their environment. They embody "neg-entropic" or "dissipative" systems that can only exist "far from equilibrium" (Prigogine 1980; Morin 2008). This implies that every life form, as a local concentration of energy, always also embodies an attractive source of energy for others, in the form of either its waste products or its body. The same goes for the specific informational and organizational *capabilities* they have developed as autopoietic beings. These capabilities can contribute *as such* to the autopoiesis of other life forms.

Rosen's notion of *anticipation* as a core feature of life itself helps to flesh out the core idea of autopoeisis (Rosen 1985, p. 7). Rosen designates living beings as anticipatory systems, that is to say: systems using (in many cases hardwired) *models* of future states of their environment as a basis for specific actions in the present. He gives the example of deciduous trees dropping their leaves in the fall before temperatures have substantially lowered:

> What is the cue for such behavior? It so happens that the cue is not the ambient temperature, but rather is day length. In other words, the tree possesses a model, which anticipates low temperature on the basis of a shortening day, regardless of what the present ambient temperature may be. (Rosen 1985, p. 7)

Rosen's notion of anticipatory behavior is related to Maturana and Varala's idea of autopoiesis, but it has the added advantage of connecting the process of autopoiesis with the development and employment of (hard-wired) *models* of possible future events in the environment. This further specification of the general notion of autopoiesis helps to clarify the possibility and importance of *interlocking* anticipatory models, on the level of both intra- and trans-species relations. In his study of the role of sign processes in nature, Hoffmeyer provides a telling example of such interlocking:

> When a brown hare spots a fox approaching in the open landscape, the hare stands bolt upright and signals its presence instead of fleeing. The explanation of this behavior, according to ethologist Anthony Holley (1993), is that a hare can easily escape from a fox simply by running – a fact that a fox seems to "know" (whether by learning or by instinct). Apparently, then, what is happening in this behavior is that the hare is telling the fox: "I have seen you" – and as a result they can both be spared the effort of running. (2008, p. XII)

Stated in terms of Rosen's concept of anticipation, the hare anticipates an unnecessary flight from the approaching fox and signals to the fox that his anticipation of a successful hunt better be cancelled. When the fox subsequently gives up his effort, they both have acted in the present on the basis of the interlocking anticipation of a *possible* future event: the hare outrunning the fox chasing him in vain.

Much more should be said about agency as a defining characteristic of living beings, but in the context of my argument these remarks suffice as an introduction to the first building block for a transdisciplinary approach to our ETPH. The importance of this building block becomes clearer when I add the next one: the shift within general biology and ethology from an aggression-centered behavioral perspective to a perspective combining aggression and peacemaking as equally constitutive for intra- and trans-species relations, as exemplified by the emergence of peace ethology.

Connecting Aggression and Peacemaking

Not so long ago, scientists perceived competition and aggression as dominant forces in nature, not only at the level of behavior but also as an ultimate cause, explaining the selection of fitness-enhancing traits in the "struggle for life." This image of "nature red in tooth and claw" has been substantially modified by a fast-growing body of literature foregrounding the role of empathy, care, and peacemaking as proximate causes, especially, but by no means exclusively, on the level of relations between higher mammals. The work of de Waal on empathy and care in primates is the pinnacle of a vast and growing body of scientific work devoted to the development of a more complex perspective on the *natural* coexistence of competition and aggression on the one hand and cooperation, peacemaking, and care on the other (de Waal 1991, 2005, 2013; Ulanowicz 1997, 2009; Fry 2013). As Verbeek succinctly summarizes this development:

> Increasing evidence suggests that in nature aggression and peace are not antithetical, but rather are linked in recurring relationships. It follows that in order to understand peace one has to understand aggression and vice versa. (Verbeek 2013, p. 63)

For the further exploration of these "recurring relationships" between aggression and peace, I follow the lead of de Waal, particularly his characterization of the human species as "the most internally conflicted animal ever to walk the earth":

> It is capable of unbelievable destruction of both its environment and its own kind, yet at the same time it possesses wells of empathy and love deeper than ever seen before. Since this animal has gained dominance over all others, it's all the more important that it takes an honest look in the mirror, so that it knows both the archenemy it faces and the ally that stands ready to help it build a better world. (de Waal 2005, p. 237)

Within de Waal's naturalistic framework, the capacity to reflect on the aggressive nature of "our inner ape" and the conscious effort to shift the balance between the two conflicting sides must be a product of our evolutionary history. De Waal presupposes here that we can *select* – up to a point – how to relate to our evolutionary heritage in view of "building a better world." The same possibility of selection between behavioral alternatives underlies the example cited above of contemporary appeals to ban specific forms of genetic modification on moral or ecological grounds. In both cases, culturally transmitted self-perceptions and moral self-evaluations of humans could act as a causal force on the future evolution of life.

I propose to designate this form of selection between behavioral alternatives as *autopoietic selection*. Obviously, such autopoietic selection is situated at the level of phenotypes and thus differs from Darwinian selection, which is based on the elimination of unsuccessful genotypes. In the next subsection, I go into new perspectives on the evolutionary significance of variation at the level of phenotypes and the possibilities for genetic assimilation of favorable traits that emerge out of this phenotypic variation.

A Broader Evolutionary Framework

The importance of the shift from a gene-centered to a development-centered evolutionary perspective for understanding our trans-species peace heritage hinges on the conceptual space this shift has created for the role of developmental plasticity in the evolution of life and for the contribution of autopoietic selection to this plasticity. According to Jablonka and Lamb:

> [M]ost biologists see heredity in terms of genes and DNA sequences, and see evolution largely in terms of changes in the frequencies of alternative genes. We doubt that this will be the situation in twenty years' time. More and more biologists are insisting that the concept of heredity that is currently being used in evolutionary thinking is far too narrow, and must be broadened to incorporate the results and ideas that are coming from molecular biology and the behavioral sciences. (Jablonka & Lamb 2005, p. 10)

West-Eberhardt's work is an influential example of this new direction in contemporary evolutionary theory (West-Eberhard 2003). On the basis of abundant evidence, she proposes shifting the focus of evolutionary theory from genotypes to phenotypes and foregrounding the importance of developmental plasticity in the evolution of life. In the general introduction to her argument, West-Eberhardt posits that the question of how developmental plasticity fits into a genetic theory of evolution remains largely unanswered:

> There is no one-to-one relation between phenotypes and genes, yet a gene mutation is often visualized as the originator of a new genotypic trait … proximate mechanisms are excluded from ultimate (evolutionary) explanations, yet a proximate mechanism (development) produces the variation screened by selection…. Here, I put the flexible phenotype first, as the product of development and the object of selection. (West-Eberhardt 2003, p. 3)

According to West-Eberhardt, recent developments in behavioral ecology and ethology have demonstrated "remarkably fine-tuned adaptively appropriate phenotypic expression of complex traits and life-history patterns" (West-Eberhardt 2003, p. 17). In her eyes, adaptive flexibility should be seen as a property of phenotypes, not of genotypes. Popular metaphors such as the genetic "blueprint" of the genetic "program," therefore, belie a fundamental misunderstanding of the role of developmental plasticity in evolution:

> The genetic program metaphor encourages the idea that each decision point in development or behavior, and each environmental input, acts as a genetic directive or a set of genetic rules, like "in condition x do A, in condition y do B." But there are no such rules in the genome, only a set of templates for molecules that will become part of the phenotype. (West-Eberhardt 2003, pp. 14–15)

The crucial point here is not only that that are no such rules in the genotype but also that there is no direct relation between genotypes and the environment. Selection works on *expressed* traits, and these are *embodied* in phenotypes. This implies that genotypes are characterized by their own developmental dynamic, which is only expressed *in part* on the level of phenotypes and, thus, exposed to natural selection.

In the context of my search for a transdisciplinary approach to our ETPH, this new evolutionary perspective is of great importance, not only because it takes a well-argued leave from mechanistic, gene-centered models of evolution, but even more so because it creates conceptual space for the role of autopoietic selection in evolution. This conceptual space opens itself most clearly in the role West-Eberhardt accords adaptive environmental *assessment*. A clear example of the causal significance of such assessment is provided by the assessment of possible mates by females and by the assessment of dominance in social insects, which influences developmental pathways leading either to worker or to queen phenotypes (West-Eberhardt 2003, p. 442). West-Eberhardt is well aware of the reluctance of many biologists to accord a form of choice (or *autopoietic selectivity*, as I would say) to nonhuman life forms, because it invokes the specter of anthropomorphism. Her counterargument is both revealing and humorous:

> Critics of anthropomorphism worry that biologists will attribute uniquely human capacities to non-human organisms. I have the opposite worry, that obsessions with human specialness will prevent us from seeing the remarkable assessment capacities of nonhuman organisms, and how these capacities relate to our own sometimes less-than-rational decision process. The fact that nonhuman organisms are known to *adaptively* switch between conditional alternatives implies that they somehow use criteria of advantageousness to assess the conditions surrounding their decisions. That many are unlikely to do this using conscious thought or a college education makes their behavior all the more intriguing. (West-Eberhardt 2003, p. 443)

Although West-Eberhardt connects this analysis in passing with the evolution of "the sophisticated device we call 'mind'" (West-Eberhardt 2003, p. 20), she refrains from analyzing the evolution of the human species and human culture. One of the reasons for my search for a transdisciplinary approach hinges, however, on my observation that changing self-perceptions and self-evaluations of humans, including insights from peace studies, can exert causal influence on our present and future relations with other species and thus modify – up to a point – the future evolution of life on our planet. For this reason, I add a fourth and last building block for a transdisciplinary approach to our ETPH. This last building block is based on Margulis' view of the social and relational character of evolutionary development.

A Social Perspective on the Evolution of Life

Whereas Neo-Darwinian theories of evolution focus almost exclusively on the role of genes, the new evolutionary perspective of West-Eberhardt and related theorists, such as Jablonka and Lamb (2005), focuses on the ongoing *integration* of environmental and genomic influences. The ensuing shift of attention to the development of concrete life forms as the "site" where this integration is brought about creates conceptual space for the role of autopoietic selectivity in evolution. Building on the work of Margulis, I argue that this role has more importance than West-Eberhardt recognizes. Autopoietic selectivity – or "adaptive environmental assessment," as she says – not only co-determines which developmental pathway will be followed at key points in the development of individual life forms. It has also influenced in decisive ways the evolution of species, by way of the interlocking of the autopoiesis of different life forms and the subsequent merging of their genotypes.

Like West-Eberhardt, Margulis argues that random variation is not the only source of evolutionary novelty. In her view, the *merging* of different genomes resulting in the emergence of new life forms is a major source of evolutionary change:

> [T]he important transmitted variation that leads to evolutionary novelty comes from the acquisition of genomes. Entire sets of genes, indeed whole organisms each with its own genome, are acquired and incorporated by others. (Margulis & Sagan 2002, p. 12)

Building on her analysis of symbiotic merging in the evolution of eukaryotic cells, Margulis designates this process as "symbiogenesis." The incorporation of photosynthetic bacteria into nucleated cells, which occurred according to Margulis some 2 billion years ago, is a prime example:

> In the final acquisition of the complex-cell generating series, oxygen breathers engulfed, ingested, but failed to digest bright green photosynthetic bacteria. The literal "incorporation" occurred only after a great struggle in which the undigested green bacteria survived and the entire merger prevailed. (Margulis 1999, p. 48)

To further clarify the process of symbiogenesis, Margulis introduces the notion of "modulated co-existence" between former predators (or pathogenic parasites) and their "hosts":

> As members of two species respond over time to each other's presence, exploitative relations may eventually become convivial, to the point where neither organism exists without the other. (Margulis & Sagan 2002, p. 12)

Margulis' views are important for my argument, because she adds social and relational dimensions to evolutionary theory. Before plants and animals appeared on earth, there was a long period in which bacteria – the first and by far most numerous life form on our planet – underwent "at least 2000 million years of chemical and *social* evolution" (Margulis 1999, p. 71; italics added). Building on Margulis' analyses, this social evolution can be specified in terms of two consecutive "social inventions": in the first place, the invention – under the pressure of selection – of exploitative relations as a behavioral alternative for destructive relations; and, in the second place, the invention of cooperative and "convivial" relations as a behavioral alternative both for destructive and for exploitative relations. Margulis' views thus provide an important addition to West-Eberhardt's analysis of "adaptive environmental assessment." The evolutionary significance of this adaptive assessment, or autopoietic selectivity, pertains not only to coordinated changes at the level of phenotypes and genotypes but also to changes in the *social pattern* of autopoietic processes that feed back into phenotypes and into the expression of specific genetic potentials. The emergence of exploitative and cooperative relations in the social evolution of bacteria was based on autopoietic selection between different behavioral alternatives corresponding with different social patterning of autopoietic processes. On one hand, "ingesting and digesting" and all forms of pathogenic exploitation occurred, leading to the (eventual) destruction of the autopoietic capabilities of the bacteria involved. On the other hand, integration of specific autopoietic capabilities of other bacteria into one's own autopoiesis occurred. When we connect these two alternatives with de Waal's analysis of empathy and cooperation (as manifested, for instance, with great clarity in the behavior of bonobos), the binary opposition between "war" and "peace" – and the related opposition between "self-preservation" and "altruism" – can be replaced by a more complex, threefold relation:

1) Incorporating or exploiting other living beings in ways that destroy their autopoietic capabilities
2) Using the autopoietic capabilities of other living beings for one's own autopoiesis in ways that leave these capabilities more or less intact or enhance them
3) Aligning one's autopoietic capabilities with that of other living beings in view of a shared and reciprocally monitored experience of well-being.

In view of its catastrophic consequences for the autopoiesis of other life forms, I designate the first behavioral pattern as *catapoiesis*. I refer to the second pattern as *ergopoiesis* to designate the incorporation of the autopoietic capabilities of different life forms in one's own autopoietic processes, while leaving these capabilities (i.e., the concrete "work" or "ergon" they perform) more or less intact. Combining the notion of dialog and autopoiesis, the term *diapoiesis* designates all relations between life forms that result in the co-creation of a reciprocally monitored experience of well-being (Kunneman 2010, 2015). The general concept of autopoiesis can thus be subdivided into four different relational patterns:

- Autopoietic relations with the environment that do not involve social interactions with other autopoietic beings, for example the photosynthetic use of sunlight by plants
- Catapoietic relations with other autopoietic beings that damage or destroy their autopoietic capabilities, such as the many forms of predation developed in the course of evolution or forms of competition resulting in the extinction of other species
- Ergopoietic relations with other autopoietic beings, ranging from the parasitic or coercive use of their autopoietic capabilities (while leaving these [more or less] intact, for example parasitic tapeworms living in the intestines of most vertebrates) to forms of cooperation that mutually enhance the autopoietic capabilities of the life forms involved, such as the cooperative relations between trees and mycorrhizal fungi or the bacteria that flourish in human intestines while contributing to our digestive processes
- Diapoietic relations, based on the co-creation of a shared and reciprocally monitored experience of well-being, such as intra- and trans-species forms of care, play, and erotic intimacy.

Using this fourfold distinction, we can analyze the autopoiesis of living beings in terms of the *relative contribution* of catapoietic, ergopoietic, and diapoietic relations with other life forms in different phases of their development. Moreover, we can analyze evolutionary development in terms of two interconnected developmental processes: (1) gradual enlargement of the *range* of intra- and trans-species social relations, starting from catapoietic relations to ergopoietic and diapoietic relations; and (2) concomitant changes in the *relational composition* of autopoietic processes at the level of intra- and trans-species relations

The possible development of more peaceful relations with other species in the near future can thus be pictured as a gradual change in the relative contribution of catapoietic, ergopoietic, and diapoietic relations with other species to the autopoiesis of the human species. In the remainder of this section, I clarify this gradual change and its significance for a deeper understanding of our ETPH in two steps. To start with, I flesh out the difference between catapoietic and ergopoietic relations by way of an instructive example: the social life of the leafcutter ants. In a second step, I elaborate the difference between ergopoietic and diapoietic relations with the help of Damasio's analysis of the nature of emotions and feelings (Damasio 1999, 2003). I focus in particular on the reciprocal modeling of feelings for the development of social relations based on the co-creation of a shared and reciprocally monitored experience of well-being. In conclusion, I argue that the transdisciplinary approach developed in this chapter contributes to a deeper understanding of our ETPH, in particular the importance of less exploitative and more cooperative *ergopoietic* relations with other species.

The Example of the Leafcutter Ants

The social life of the leafcutter ants provides an instructive example of a network of ergopoietic relations that have sustained their autopoiesis for approximately 20 million years. I focus on the *Atta*, a tribe belonging to the fungus-growing termites of the subfamily of *Macrotermitinae*. According to Hölldobler and Wilson (2011), visitors from another star system a million years ago might have concluded that:

> leafcutter colonies were the most advanced societies this planet would ever be able to produce.... Because they possess one of the most complex communication systems known in animals, as well as the most elaborate caste systems, air-conditioned nest architecture, and populations into the millions, they deserve recognition as Earth's ultimate superorganisms. (Hölldobler & Wilson 2011, pp. 3–4)

The specific ergopoietic repertoire of the Attine leafcutters has developed from an earlier form of mutually beneficial ergopoiesis between basal Attine species, such as the *Mycetarotes*, and a specific family of fungus, the *Lepiotaceae*. These ants made the decisive transition from a hunter-gatherer existence to a form of agriculture some 50 to 60 million years before the human species (Hölldobler & Wilson 2011, p. 11). The first ant species to develop the culture of fungi assembled pieces of dead leaves and other debris to feed the fungi, which they cultivated inside their nests. The hyphae sprouting from these fungi provided food for their larvae. Their colonies were quite small, on average a few hundred workers. In contrast, the Attine ants made a big leap in the size, social complexity, and stability of their colonies, based on a new "adaptive environmental assessment," as West-Eberhardt would say, namely the invention of leafcutting. Hölldobler and Wilson describe the concomitant reorganization and refinement of the internal division of labor between foragers, gardeners, and caretakers as follows:

> The Atta workers organize the gardening operation in the form of an assembly line. The most frequent size group among foragers, at the start of the line, consists of workers with a head of 2.0 to 2.2 millimeters. At the end of the line, the care of the delicate fungal hyphae requires very small workers, a task filled within

the nest by workers with a head width of predominantly 0.8 millimeters. The intervening steps in gardening are conducted by workers of graded intermediate size. (Hölldobler & Wilson 2011, p. 53)

The autopoiesis of the colony is thus realized through a high degree of differentiation between individual ants, allowing for the combination of the concomitant specialized capabilities in a fine-tuned cooperative process. The combined autopoiesis of the individual ants contributes to and is supported by the autopoiesis of the colony as a whole.

The remarkable stability of the ergopoietic form of autopoiesis developed by the leafcutter ants rests, however, only in part on their high degree of intra-species cooperation. The characteristic features of this cooperation have evolved in direct connection with the development and stabilization of cooperative *trans-species* relations. This goes in particular for their ergopoietic relations with the *Lepiotaceae* fungi. The division of labor between foragers, gardeners, and caretakers is geared completely to the flourishing of these fungi. And the sophisticated forms of communication that support their internal cooperation are complemented by a form of communication with the fungi, allowing the ants to detect distress signals emitted by them. This conclusion follows from experiments by researchers who contaminated the leaves fed to the fungi by Attine workers with a substance the workers could not detect. The fungi effectively communicated the presence of these contaminants to the ants, as demonstrated by the fact that the ants stopped feeding "their" fungi with the contaminated leaves within 6 to 10 hours and refrained from doing so for a period of weeks (Hölldobler & Wilson 2011, pp. 89–91).

Following from my definition of ergopoietic relations, these relations range from parasitic and exploitative relations with other species to forms of mutually beneficial cooperation. The autopoiesis of the leafcutter ants provides a beautiful illustration of this range. To start with, their relationship with the trees that grow the leaves they harvest is clearly exploitative but not destructive. Moreover, the large amounts of food assembled in their huge colonies attract other life forms that realize their autopoiesis by way of exploitative or parasitic forms of ergopoiesis. Smaller ant species, for example, inhabit the colonies of the leafcutters on a regular basis and feed on the cultivated fungi without contributing to the work necessary for their flourishing. Apparently, the Atta can afford to let them cohabit with their colony for free, because they do not try to block or remove them. The same does not go for the pathogenic parasites that feed on their fungi and embody a threat to their health. In the ongoing struggle with these parasites, the Atta ants have developed an important ergopoietic alliance with a small bacterium of the genus *Pseudonocardia*. These bacteria produce antibiotics that suppress the spread of the pathogenic parasites. They live in a cavity adapted for this purpose inside the body of the small workers that take care of the fungi (Hölldobler & Wilson 2011, p. 99).

Significantly, the Attine leafcutters stay almost completely within the range of ergopoietic trans-species relations. Although they have developed a "warrior caste" of large specimens with strong jaws, these are employed mainly when enemies threaten to destroy the colony. They are only employed for aggressive purposes in the case of an emergency. When pathogenic parasites destroy the fungi inside the nest, the ants will try to steal fungi from another colony, or even raid it for that purpose, so as to start anew their cooperation with the *Lepiotaceae* fungi.

Diapoietic Relations

I have described the example of the leafcutter ants in some detail because their way of life is of special interest for my argument. To start with, in the course of their development, they have explored the full range of ergopoietic relations. On the one hand, close to the border with catapoietic relations, their leaf cutting forcefully exploits the "ergon" brought into the world by trees. Moreover, in the case of an emergency, they also steal fungi or even raid neighboring ants. On the other hand, they have developed very stable, mutually beneficial forms of cooperation with their fungi and with the bacteria that produce antibiotics for them. The ergopoietic lines along which they have organized their autopoiesis are also instructive for a different reason, however. Their sensitivity to distress signals emitted by the *Lepiotaceae* fungi when these are contaminated comes very close to the border between ergopoietic and diapoietic relations, but they have never "crossed the border" toward the further exploration of such relations.

As defined in this chapter, diapoietic relations are based on the co-creation of a shared and reciprocally monitored experience of well-being. The development of diapoietic relations thus presupposes the autopoietic capability of experiencing well-being and the lack of it oneself, that is, the capability to "emote" (Damasio 1999). Moreover, they presuppose the shared capability of the life forms involved to perceive, be concerned about, and *enjoy* up to a point the well-being of other life forms. This concern and enjoyment presuppose the ability not only to *feel* one's own emotions and feelings, but also to feel the emotions of others and infer their feelings from feeling their emotions. In contrast to ergopoietic relations, which only presuppose emotions, diapoietic relations presuppose emotions, feelings, and consciousness on *both* sides.

This articulation of the dividing line between ergopoietic and diapoietic relations builds on the complex, neurobiological perspective on the relations between emotions, feeling, and consciousness developed by Damasio (1999, 2003). According to Damasio, emotions should be considered as *collections* of *responses* to certain *stimuli*, especially stimuli that are potentially useful or dangerous:

> [A]t their most basic, emotions are part of homeostatic regulation and are poised to avoid the loss of integrity that is a harbinger of death or death itself, as well as to endorse a source of energy, shelter or sex…. Emotions are inseparable from the idea of reward or punishment, of pleasure or pain, of approach or withdrawal, of advantage and disadvantage. Inevitably, emotions are inseparable from the idea of good and evil. (Damasio 1999, p. 54)

Damasio proposes to distinguish background emotions from primary ("basic") emotions. Background emotions concern the *state of being* of life forms, "good, bad, or somewhere in-between" as he says (Damasio 2003, p. 44). He considers primary emotions as modulations of this good or bad condition in the form of specific bodily states and concomitant patterns of behavior. He argues that background emotions and primary emotions should be clearly distinguished from feelings. The distinction between emotions and feelings is perhaps his most important contribution to (neuro-)biology. In Damasio's eyes, feelings are based on "sensory maps of the transformations that occur in the body when emotive reactions take place" (Damasio

2003, p. 51). The ability to create such maps depends on the development of brains complex enough to create mental maps of sensory maps of different emotive reactions. Damasio designates feelings also as "portrays," as mental images created by brains of their "own" sensory maps of bodily transformations connected with specific emotions, such as an increased heart rate or arousal of sexual organs. According to Damasio, this ability to create mental maps of sensory maps is of decisive importance for the emergence of consciousness, and consciousness hinges on the ability to *know* the feelings caused by emotions – the ability to *feel feelings* – and develops further when this ability gets connected with memory and with foresight (Damasio 1999, p. 31).

The emergence of this ability to feel one's own feelings opens the door for the further exploration of the space of diapoietic relations. This door opens when conscious beings develop the ability to feel not only their own feelings but also the feelings of other life forms, as inferred from the emotions they *show* when feeling them. This inference builds upon older forms of "emotional contagion," as de Waal says, in particular distress caused by another's distress (de Waal 2005, p. 178). The ability not only to resonate empathically with the emotions of other life forms but also to concurrently *feel* these emotions makes a big difference, because then the "portrays" involved in felt feelings can be connected with past memories and with imagination and foresight. The role of anticipatory modeling (as Rosen says) in the autopoiesis of life forms now also starts to involve the anticipatory modeling of the felt feelings of other life forms. In the wake of this development, new forms of diapoietic sociality emerge, connected with the reciprocal modeling of feelings of well-being, in particular social play, enjoyed care, and erotic intimacy. The *co-creation* of a *shared* and reciprocally *monitored* experience of well-being characterizes all three.

Burghardt's (2005) analysis of animal play and its evolution inspires this characterization. Burghardt proposes five criteria for distinguishing animal play from other forms of behavior. His first criterion for recognizing animal play is based on its direction toward stimuli "that do not contribute to current survival" (Burghardt 2005, § 3.7.1). This criterion is closely related to his second and fifth criteria, which are worth quoting in full:

> The second criterion for recognizing play is that the behavior is spontaneous, voluntary, intentional, pleasurable, rewarding, reinforcing or autotelic ("done for its own sake").... The fifth criterion for play is that the behavior is initiated when an animal is adequately fed, healthy and free from stress (e.g. predator threat, harsh microclimate, social instability), or intense competing systems (e.g. feeding, mating, predator avoidance). In other words, the animal is in a "relaxed field." (Burghardt 2005, §§ 3.7.2 and 3.7.5)

These definitions point to the specific forms of diapoietic sociality that can emerge from welcomed and accepted efforts to bring *another* life form in a "relaxed field." The relational pattern involved here is connected with feelings of pleasure and relaxation that are *shared* by the life forms that co-create this experience. We can speak of co-creation here, because such diapoietic interactions are sustained by the recognition of the felt intention to contribute to a "relaxed field" for the other and the complementary recognition of the welcoming acceptance of this intention.

Social play by animals provides a good example of this diapoietic sociality, as testified by Bekoff's (2008) analysis of social play among animals cited earlier in this chapter. In his example, a chimpanzee *offers* to play with another. This offer has to be accepted as leading to pleasure and relaxation. This acceptance in turn has to be recognized and acknowledged for the play to continue. During the play itself, signals are exchanged to communicate that diapoietic intentions and their acceptance continue to shape the relation.

Social play is not the only form of diapoiesis, however. Enjoyed care and erotic intimacy share the characteristics of diapoietic sociality just mentioned. Hrdy (2009) argues that the attractiveness of babies, their delicious smell, their endearing smiles, and their sounds of contentment and well-being have developed as an evolutionary answer to the problem posed by the long period of complete dependence of small children on care by others. The invention of "co-parenting" provided a solution to this problem. The motivation of different relatives to take care of very young children was greatly enhanced in her eyes by their increasing attractiveness and their ability to communicate and share their feelings of well-being (Hrdy 2009). Compared with the "burdensome care" of the Atta leafcutters for their own larvae and their fungi, this form of care brings in elements of enjoyment and relaxation, and a fine-grained attunement of the feelings of well-being felt by the parties involved.

The same goes in my eyes for the different forms of erotic intimacy that have evolved on the basis of the reciprocal modeling of feelings. This notion is meant to designate the experience of deep well-being arising from the combination of feeling completely secure with another and experiencing the body of the other as attractive, pleasurable to touch, and overall emotionally "touching." Erotic intimacy can emerge and be renewed when the participants in this diapoiesis share this experience of feeling safe with and being attracted to the other, reciprocally monitor and confirm this experience, and try to repair it when it threatens to disappear. Erotic intimacy is frequently connected with sexual desire. But it is also a robust characteristic of relations not involving sexual desire and sexual arousal, such as diapoietic relations between relatives and friends who enjoy each other's company, find it pleasurable to touch and be touched by the other, and reciprocally see beauty in the other.

Conclusion: The Moral Import of Our Trans-Species Evolutionary Heritage

Looking at the social patterns dominating the relations between humans and other species at the start of the 21st century, it is clear that these relations can hardly be called peaceful. With a small number of "elected species" – our pets and "house animals" – we sustain more or less diapoietic relations, but the present dominance of the human race has well-documented catastrophic consequences for a great number of other species. Moreover, we have developed a still-expanding network of exploitative relations with staggering numbers of cows, pigs, chickens, fish, and other animals, whose autopoiesis is forcefully bent to maximal human usefulness. This exploitation also has destructive consequences for other species, as witnessed for example by the rapid diminishment of tropical forests and the many life forms harbored by them, due in part to the steady increase of the area where humans cultivate fodder for "meat-producing" animals (Hosonuma *et al.* 2012).

This situation points to the urgency of a deeper understanding of the evolutionary resources we humans could use to further more peaceful relations with other species. On the basis of my foregoing analysis, I will argue in conclusion that the moral import of our ETPH should be located primarily in the domain of ergopoietic relations. Instead of looking for moral resources for trans-species peacemaking primarily in the domain of diapoietic relations, the most promising resources contained in our ETPH can be found in the ergopoietic domain. The following considerations support this conclusion.

To start with, my analysis strongly suggests that catapoietic and exploitative relations are part and parcel of our evolutionary heritage and are there to stay. There is no moral teleology in evolution, only an increase in the complexity of social patterns. The emergence of ergopoietic relations in the evolution of life proved to be a viable addition to, but certainly not a replacement of, catapoietic relations. The same goes for the emergence of diapoietic relations on the basis of the reciprocal modeling of feelings. This addition no doubt has enhanced the complexity of the social patterns available to specific species, allowing for new combinations and interference between them. But there is no "most basic" or "most fundamental" pattern in our evolutionary heritage that in the end would determine the overall structure of our social relations with other species. In our encounters, one of them can be in the foreground and function as a primary social pattern structuring the actual relations between the participants, but the other two will always be there in the background.

The second consideration supporting the importance of the ergopoietic domain for developing more peaceful relations with other species flows directly from the first. When the social relations between humans and other species are indeed characterized by a structural complexity, it is probably asking too much of us to singlehandedly bring about a permanent shift toward exclusively diapoietic relations with other species. Also, in view of the fact that only a relatively small number of life forms have actually incorporated diapoietic relations into their autopoietic repertory, the potential for developing more peaceful relations with other species can be found primarily in the ergopoietic domain. The majority of life forms on our planet realize their autopoiesis by way of a mixture of catapoietic and ergopoietic relations with other species, and for a great number of them these relations are situated primarily in the ergopoietic domain. So it is here that the chances for us to develop more peaceful relations with other species are by far the greatest, because here we are not the only stakeholders; many other species are de facto interested in and capable of co-creating less destructive, more peaceful relations with us. A simple but instructive example of this is the strong increase in Europe and the United States of infectious bacteria that have become resistant to antibiotics. We release antibiotics on them in massive doses in order to destroy them. They "retaliate" in kind by a strong increase in the genetic expression of alternative morphological possibilities, leading to the emergence of "resistant variants." Interestingly, however, in countries where fewer antibiotics are used, the prevalence of resistant infectious bacteria in hospitals is much lower (de Boer *et al.* 2011, p. 43). Apparently, when the conflict between humans and infectious bacteria is approached differently and we replace all-out "war" with a limited offer of co-existence from our side, the other side also shapes its autopoiesis along different, less belligerent lines. As Verbeek (2013, p. 60) reminds us: "we go to war not because we are naturally driven to do so, but because we choose to do so." This choice is moreover relational and co-created up to a point between species. The "relational offer" of the one influences perceived possibilities of the other to either move or not move in the direction of less destructive, ergopoietic relations. In this respect, the capabilities for conflict management and reconciliation that we share with many other species according to peace ethologists such as Verbeek are of the utmost importance. The same applies at the other side of the range of ergopoietic relations, where they border on diapoietic relations. There is a vast and mostly unexplored potential here for the co-creation of new, mutually beneficial cooperative relations between humans and other species. Our diapoietic capabilities can be of help here, insofar as they enable us to enjoy the flourishing of other living beings as a contribution to our own feelings of well-being, but my argument suggests that we should be careful not to rely first and foremost on our diapoietic capabilities. The development of less catapoietic and more ergopoietic relations with the majority of life forms on this planet would already be a vast and more feasible step toward a better use of our evolutionary trans-species peace heritage.

References

Bekoff, M. (2008). *Animals at play: Rules of the game*. Philadelphia: Temple University Press.

Burghardt, G. (2005). *The genesis of animal play*. Cambridge: MIT Press.

Damasio, A. (1999). *The feeling of what happens: Body and emotion in the making of consciousness*. New York: Harcourt Brace.

Damasio, A. (2003). *Looking for Spinoza: Joy, sorrow and the feeling brain*. London: Vintage.

Dawkins, R. (1976). *The selfish gene*. Oxford: Oxford University Press.

Dawkins, R. (1986). *The blind watchmaker*. Harlow: Longman Scientific & Technical.

de Boer, C.S., van Vliet, J.A., & Coutinho, R.A. (2011). *Strategic Policy Plan RIVM-Centre for Infectious Disease Control 2011–2015*. The Hague: Ministry of Health, Welfare and Sports.

de Waal, F. (1991). *Peacemaking among primates*. London: Penguin.

de Waal, F. (2005). *Our inner ape: The best and worst of human nature*. London: Granta Books.

de Waal, F. (2013). *The bonobo and the atheist: In search of humanism among the primates*. New York: W.W. Norton & Company.

Habermas, J. (1987). *The theory of communicative action, vol. 2: The critique of functionalist reason*. Boston: Beacon Press.

Hoffmeyer, J. (2008). *Biosemiotics: An examination into the signs of life and the life of signs*. Scranton, PA: University of Scranton Press.

Hölldobler, B., & Wilson, E.O. (2001). *The leafcutter ants: Civilization by instinct*. New York: W.W. Norton.

Hosonuma, N., Herold, M., De Sy, V., et al. (2012). An assessment of deforestation and forest degradation drivers in developing countries. *Environmental Research Letters*, 7(4), 4009.

Hrdy, S.B. (2009). *Mothers and others: The evolutionary origins of mutual understanding*. Cambridge, MA: Harvard University Press.

Jablonka, E., & Lamb, M. (2005). *Evolution in four dimensions: Genetic, epigenetic, behavioral and symbolic variation in the history of life*. Cambridge, MA: MIT Press.

Kauffman, S. (1995). *At home in the universe*. Oxford: Oxford University Press.

Kauffman, S. (2000). *Investigations*. Oxford: Oxford University Press.

Kauffman, S.A. (2008). *Reinventing the sacred: A new view of science, reason, and religion*. New York: Basic Books.

Kunneman, H.P. (2010). Ethical complexity. In P. Cilliers & R. Allen (Eds.), *Complexity, difference and identity* (pp. 131–164). Dordrecht: Springer.

Kunneman, H. (2015). The political importance of voluntary work. *Foundations of Science*, *20*(1). doi:10.1007/s10699-014-9403-x

Margulis, L. (1999). *The symbiotic planet: A new look at evolution*. London: Phoenix.

Margulis, L., & Sagan, D. (2002). *Acquiring genomes: A theory of the origins of species*. New York: Basic Books.

Maturana, H.R., & Varela, F.J. (1980). *Autopoiesis and cognition: The realization of the living (Boston Studies in the Philosophy of Science*, Vol. *42*). Dordrecht: D. Reidel.

Maturana, H.R., & Varela, F.J. (1992). *The tree of knowledge: The biological roots of human understanding* (Rev. ed.). Boston: Shambhala.

Morin, E. (2008). *On complexity*. New York: Perfect Paperbacks.

Prigogine, I. (1980). *From being to becoming: Time and complexity in the physical sciences*. San Francisco: W.H. Freeman.

Rawls, J.A. (1971). *A theory of justice*. Cambridge, MA: Harvard University Press.

Ricoeur, P. (1992). *Oneself as another*. Chicago: The University of Chicago Press.

Rosen, R. (1991). *Life itself*. New York: Columbia University Press.

Rosen, R. (1985). *Anticipatory systems: Philosophical, mathematical and methodological foundations*. Oxford: Pergamon Press.

Rosen, R. (2000). *Essays on life itself*. New York: Columbia University Press.

Ulanowicz, R.E. (1997). *Ecology, the ascendent perspective*. New York: Columbia University Press.

Ulanowicz, R.E. (2009). *A third window: Natural life beyond Newton and Darwin*. Templeton Press.

Verbeek, P. (2008). Peace ethology. *Behaviour*, *145*, 1497–1524.

Verbeek, P. (2013). An ethological perspective on war and peace. In D.J. Fry (Ed.), *War, peace and human nature: The convergence of evolutionary and cultural views*. Oxford: Oxford University Press.

West-Eberhard, M.J. (2003). *Developmental plasticity and evolution*. New York: Oxford University Press.

Wilson, E.O. (2000). *Sociobiology: The new synthesis*. Cambridge, MA: Harvard University Press. (Original work published in 1975)

Wilson, E.O. (2014). *The meaning of human existence*. New York: Liveright.

16

Natural Peace

Peter Verbeek

The Peace of Wild Things

When despair for the world grows in me
and I wake in the night at the least sound
in fear of what my life and my children's lives may be,
I go and lie down where the wood drake
rests in his beauty on the water, and the great heron feeds.
I come into the peace of wild things
who do not tax their lives with forethought
of grief. I come into the presence of still water.
And I feel above me the day-blind stars
waiting with their light. For a time
I rest in the grace of the world, and am free.

Copyright © 2012 by Wendell Berry, from *New Collected Poems*.
Reproduced with permission of Counterpoint.

When nature draws us in, she does not leave us cold. As beautifully expressed by the conservationist and poet Wendell Berry, nature can bring us feelings of peace when our human world brings us feelings of despair. In this chapter, I show that turning to nature to find peace is not only an emotional experience but also a scientific endeavor that can help fill important gaps in our knowledge of peace.

Looking to nature to understand peace is still far from commonplace in Western culture. To wit, *natural peace* (or even the word *nature*) is absent from among the 40 types and components of peace extracted from a literature search on recent academic articles with *peace* in their title (Coleman 2012, p. 354). A comprehensive survey of predominantly male (76%), primarily Western-based (91%) scholars in disciplines dealing with peace and conflict makes no mention of nature as a meaningful factor in peace (Advanced Consortium on Cooperation, Conflict and Complexity [AC⁴] 2015). In fact, the word *species* was excised from the process definition of peace that I contributed to

Peace Ethology: Behavioral Processes and Systems of Peace, First Edition.
Edited by Peter Verbeek and Benjamin A. Peters.
© 2018 John Wiley & Sons Ltd. Published 2018 by John Wiley & Sons Ltd.

the survey (AC[4] 2015, p. 10). Indeed, persistent traditional Western views of animal life as unmitigated violent competition, as in Tennyson's "nature red in tooth and claw" and Huxley's "gladiator show," seem to suggest that there is no peace in nature. Traditional conclusions drawn from life science tend to reinforce this impression by suggesting that the essence of life consists of the replication of genes, and that the "gene's eye view of the world" is inherently "selfish" (Dawkins 1976). Taken uncritically, such assumptions lead one to believe that peace, if it exists at all, must be a cultural phenomenon and should be left to philosophers and social scientists to contemplate. And yet, ample findings of peaceful behavior among nonhuman primates, our closest evolutionary relatives, suggest that our evolved inner ape is not merely selfish but rather as predisposed to peace as to aggression (de Waal 2006). And the primates are not alone in these natural predispositions. Peace, like aggression, drives the life processes of numerous creatures, and through their lives it drives life itself.

Research on the evolution of our social behavior tends to focus on primate comparison, yet we share ancestry with all other mammals and vertebrates. Complementing the findings on primates, research of recent decades on other vertebrates shows that peaceful behaviors, such as non-aggressive cooperation, helping and sharing, and behaviors that keep aggression in check or reestablish relations and tolerance following aggression, appear in a wide range of species (Verbeek 2013). When we focus on microbes, which are some of the smallest organisms that help sustain life on earth, we find that cooperation is widespread among them (Nadell *et al.* 2016). Taken together, these findings show that peaceful behavior, like aggression, is an integral part of the behavioral repertoire of organisms big and small.

While still largely absent from peace research, peace in nature needs to be understood, as leaving nature out of the study of peace and peace out of the study of nature will keep us not only from truly understanding peace but also from understanding nature and ourselves. This chapter offers an introduction to the study of natural peace through the lens of peace ethology. The first two sections of the chapter, "Concepts and Contexts" and "Situating Human Nature," provide a brief historical, conceptual, and contextual backdrop for the third main section, "Situating Natural Peace," which presents selected findings of peaceful behavior in nature organized along peace ethology's three social dimensions of peace: *direct*, *structural*, and *sociative* peace (Verbeek & Peters this volume, Chapter 1).

As a single chapter, the discussion here is necessarily limited and offers neither a comprehensive review nor a developed theory. Instead, it points to opportunities to measure peace in the lives of human and nonhuman animals and shows how the methods of ethology, the biological study of behavior, can be put to use for that task. To obtain a deeper understanding of the contents of this chapter, I encourage the reader to investigate further the literature referenced throughout. Like Theodore Lentz did in his pioneering book on peace science (1955, p. 194; and see the following section), I encourage readers to contact me with thoughts and ideas: peace and peace science flourish through dialogue and wither through monologue.

Concepts and Contexts

Defining Peace …

By *peace*, we mean the

Behavioral processes and systems through which species, individuals, families, groups, and communities negate direct and structural violence (*direct peace; structural peace*), keep aggression in check or restore tolerance in its aftermath (*sociative peace*), maintain just institutions and equity (*structural peace*), and engage in reciprocally beneficial and harmonious interactions (*sociative peace*). (Verbeek & Peters this volume, Chapter 1; Table 16.1)

… And Behavioral Science of Peace

In a 1962 *Science* article entitled "The Behavioral Sciences and War and Peace," US National Institute of Mental Health psychiatrist William Pollin laments,

> in the entire nation there are less than 100 individuals working full time at projects which might be thought of as aimed at the development of a social and behavioral science of peace.

Now, more than five decades later, hundreds of scientists from around the globe and across disciplines are contributing findings that together can form the foundation for a veritable behavioral science of peace (e.g., Fry 2013 and this volume, Chapter 14; Verbeek & Palagi 2016). This sounds like news that would cheer up the late William Pollin were it not for the fact that many of these scientists may be unaware of work on peace outside of their own research areas or of how such work in other disciplines fits with their own work. What is more, they may not interpret their own work as work on peace or may shy away from labeling it as such for fear of not being able to obtain funding for it or publish it in influential journals (Verbeek 2013). What is needed is an acceptance of the fact that peace is behavior that can be operationally defined and studied like any other behavior. In addition, what is needed is a conceptual and methodological framework that can integrate existing findings that are relevant to peace and foster collaboration among researchers across scientific disciplines. As this multidisciplinary volume shows, peace ethology provides such a framework. The following historical sketch of research on peace puts this further into perspective.

Peace Research

It is remarkable that in view of its deep philosophical roots (e.g., Kant 1795/1991) and profound significance for life on Earth, peace has only been researched scientifically for a little over five decades (Gleditsch *et al.* 2014). Contemporary social science research on peace goes by various names, ranging from *peace and conflict research* to *peace studies, peace and justice studies*, and, simply, *peace research* (Gleditsch *et al.* 2014). Much of the work focuses on violence, in particular organized violence such as war (Gleditsch *et al.* 2014; Diehl 2016; cf. Fry 2013). This approach presents a "negative" notion of peace as the absence of violence (Galtung 1969), and finding ways to quell violence is the driving force behind much of this research. Chroniclers of peace research tend to agree that a desire never again to repeat the atrocities of World War II inspired much of the contemporary peace research (e.g., Kelman 2010). Driven by feelings of urgency to understand the roots of that horrific violence, fields like psychology built new subfields on peace (e.g., *peace psychology*) with their own dedicated journals (Wessells *et al.* 2010) or developed thematic strands in existing subfields (e.g., in social psychology: Cohrs &

Boehnke 2008; Vollhardt & Bilali 2008; Kelman 2012; and in cultural and biological anthropology: e.g., Fuentes 2004; Fry 2006; Fry & Souillac 2017). Going back in time even further to the pre- and post-WWI era, we find that natural science had already conceived of its own approach to peace through "peace biology," a controversial and ultimately doomed endeavor with eugenic overtones (Crook 1994). Despite all this activity, efforts to build a comprehensive, interdisciplinary behavioral science of peace have been few and far between (but see Aureli & de Waal 2000; Verbeek 2008; Fry 2013; AC^4 2015; Verbeek & Palagi 2016).

Peace Science

An inspiring exception to the segregated approach that has characterized much of the research on peace thus far is the work of psychologist Theodore Lentz (1888–1976). Initially focused on attitude research, Lentz devoted the last three decades of his life to promoting research on peace as a comprehensive *science*. In his seminal book *Towards a Science of Peace* (1955), Lentz outlines a clear vision for a science of peace:

> Stated bluntly and simply, the positive opportunity is to ascertain not what causes war, but what can cause peace. (p. 5)

And:

> Peace is process. War is the absence of the processes, which are unifying from the point of view of all the nations of the world. What these processes are – finding them out – is the job of the science of peace. (p. 5)

Following Lentz' death, one of his former students commented that although Lentz was an expert on human attitudes, "he was not concerned with converting the masses to peace-mindedness." Rather, Lentz stressed the importance of redirecting science away from preparing for war to developing experts in the science and technology of peace (Parker 1977). While most anthologies of peace research credit Lentz for his early insights (e.g., Kelman 2010), his groundbreaking call for a comprehensive behavioral science of peace was not heeded in his lifetime.

Looking at it today, Lentz' invitation to the biologist Julian Huxley to write the foreword to *Towards a Science of Peace* was an inspired move underscoring the need for natural science and the social sciences to join forces on peace research. Huxley's genuine interest in Lentz' ideas for a science of peace anticipates Niko Tinbergen's call for an ethology of peace (Tinbergen 1968; cf. Verbeek 2008, 2013; Verbeek & Peters this volume, Chapter 1), as discussed later in this chapter. Concluding his foreword to Lentz' magnum opus on peace science, Huxley (1955) writes,

> There are still a number of natural scientists who are opposed to the extension of science in human affairs, often because they fear that it will then lose the accuracy and razor-edged certitude that is has achieved in the physical world. And there are many humanists and idealists who are equally opposed to such an extension, often because they are afraid of human values being denatured, so to speak, by what they regard as the cold and impersonal methods of science. Such fears are, I am sure, groundless: in the long run nothing but good can come from the marriage of the scientific method with human values and ideals. (p. vii)

Tinbergen, co-founder of modern ethology, admired Julian Huxley and credited him for inspiring three of the four principal questions of contemporary ethology. Tinbergen writes,

> The biological method is characterized by the general scientific method, and in addition by the kinds of questions we ask, which are the same throughout biology and some of which are peculiar to it. Huxley likes to speak of "the three major problems of biology: that of causation, that of survival value, and that of evolution" – to which I would like to add a fourth, that of ontogeny. (Tinbergen cited in Kruuk 2004, p. 219; for a detailed account of Huxley's influences on ethology, see Burkhardt 2005, pp. 103–126)

Peace ethology. In his 1966 inaugural lecture at Oxford University as a new Professor of Animal Behavior, Tinbergen issued a call to apply the methods and four questions of ethology to the study of "war and peace in animals and men" (Tinbergen 1968; cf. Burkhardt 2005; Verbeek 2008; Verbeek & Peters this volume, Chapter 1).

The emerging field of peace ethology sets out to answer Tinbergen's call by focusing on peaceful behavior as it occurs naturally in a species, be it in the Tanzanian woodlands (e.g., *post-aggression consolation among adult chimpanzees*: Kutsukake & Castles 2004) or the streets of Amsterdam (e.g., *post-aggression consolation among adult humans*: Lindegaard *et al.* 2017). Peace ethology seeks answers to the four guiding questions of ethology as they pertain to peaceful behavior. In terms of *proximate causation*, peace ethology asks what biological, psychological, political, cultural, and environmental factors make peaceful behavior happen at any given time, and how learning and experience modify it (e.g., Jaeggi *et al.*, 2016; Romero *et al.*, 2016; Ali & Walters this volume, Chapter 5; Otten *et al.* this volume, Chapter 3; Romero this volume, Chapter 4; Shnabel this volume, Chapter 2).

Regarding *development*, peace ethology asks when and how peaceful behavior first emerges in the behavioral repertoire of species, individuals, groups, communities, and cultures. It also investigates the capacity for change or transformation of peaceful behavior within these developmental domains in response to different environmental conditions. (e.g., Cordoni *et al.* 2016; Himmler *et al.* 2016; Webb & Verbeek 2016; Furnari this volume, Chapter 8; Narvaez this volume, Chapter 6; Roseth this volume, Chapter 7; Wessells & Kostelny this volume, Chapter 9).

Concerning *function*, peace ethology asks what the immediate and delayed benefits of peaceful behaviors are and how they affect the survival, well-being, and lifetime success of individuals, groups, communities, and cultures (Ikkatai *et al.*, 2016; Palagi *et al.*, 2016; Pallante *et al.*, 2016; Adang *et al.* this volume, Chapter 10; Evans Pim this volume, Chapter 12; Hyslop & Morgan this volume, Chapter 13; Peters this volume, Chapter 11).

Finally, with regard to *evolution*, peace ethology asks why and how the ability to engage in peaceful behavior evolves over generations and evolutionary time in species, individuals, groups, communities, and cultures. In addition, it asks how peaceful behaviors compare across extant species, communities, and cultures (e.g., Czárán & Aanen 2016; Brosnan & Bshary 2016; Clay *et al.* 2016a; Rapaport *et al.* 2016; Fry this volume, Chapter 14; Kunneman this volume, Chapter 15).

The chapters in this volume, together with the recent journal articles cited in this chapter as examples of research on each of the four questions of peace ethology, document research in anthropology, sociology, political science, comparative literature, ethology, evolutionary biology, environmental science, genetics, mathematics, primatology, and psychology, illustrating the fact that answers to peace ethology's four questions are being pursued across multiple disciplines. What peace ethology offers is the conceptual framework that can inspire multidisciplinary efforts and interdisciplinary collaboration toward a truly comprehensive behavioral science of peace (Verbeek & Palagi 2016).

Zoologist cum ethologist Frans de Waal was among the first to operationalize Tinbergen's call for action on aggression and peace through his groundbreaking work on post-aggression reconciliation and consolation in nonhuman primates (de Waal & van Roosmalen 1979; de Waal & Yoshihara 1983). De Waal's Relational Model of aggressive conflict predicts that opponents are likely to resume peaceful interactions shortly after their confrontation when they share a mutually beneficial relationship (de Waal 2000a, 2000b). According to the Relational Model, post-aggression conciliatory behavior serves to repair the damage done to the relationship by aggression, and successive rounds of reconciliation may actually strengthen an existing relationship (cf. Adang *et al.* this volume, Chapter 10). Building on his early work on peacemaking in chimpanzees (*Pan troglodytes*), de Waal collaborated with ethologist Filippo Aureli in putting together the first multidisciplinary volume on post-aggression peaceful behaviors in primates and other mammals, *Natural Conflict Resolution* (Aureli & de Waal 2000). The edited volume inspired new research on peaceful behavior that optimizes kin and non-kin relations in species ranging from mammals to birds and from fish to insects (cf. Verbeek 2008, 2013; Verbeek & Peters this volume, Chapter 1; and see below).

In science and art, ideas emerge at times that transcend boundaries and help set a course for fresh visions and new fields. The work and ideas of Lentz, Huxley, Tinbergen, and de Waal, among others, helped inspire a new vision for how to study peace. This new vision is characterized by what anthropologist Douglas Fry has identified as a convergence of evolutionary and cultural views (Fry 2013), as evidenced by a recent flurry of innovative and wide-ranging publications in this general area (e.g., Christie 2012; Coleman & Deutsch 2012; Fry 2006, 2012). The comparative aspect of this research is important as understanding the role of peaceful behavior in the survival and propagation of nonhuman animal life has significance for improving understanding of our own evolved abilities for peace. Peace ethology exemplifies a new vision of natural peace. The next sections of the chapter present a theoretical and conceptual backdrop for peace ethology by situating human nature within nature as a whole.

Situating Human Nature

A traditional and persuasive view of behavior in humans and nonhuman animals portrays it as a function of stimulus and response. The role of the brain or mind in the endeavor is to construct a model of the world that optimizes stimulus and response patterns (Craik 1943, cited in Reed 1996). This mechanistic view ignores the fact that the nervous system evolved within populations of organisms already quite active in encountering nature (Passano 1963, cited in Reed 1996; Mackie 1970). In fact, like other animals, humans evolved as autonomous agents that continuously explore what nature affords and seek to extract value from it (cf. Reed 1996, pp. 10; Verbeek & de Waal 2002). The nervous system does not need to be switched on like a machine or activated like a software application. Our evolved nervous system is always "on" as we actively seek out and encounter the world. It follows that to understand ourselves, we first and foremost need to focus on how we make our way in the world, not on how a world is made inside of us (Reed 1996, p. 11). Peace ethology mirrors the empirical stance of ecological psychology by emphasizing that to understand peace, we first and foremost need to understand how we engage in it, not how peace is made inside of us (cf. Lentz 1955). It is through this empirical focus on behavior that peace ethology differentiates itself from philosophical and religious analyses of peace that center on transformative interpersonal processes (cf. Galtung & Ikeda 1995).

Happiness and Well-Being

The human species is a manifestation of nature; we have not been parachuted in. It follows that *human nature*, our species-typical being, is the result of natural, evolutionary processes. This biological truth directs medical science, which squarely relies on the fact that we share many of our evolved biological systems with other animals. While medical science could not exist without turning to nature, behavioral science, as practiced in the social sciences, exists in great part without doing so. Unfortunately, by neglecting, misinterpreting, and at times outright rejecting the nature in human nature, behavioral science has limited itself and, as a result, has yet to match medical science in its relevance and significance for human happiness and well-being. A new focus on natural peace provides an opportunity for behavioral science to catch up with medical science in terms of its contribution to the greater good.

Human happiness and well-being rest on two pillars, *health* and *peace*, as a deficit in either diminishes them. Both health and peace are best understood as processes, not as states. Health can be defined as "a process leading to physical, mental, social, and spiritual well-being," and "a resource for the full realization of the human potential" (Simonelli *et al.* 2014). Correspondingly, and as defined at the beginning of this chapter, peace is a *behavioral process aimed at negating violence, keeping aggression in check, and maintaining just and mutually beneficial and harmonious interactions and relationships.*

Aggression, in this context, is defined as "Behavior through which species, individuals, families, groups, and communities pursue active control of resources and the social environment at the expense of others" (Verbeek & Peters this volume, Chapter 1; cf. de Boer in Kruk & Kruk-de Bruin 2010). *Species-typical* aggression is aggression common for the species, while *species-atypical* aggression is aggression that is less frequently shown by members of the species (Chapter 1; de Boer in Kruk & Kruk-de Bruin 2010). *Violence*, in this context, is defined as "escalated aggressive behavior that is out of inhibitory control" (Verbeek & Peters, this volume, Chapter 1; cf. de Boer *et al.* 2009) and as such appears to be a species-atypical form of aggression. Violence can be physical as well as structural. Examples of *structural violence* include social injustice and inequity, discrimination, prejudice, social or moral exclusion, poverty linked to these conditions, and their intended or unintended cultural justifications (Verbeek & Peters this volume,

Chapter 1; cf. Galtung 1996). Peace ethology aims to understand the natural balance between species-typical aggression and peace and the deviations thereof as manifested through species-atypical aggressive and peaceful behavior.

Human Relationship with Nature

The connections between health, peace, happiness, and well-being are evident in our multifaceted relationship with nature. For example, health and peace – and, by extension, happiness and well-being – are at risk from global environmental threats such as climate change and the accelerating loss of biodiversity. Human cultural behavior is a major causal factor in these environmental threats, and its impact is so large that Earth itself is undergoing changes in the functioning of key life-sustaining processes such as the cycles of water, nitrogen, and carbon (Sachs 2015, p. 5). Mindful of natural interconnections and interdependencies, renowned economist and leader in sustainable development Jeffrey Sachs clearly positions the science and practice of sustainable development within a science of peace:

> Sustainable development is also a normative outlook on the world, meaning that it recommends a set of goals to which the world should aspire. The world's nations adopted Sustainable Development Goals (SDGs) precisely to help guide the future course of economic and social development on the planet. In this normative (or ethical) sense, sustainable development calls for a world in which economic progress is widespread; extreme poverty is eliminated; social trust is encouraged through policies that strengthen the community; and the environment is protected from human-induced degradation…. SDGs call for socially inclusive and environmentally sustainable economic growth. (Sachs 2015, p. 3)

The ecological domain sets the stage for the social and economic domains rather than the other way around (World Wide Fund for Nature 2014). Life itself, as well as the entire human economy, depends on goods and services provided by Earth's natural economies (i.e., ecosystems; Daily 1997), and sustaining the natural conditions that are essential for the functioning of the natural economies on which our lives depend sustains peace. Moreover, human-caused environmental degradation that causes human and nonhuman suffering is structural violence (cf. O'Brien 2017); thus, working to negate this form of structural violence through sustainable development is working for peace.

Humans also commit violence against other species, for example through actions resulting in unnatural rates of extinction, and, in a related way, against life itself, through environmental destruction that threatens the planetary boundaries on life as we know it (Sachs 2015; cf. Verbeek 2017a). Our violence against nature is coming back to haunt us as studies from multiple disciplines identify threats to both our physical and mental health posed by biodiversity loss (Chivian & Bernstein 2008; Fritze *et al.* 2008, cited in Verbeek 2009). Finding ways to end our violence against nature is finding ways for peace.

The connections among health, peace, happiness, and well-being are also evident in our inherent attraction to nature. We inherited an affinity for life that undoubtedly served our early ancestors well (Verbeek 2009). Evolutionary biologist and conservationist Edward Wilson coined the term *biophilia* (Wilson 1984) to describe our attraction to nature, linking it to our evolutionary history of adapting and learning from our ancestral natural environment (Kellert & Wilson 1993). Our attraction to nature is multimodal, comprising symbiotic, cognitive, affective, and experiential traits (Zylstra *et al.* 2014); is reflected in specific neurobiological substrates (Lengen & Kistemann 2012); is present early in life (Kahn 1999) and shared with our extant primate kin (Verbeek & de Waal 2002); informs our universal values (Kellert 1996); and correlates positively with health and well-being (Howell *et al.* 2012; Passmore 2014).

As Paul Shephard reminds us, "when young, we delight in the play space that nature affords: trees, shrubs, places to hide and climb, fields to wonder, and rivers to swim and fish in" (Shepard 1982, cited in Verbeek 2009). Unfortunately, new generations of young people are increasingly deprived of opportunities to act on their natural attraction to nature because natural areas are lacking or degraded in the places where they grow up. People are thought to construct a conception of what is environmentally normal based on the natural world they encountered in childhood. With each ensuing generation, environmental degradation can increase, and each generation tends to take that degraded condition as the normal experience of nature, a problem that developmental psychologist Peter Kahn (2002) has referred to as *environmental generational amnesia*.

In his 2012 memoir *Birthright: People and Nature in the Modern World*, social ecologist Stephen Kellert, who with Wilson was a pioneer in the study of biophilia, warned, "We may construct and create our world through learning and the exercise of free will, but to be successful, we must remain true to our biology, which is rooted in nature." He added, "If we stray too far from our inherited dependence on the natural world, we do so at our own peril" (Kellert 2012). Creating opportunities for children to act on their natural attraction to nature is an essential component of what developmental psychologist Darcia Narvaez identifies as our evolved developmental niche for peace (Narvaez this volume, Chapter 6). From the perspectives of evolutionary and individual development, we can say that we derive our species-typical profile, our humanity, from our attraction to nature and our adaptation for peace. Like environmental generational amnesia can result in the interpretation of impoverished environmental conditions as the normal state, obstacles to our natural development for peace may lead to the interpretation of the absence of peace as the normal state.

In social science, happiness and well-being are generally seen as uniquely human constructs and goals, and their interdependencies are seen as a function of cultural and economic processes. In a treatise on comparative (ecological) economy, evolutionary biologist and paleontologist Geerat Vermeij notes in this context,

> [A]t the heart of economic theory and practice in the human realm is the idea that humans and their societies develop stable tastes and preferences, which determine economic decisions. (Vermeij 2009, p. 119)

He adds,

> These preferences reflect goals that other life forms are thought not to have: health, happiness, affluence, influence, reputation, and the like.

Vermeij cites Mokyr (2006 p. 1011) in this context, who proposes that economies are not like ecologies in that the main purpose of life seems to be life itself.... There is no real analog in biology to the economist's concept of utility.

Vermeij disagrees and believes that the reality is more nuanced than that. Vermeij's view reflects work in peace ethology that suggests that the difference between humans and other animals in the concept of utility is likely one of degree rather than of kind (cf. Bekoff 2007; Bekoff & Pierce 2009). Vermeij captures this as he refers to the work on natural peace by de Waal:

> With the evolution and elaboration of nervous systems and social organization, some animal lineages became increasingly endowed with emergent emotions, meaning, purpose, and intentional preferences for happiness, wealth, leisure, status, honor, peace, reputation, arousal, dignity, and pleasure. In our social species, shared culture-music, religion, fashion, and the pursuit of knowledge-provides the context in which these diverse forms of utility flourished. (Vermeij 2009, p. 119; cf. de Waal 1996)

Peace ethology's focus on natural peace offers the kind of hope that springs from a deeper understanding of nature's ways. To paraphrase the conservation biologists Swaisgood and Shephard (2010), it is unlikely that knowledge gained through peace ethology can help avert the expected worsening of the current environmental crises. However, by turning our attention to natural peace, we can find meaning in nature even in a dramatically altered world. The remainder of this chapter sets out to delineate such meaning through documented examples of natural peace.

Situating Natural Peace

The potential passion for peace can be as deep as the desire for survival which is potentially as real as the total passion for life. It is buttressed by the urge of evolution itself.

Lentz (1955, p. 187)

Lentz clearly saw the roots of peace in nature, linking it to "the urge of evolution itself." Peace is by definition a social process, and Lentz' vision reflects the evolution of Earth's vast number of social organisms. In essence, a social organism is an organism that behaves in a way that affects and is affected by the behavior of another organism. We can call such behavioral interaction *sociality*. Some authors restrict sociality to members of the same species (e.g., Robinson *et al.* 2008), but here sociality is seen as both within species (intraspecific) as well as between species (interspecific). So defined, sociality is so widespread that it has been argued that all of life on Earth is social (Frank 2007; Ghoul *et al.* 2017).

Sociality involves interactions between actors with potentially diverging interests. How will they work out inevitable conflicts of interest? Do they deal with them aggressively or peacefully? And what affects their choice of strategy? How do peaceful choices contribute to the propagation of life? Seeking answers to these fundamental questions requires peace ethology's fourfold focus on the immediate causation,

Table 16.1 Social dimensions of natural peace.

Sociative	Structural	Direct
Symbiosis / mutualism	Social norms	Behavioral inhibition
Tolerance / habitat sharing	Inequity aversion	Social inhibition
Impartial intervention		
Postconflict behavior		

development, function, and evolution of behavior. To sum up peace ethology's approach, we can say that to uncover the *proximate nature* of peaceful behavior, we must determine what triggers the behavior and how *environmental, contextual, developmental, social, emotional, cognitive,* and *physiological* mechanisms affect it. To uncover the *ultimate nature* of peaceful behavior, we must determine how the peaceful behavior contributes to *inclusive fitness,* where *inclusive fitness = direct fitness + indirect fitness.* In evolutionary terms, *direct fitness* means individual survival and reproductive success, and *indirect fitness* means the survival and reproductive success of closely related individuals. Simply put, peaceful behavior can evolve if it benefits the fitness of the actor and/or benefits the fitness of individuals who share some of the same genes as the actor.

The first chapter in this volume identifies three social dimensions of peace: *direct, structural,* and *sociative* (Verbeek & Peters this volume, Chapter 1). *Sociative peace* is behavior associated with the preservation of harmony in relations through the pursuit, establishment, or deepening of mutual or reciprocal interests; helping and sharing; and the active avoidance of aggressive confrontations and restoration of peaceful relations following aggression (Ali & Walters this volume, Chapter 5; Evans Pim this volume, Chapter 12; Fry this volume, Chapter 14; Kunneman this volume, Chapter 15; Narvaez this volume, Chapter 6; Romero this volume, Chapter 4; Roseth this volume, Chapter 7).

Direct peace involves behavior that negates or overcomes physical violence (e.g., Adang *et al.* this volume, Chapter 10; Furnari this volume, Chapter 8; Peters this volume, Chapter 11; Shnabel this volume, Chapter 2; Wessells & Kostelny this volume, Chapter 9), while *structural peace* involves behavior aimed at negating or overcoming structural violence (e.g., Otten *et al.* this volume, Chapter 3; Hyslop & Morgan this volume, Chapter 13).

In the remainder of this section, I explore whether and how these social dimensions of peace apply to the lives of nonhuman organisms and how peace ethology investigates this. Table 16.1 presents areas of behavior that correspond to each social dimension of peace. Like the social dimensions, these aspects of behavior are not mutually exclusive, and there are other aspects of peaceful behavior such as cooperation that can co-occur with any or all of them (cf. Brosnan & Bshary 2010).

Sociative Peace

Symbiosis and Mutualism

To understand the function and evolution of peaceful behavior, we need to understand the interplay between *genotypes* and *phenotypes.* Genotypes are genes included in the full genetic makeup of an organism (the *genome*) that are associated with an organism's traits. A *phenotype* is the expression of a *genotype* as mediated by the organism's developmental history (West-Eberhard 2003) and active engagement with the natural world (Piersma & van Gils 2011).

While genes are considered "selfish," when expressed as social phenotypes they can transcend selfishness and become important for all living systems. *Social phenotypes are* phenotypes that exert an effect (either positive or negative) on the fitness of other individuals. Social phenotypes evolved, in part, because of this fitness effect that they exert (Nadell *et al.* 2016). Social phenotypes can be found at multiple levels of organismal complexity, from microbes to vertebrates and from mice to men. Here, social phenotypes that exert positive effects are referred to as *peaceful phenotypes*. Delineating peaceful phenotypes and their interactions within and across species and mapping their development and constraints are the job of peace ethology.

Insights gained from research on microbes show that peaceful phenotypes, including cooperative phenotypes, may be linked to *pleiotropy*, which is the influence of one gene on two or more seemingly unrelated phenotypic traits. The life cycle of the social slime mold, *Dictyostelium discoideum*, is an interesting example of this. In this social species, a single gene, *dimA*, links altruistic with reproductive behavior. Hatching from spores, *D. discoideum* starts its life as a unicellular organism feeding on bacteria and dividing by cell division (mitosis). As part of the species' life cycle, some cells aggregate in a stalk that holds up other cells as reproductive spores. Up to 20% of the cooperative cells die in the process of forming the species-sustaining stalk. The fact that both the costly cooperative trait of stalk production and the beneficial trait of spore production are linked to the functioning of a single gene limits the range of possible cheating strategies (Foster 2011). Cheating in a biological context refers to reaping the benefits of cooperation without making any contribution to it. If cooperative traits are genetically linked to traits that provide a benefit to the individual cells, then this will constrain mutations that produce noncooperative phenotypes because the mutation will also remove the individual benefit. Pleiotropic links between costly cooperative traits and individual beneficial traits may be a frequent property in the genetics underlying cooperative traits, which helps to stabilize them over evolutionary time (Foster 2011).

Microorganisms often live in diverse social communities bound together by a secreted polymer matrix called a *biofilm*. Biofilms are essential for the healthy functioning of the microbiota of multicellular organisms such as the microbiota in our gut that aid in our digestion of food. Biofilms also cause antibiotic-tolerant infections and so are of great concern in medical settings (Nadell *et al.* 2016). Often, the cells in biofilms positively influence each other's fitness through the secretion and exchange of products that are mutually beneficial, such as digestive enzymes necessary for the digestion of nutrients. Microorganisms are thus fundamentally social organisms.

By affecting the fitness of neighboring cells in a biofilm, social phenotypes can alter the composition and structure of microbial communities and their function (and, in the case of pathogens, their virulence). To understand the sociality of microorganisms, it is thus important to understand the balance between competition and cooperation in biofilms (Nadell *et al.* 2016). A rapidly developing new field in microbiology, *sociogenomics*, focuses on this question. Life on Earth has transitioned from the evolution of cells to multicellular organisms to the organization of these organisms into societies (Robinson *et al.* 2005), and research in sociogenomics helps to situate sociative peace at basic natural levels. The specific goal of sociogenomics is to uncover genes and pathways that regulate the aspects of development, physiology, and behavior that influence sociality, and the ways all of these are influenced by social life and social evolution (Robinson *et al.* 2005). This is a task that transcends disciplinary boundaries, and peace ethology's fourfold approach is crucial for its success.

The traditional view of nature as a combat scene had to yield for a field like sociogenomics to emerge. Microbiologist Lynn Margulis is one of the pioneers whose work precipitated this, as she famously argued, "life did not take over the globe by combat but by networking" (Margulis & Sagan 2001). She referred, in part, to endosymbiotic theory (Sagan 1967; Margulis 2010), her life's work on the symbiotic origins of the eukaryotic cell. Building on earlier work by the botanist Boris Kozo-Polyansky, Margulis' work showed that some of the organelles that distinguish eukaryotic cells from prokaryotic cells were once free-living single-celled prokaryotes that merged with an ancestral host cell. Specifically, the origin of the mitochondrion, the organelle that provides eukaryotic cells with their chemical energy, is linked to an initially free-living oxygen-dependent (aerobic) bacterium, while the origin of the chloroplast, the organelle that conducts photosynthesis in plant and algal cells, is linked to an initially free-living photosynthesizing cyanobacterium.

Fifty years after being introduced amidst much controversy, endosymbiotic theory has matured into symbiogenesis (Margulis 2010; Kozo-Polyansky 2010), a universally accepted biological principle that sheds light on basic interdependencies of life on Earth and adds a networking challenge to the traditional focus on evolutionary novelty through random genetic mutations (Gray 2017; López-García *et al.* 2017; O'Malley 2017). The work of Margulis and others puts a spotlight on how interdependencies within and across species play an important role in the evolution and propagation of life on Earth. Thanks to this research focus, we now know that no multicellular eukaryotic organism is capable of surviving and reproducing using only its nuclear genes and the gene products it makes (Thompson 2006). It follows that situating natural peace is a holistic enterprise that must consider multiple forms of interdependencies, including symbiosis.

Symbiosis is the living together in close proximity of different organisms that confers mutual or unilateral benefits to them. Like in the case of the eukaryotic cell, symbiosis may involve the merging of previously independently living organisms. Symbiosis that confers mutual benefits is called *mutualism*, while symbiosis with unilateral benefits is distinguished by either the absence (*commensalism*) or presence (*parasitism*) of negative effects on the non-benefiting organism(s).

There are ample examples of symbiotic mutualisms that help sustain life. For example, many animals, including our own species, host numerous species of microbes in their cells and bodies, and these *symbionts* are functional in their host's metabolic pathways and other physiological functions (Gilbert *et al.* 2012). Symbionts can also mediate the development of their host. In the developing guts of mice and zebrafish, for example, hundreds of genes are activated by bacterial symbionts (Hooper *et al.* 2001; Rawls *et al.* 2004; cited in Gilbert *et al.* 2012). Algal symbionts provide nutrients that coral needs for survival. When rising sea-surface temperatures kill their algal symbionts, corals "bleach" and die (Gilbert *et al.* 2012).

The immune system has long been seen as a defense network against a hostile exterior world that rejects anything that is not "self." We now know that the function of the immune system is significantly reduced when symbiotic microbes are absent in the gut

(Mazmanian *et al.* 2008; Round *et al.* 2010; cited in Gilbert *et al.* 2012). The more we learn about symbiotic relationships in nature, the more the boundaries between self and other that have characterized Western thought are becoming blurred (Tauber 2008, 2009; cited in Gilbert *et al.* 2012). At this level of analysis, peace, defined as mutually beneficial interactions (*sociative peace*), is a process that transcends traditional theoretical boundaries of self and species. Exciting new areas of inquiry spawned from symbiogenesis and sociogenomics that synthesize phenotypic and genomic analyses (e.g., Ghoul *et al.* 2017; Kasper *et al.* 2017) are set to offer new insights into social evolution and natural peace as a crucial aspect of it (Verbeek 2016).

Tolerance and Habitat Sharing

> But the question now is what we as biology teachers should do so people realize that mutual cooperation between different kinds of organisms is just as important in helping each other to survive. First, as biology teachers, we need to explain and illustrate symbiosis mechanism not only through microscopic organisms but throughout all the levels of biological and social organizations particularly if we want to replace the image of mankind as the conqueror with the image of mankind as an integral part of the natural world.
>
> Cherif (1990, p. 207)

Multicellular organisms such as vertebrates show multiple forms of sociative peace. For example, there are numerous related vertebrate species (*related* meaning grouped together taxonomically in a genus, family, or order) that show tolerance toward each other or travel and forage together within their shared habitat. Mixed-species flocks of birds are a good example of this. What attracts these bird species to each other, and how are their mixed flocks maintained? What is the function of these mixed-species flocks? A large-scale, multiyear study covering 30 bird families from 17 sites worldwide showed that vulnerability to predation is an important factor in a species' tendency to join a mixed-species flock. Small insect-eating species that forage in trees are especially vulnerable to predation and often join mixed flocks (Sridhar *et al.* 2009).

Mixed-species flocks include flock leaders and flock followers, and the former are commonly cooperatively breeding species that forage in kin groups. Members of such kin groups are likely to have well-developed intraspecific communication and alarm call systems, and flock followers that join cooperative breeders might be able to exploit this anti-predator system (Sridhar *et al.* 2009). Large aggregations of birds may flush more insects or other prey than smaller ones do, and this benefits flock leaders and flock followers alike. The presence of many eyes in a mixed-species flock reduces the amount of time individuals need to spend on vigilance, which allows for more time for feeding (Sridhar *et al.* 2009).

A field experiment in Japan showed that willow tits (*Poecile montanus*) use a specific "tää" call to attract birds of their own and other species when they arrive at a food patch alone. Foraging in mixed-species flocks provides better predator vigilance and avoidance for the willow tit compared to foraging alone (Suzuki & Kutsukake 2017). However, the willow tit has two distinct foraging behaviors; one is eating food on the spot, and the other is caching it for later consumption. While foraging with other birds is advantageous in terms of predator protection, willow tits likely do not want others around when they cache food to avoid the risk of having it stolen. In a cleverly designed field experiment using artificial foraging patches and playback of "tää" recordings, it was found that willow tits use their recruiting call only when they set out to eat food directly and not when they cache food (Suzuki & Kutsukake 2017). This study and others like it show that individuals within mixed-species flocks actively coordinate peaceful associations with other species through interspecific signaling.

Mixed-species groups also occur in mammals, in particular among primates, ungulates, cetaceans, and certain carnivores (Stensland *et al.* 2003). Mammalian mixed-species groups are made up of social species that live in loosely formed rather than stable social groups. Direct benefits for peaceful association with other species include detection and utilization of food, and predator detection, avoidance, or deterrence. Social advantages include home range increase, territory defense, as well as practicing social behaviors (Stensland *et al.* 2003).

Similar to birds, mammals coordinate peaceful association through signaling. The case of the plains zebra (*Equus quagga*) is an interesting example. Zebra occur in mixed-species groups more often than expected by chance and associate with a range of species, including blue wildebeest (*Connochaetes taurinus*), impala (*Aepyceros melampus*), giraffe (*Giraffa camelopardalis*), cape buffalo (*Syncerus caffer*), oryx (*Oryx beisa/gazelle*), common eland (*Taurotragus oryx*), Grant's gazelle (*Nanger granti*), and Thomson's gazelle (*Eudorcas thomsonii*) (Ireland & Ruxton 2017).

Zebra are hind-gut fermenters, while their bovid associates are ruminants. Zebra require larger quantities of low-nutrition foliage, while bovids require small quantities of high-nutrition short grass. Zebra have opposing incisors and can eat woody stems, while ruminants have only lower incisors and cannot, so zebra can clear rank grass and overgrown vegetation for ruminants (Ireland & Ruxton 2017). For zebra, the benefit of peaceful association with bovid species is reduced predation risk, particularly the risk of being caught by lions (*Panthera leo*). Zebra can outrun lions, and so for zebra, spotting lions is as important as avoiding being spotted by them, and the extra senses and alarm calls of their bovid associates help.

Bovids have poor daytime visual acuity, and it is possible that zebra stripes evolved to signal their identity to potential bovid associates (Ireland & Ruxton 2017). Zebra, in fact, appear to be the leader species that attracts others to the mixed-species group. In mixed-species bird flocks, leader species also attract other species to the flock and maintain cohesion through calls, behavior, or coloration (Ireland & Ruxton 2017). Like zebra, leader species in bird flocks tend to be at risk for predation, tend to be feeding generalists that accommodate other species, and tend to alarm call more than follower species (Ireland & Ruxton 2017).

Peaceful mixed-species aggregation also confers advantages to many nonhuman primates. For example, in eastern Brazilian Amazonia, peaceful associations between squirrel monkeys (*Saimiri sciureus*) and brown capuchin monkeys (*Sapajus apella*) may last from a few hours to several weeks (Pinheiro *et al.* 2011). Besides enjoying better predator defenses, squirrel monkeys benefit from the capuchin monkeys' ability to get at hard-cased foods such as palm seeds, while the capuchin monkeys can capture more arthropod prey disturbed by the squirrel monkeys. Long-term field observations revealed very few aggressive interactions between individuals of the two species, and when threats do occur they are commonly followed by dispersal rather than physical confrontation (Pinheiro *et al.* 2011; for more on mixed-species groups in nonhuman primates, see Heymann 2011; Verbeek 2006).

A widespread aspect of *sociative peace* is the formation of dominance relationships among social animals. In many social species, individuals gain information from interactions with other individuals and recognize previously encountered individuals. Often, individuals that lost a fight are likely to subsequently defer to the winner, and winners may show less or no aggression toward individuals that they won a previous fight against. Aggression is costly behavior with the potential of fatal injury, and the establishment of dominance relationships, often managed through specific dominance or submissive behavioral signals, is one of nature's ways to keep aggression in check (Chase 1985).

While in many social species aggressive familiarity limits subsequent aggression, nonaggressive familiarity can also reduce aggression among individuals. Hamsters (*Mesocricetus* spp.) are favorite subjects for the aggression laboratory as males tend to show intense aggression toward each other in confined experimental settings (delBarco-Trillo *et al.* 2009). In one laboratory study, male Turkish hamsters (*Mesocricetus brandti*) were housed together separated by a wire mesh for 10-minute periods over 10 days. When they were exposed to each other 10 days after they last saw each other, they showed a decrease in the number of fights, the percentage of time fighting, as well as latency to fight when they were compared to males who were not familiar with each other. The authors concluded that the decrease in aggression linked to the experimentally induced nonaggressive familiarity resembles what is called the *dear enemy effect* in free-ranging territorial animals (delBarco-Trillo *et al.* 2009).

Some territorial animals direct less or no aggression to established neighbors while directing high levels of aggression to intruding strangers. This *dear enemy effect* depends on the ability to discriminate between familiar neighbors and strangers. In songbirds, for example, males learn to distinguish the vocal signatures of neighbors from those of strangers. Other species use vocal signature discrimination in this context as well. A field experiment in Taiwan using playback tests showed that territorial males of the olive frog (*Babina adenopleura*) had higher thresholds for producing aggressive calls in response to the advertisement calls of their nearby neighbors compared with those of strangers (Fen-Chuang *et al.* 2017).

Dear enemy–like effects can also function to facilitate peaceful coexistence at the species community level. Field experiments in Nicaragua on cichlid fish of the genus *Amphilophus* showed that *Amphilophus zaliosus* territorial males direct less aggression to breeding neighbors of the related species *Amphilophus astorquii* compared to nonbreeding *A. astorquii* neighbors (Lehtonen *et al.* 2010). Nonbreeding individuals of *Amphilophus* species have been shown to prey on each other's offspring, while breeding individuals are too preoccupied with raising their young to do the same. Follow-up field experiments in the same natural habitat showed that territorial aggression in both males and females of the bi-parental moga cichlid (*Hypsophrys nicaraguensis*) decreased quicker when they were sequentially presented with the same heterospecific intruder stimulus (*Amphilophus sagittae*, *Amphilophus xiloaensis*, and *Parachromis managuensis*) than when they were presented on each round with a different stimulus. Males showed higher levels of aggression, and the decrease in aggression was quicker in males than in females. This sex difference may relate to the different roles of the sexes in territory defense and parental care (Lehtonen & Wong 2017).

A recent laboratory study provides insights into the proximate mechanisms of dear enemy effects in cichlid fish. Androgen hormone levels are known to increase in response to territorial intrusions, and in a study on Mozambique tilapia (*Oreochromis mossambicus*), neighbor intruders elicited lower aggression and a weaker androgen response than strangers on first intrusion (Aires *et al.* 2015).

Intervention in Aggression

Nonhuman primate bystanders may intervene in the aggressive interactions of other group members with the apparent goal of ending the aggression. Such impartial interventions have been observed in a number of macaque species as well as in chimpanzees, bonobos, orangutans (*Pongo pygmaeus*), golden snub-nosed monkeys (*Rhinopithecus roxellanae roxellanae*), and hamadryas baboons (*Papio hamadryas*; all cited in Rudolf von Rohr *et al.* 2012).

In chimpanzees, impartial interventions are commonly conducted by high-ranking individuals and are aimed at all sex-dyad combinations (de Waal 1982, 1992). Impartial interventions often involve aggression among multiple actors and occur more often during periods of social instability in a group, for example following the arrival of new group members. As such, impartial interventions may function as a group stabilizer by preventing aggression from escalating and further disrupting group stability (de Waal 1982, 1992; Rudolf von Rohr *et al.* 2012). Both in the natural habitat (Boehm 1992, 1994; Nishida & Hosaka 1996; Pusey *et al.* 1997) and in captivity (Rudolf von Rohr *et al.* 2012), maintaining social stability appears to be a shared concern among chimpanzees (cf. de Waal 2014).

A recent study on seven large captive groups of rhesus macaques (*Macaca mulatta*) identified various behaviors aimed at restoring group stability. Across groups, bystanders were shown to engage in impartial interventions and support subordinates in both *dyadic* (two opponents) and *polyadic* (more than two opponents) fights (Beisner & McCowan 2013). Supporting one of the actors in a fight is commonly seen as a behavior that benefits the actor, for example by elevating or solidifying dominance rank or by facilitating mating opportunities. In this large study, however, no evidence for direct benefits of partial interventions was found, and the authors suggest that in rhesus macaques, both impartial and partial interventions can serve group stability that benefits most if not all members of the group.

Postconflict Behavior

> [I]t happened one day when the colony was locked indoors in one of its winter halls. In the course of a charging display, the highest-ranking male fiercely attacked a female. This caused great commotion as other apes came to her defense. After the group had calmed down, an unusual silence followed, as if the apes were waiting for something to happen. This took a couple of minutes. Suddenly the entire colony burst out hooting, and one male produced rhythmic noise on metal drums stacked up in the corner of the hall. In the midst of this pandemonium two chimpanzees kissed and embraced. (de Waal 2000a, p. 16; cf. de Waal 1989)

The two chimpanzees that kissed and embraced in de Waal's 1970s groundbreaking observation were the male and female who just minutes earlier were engaged in a fierce physical fight (de Waal & van Roosmalen 1979). As de Waal writes, once recognized, this kind of peaceful postconflict behavior was seen on a daily basis, and it was hard to imagine that it had gone unnoticed for so long (de Waal 2000a). These peaceful reunions between former opponents were labeled *reconciliation* on the assumption that they have a lasting effect on their relationship.

De Waal also noticed another peaceful postconflict behavior in which uninvolved bystanders contacted recipients of aggression in a friendly manner shortly after a fight. As this bystander behavior was shown to comfort and reassure the often still visibly upset targets, it was labeled *consolation* (de Waal & van Roosmalen 1979). A controlled observation method was developed to systematically test the selective occurrence of these postconflict behaviors against chance (i.e., the postconflict matched control [PC-MC] method; de Waal & Yoshihara 1983; Veenema *et al.* 1994). With the use of this standardized observation method, postconflict reconciliation has been demonstrated in more than 40 Old and New World nonhuman primate species, in both natural and captive settings, as well as in a number of other social mammals, ranging from wolves (*Canis lupus*: Cordoni & Palagi 2008) and their domesticated cousin the dog (*Canis lupus familiaris*: Cools *et al.* 2008) to red-necked wallaby (*Macropus rufogriseus*: Cordoni & Norscia 2014) and bottlenose dolphin (*Tursiops truncates*: Yamamoto *et al.* 2016), and also in birds, including raven (*Corvus corax*: Fraser & Bugnyar 2010, 2011) and budgerigars (*Melopsittacus undulates*: Ikkatai *et al.* 2016), and fish (*Labroides dimidiatus*: Bshary & Würth 2001) (see Verbeek 2013 for a more detailed listing of reconciling species).

Research on reconciliation in nonhuman primates inspired similar research in young children of various cultures using the same observation method (e.g., Verbeek & de Waal 2001; Fujisawa *et al.* 2006). Verbeek (2008) compared postconflict behavior in nonhuman primates and children and found significant similarities in rates and timing, function, and proximate causation of reconciliation. Across cultures, young children tend to make peace following peer aggression and conflict on average four times out of ten. Studies from countries with different social systems, including Japan and the United States, show that teacher intervention in peer conflict decreases the likelihood that the opponents make peace (Verbeek 2008). Taken together, the comparative evidence suggests that a peacemaking tendency is a universal feature of early childhood and is molded by culture as children grow older (Butovskaya *et al.* 2000; Verbeek 2008; Roseth this volume, Chapter 7; cf. Verbeek 2013).

As defined in this chapter, consolation is postconflict behavior in which an uninvolved bystander initiates friendly contact with a recipient of aggression. First observed in chimpanzees, consolation has been a highly replicable finding in wild and captive chimpanzee groups and has also been documented in bonobos, mountain gorillas (*Gorilla beringei beringei*), Western lowland gorillas (*Gorilla gorilla gorilla*), and Tonkean macaques (*Macaca tonkeana*) (all cited in Lindegaard *et al.* 2017). In monkeys, bystanders have also been observed to offer friendly contact when solicited to do so by victims (e.g., *Sajapus* [formerly *Cebus*] *apella*: Verbeek & de Waal 1996). The distinction between (1) consolation initiated by the bystander ("consolation proper") and (2) friendly bystander contact solicited by recipients of aggression is important in terms of understanding the function as well as the cognitive and emotional causal factors of the behavior (Verbeek & de Waal 1996; Romero this volume, Chapter 4). Indicative of convergent evolution, consolation has now also been observed in nonprimate species such as canids, corvids, rodents, and elephants (all cited in Lindegaard *et al.* 2017). Empathy, the ability to relate to the emotional state of others, is seen as a major driver of this particular aspect of sociative peace in nonhuman and human animals alike (Preston & de Waal 2002; de Waal & Preston 2017).

Genus *Pan*

Social conflicts are inevitable in social groups, and postconflict reconciliation functions to signal an end to aggression and to restore potentially valuable social interactions as well as established social relationships. Chimpanzees and bonobos are our closest living evolutionary relatives, and finding out how adults of both species balance aggressive and peaceful behavior in their social life is of significant interest in terms of understanding natural peace in our own species. Not unlike human society, both chimpanzees and bonobos live in a fission–fusion society in which small groups alternate between splitting off from the larger community (fission) and merging back in it again (fusion). While postconflict reconciliation functions to repair relations in the aftermath of aggression in both species, chimpanzee society is characterized by frequent aggression within and between communities, whereas bonobo society is built on more peaceful relations, both within and between communities (Clay *et al.* 2016a).

Male–male and female–male relations are important factors with regard to the difference in overall peacefulness between chimpanzee and bonobo society. In both species, females show the periodic swelling of the sexual skin that is common in nonhuman primates. In chimpanzee females, further increases in sexual swelling occur as the time of ovulation approaches, and the interest of dominant males in the female, as indicated by association, copulations, and the rate of intervention in copulation attempts by lower ranking males, closely matches these subtle changes in swelling size (Deschner *et al.* 2004). In contrast, in bonobo females, the onset and duration of the maximum swelling phase are highly variable and only weakly predictive of ovulation (Douglas *et al.* 2016). As swelling cycles also occur without ovulation, the sexual swellings of bonobo females are significantly less reliable indicators of ovulation than those of chimpanzee females (Douglas *et al.* 2016). These different patterns of sexual signaling in chimpanzee and bonobo females correlate with a reduced value of intermale aggression in bonobos compared to chimpanzees and with an increased value of close male–female relations in bonobos (Clay *et al.* 2016a; Surbeck *et al.* 2017).

While chimpanzee males are more likely to fight each other than bonobo males, they are also more likely to reconcile after their fights and to maintain close relationships and groom each other compared to bonobo males (Surbeck *et al.* 2017). Compared to chimpanzee society, the more peaceful society of bonobos is characterized by less sexual aggression by males; greater initiative of females in social, sexual, and ranging behaviors; and ecological factors, such as greater food abundance and shorter travel distance between food patches (Clay *et al.* 2016a).

Structural Peace

In peace ethology's conceptual framework, direct peace is associated with physical violence, in particular lethal violence, while structural peace is linked to social and psychological violence. In this context, and as we saw in this chapter, violence is seen as aggression out of inhibitory control. Physical violence violates life itself, while structural violence violates well-being and social norms that sustain peaceful life, such as norms of fairness and equity and the protection of the weakest members of society. Actions that oppose physical violence and aim to undo the conditions that bring it about are direct peace processes. Actions that oppose social norm violations and aim to undo the conditions that bring them about are structural peace processes. Here, I explore behavioral processes indicative of structural peace in nonhuman animals.

Norm Violations

In human and nonhuman animal societies, infants and juveniles commonly occupy a protected position, and violence directed at these vulnerable members of society violates the putative norm of protecting them. In human society, the social norm of protecting children is formalized locally and globally in civil and human rights law (Ruck *et al.* 2017). Opposing the violation of the basic norm of protecting the young is a structural peace process that can range from communicating distress about the violation, attempting to stop it, and consoling the victim to taking civil or legal action against it.

In chimpanzee society, infants generally enjoy high levels of tolerance (Rudolf von Rohr *et al.* 2011), but they are also at risk during severe aggression both within and between groups (Rudolf von Rohr *et al.* 2015). At several field sites, lethal violence against infants has been observed to provoke strong reactions in bystanders, ranging from high arousal and screaming to attempts to intervene and coalitions in defense of the mother–infant pair (Townsend *et al.* 2007; Rudolf von Rohr *et al.* 2015).

Following the characteristic ethological method of experimentally verifying naturalistic findings, Rudolf von Rohr and colleagues showed two captive groups of chimpanzees video depicting violence against infants, including infanticide, as well as other intense social scenes including hunting and killing of a colobus monkey and severe aggression among adults (Rudolf von Rohr *et al.* 2015). Chimpanzees in both groups spent four times as much time looking at the infanticide scenes compared to the other social scenes. Effectively ruling out other explanations, the researchers concluded that the longer looking times were consistent with the idea that violence against infants did not match chimpanzees' social expectation of tolerance afforded to infants. Being confronted with something they did not expect to see inspired the intense concentration the chimpanzees in both groups showed. Taken together, these studies show that structural peace processes surrounding the normative treatment of infants, including communicating distress and acting to oppose the norm violation, are present in chimpanzees.

Species-Typical and Species-Atypical Aggression

Recent research shows that bonobos, like chimpanzees, have specific expectations about interspecific aggression. The research conducted in a sanctuary located in the natural habitat of the species showed that when confronted with aggression that occurred without provocation or warning, bonobos would "protest vocally" using screams that were acoustically and temporally significantly different from screams associated with aggression associated with inter-individual competition (Clay *et al.* 2016b). The bonobos appeared to have expectations about what constitutes "normal" (i.e., species-typical) aggression and protested against unexpected (species-atypical) aggression that violated these expectations (Clay *et al.* 2016b).

Inequity Aversion

The examples in this chapter from chimpanzee and bonobo research show that the violation of social expectations can provoke strong emotions in nonhuman animals. Captive and experimental studies across vertebrates also suggest that the violation of expectation can provoke aggression as well. Atlantic salmon (*Salmo salar*), for example, showed increased aggression when they did not receive a food reward that they had been trained to expect (Vindas *et al.* 2012). For our understanding of natural peace, reactions to the violation of social norms are of particular interest as social norms are instrumental in getting along peacefully in the social group. Negative reactions to the violation of social norms reflect a shared expectation of how social norms ought to preserve peaceful behavior (de Waal 2014; cf. Vermeij 2009; Andrews 2013). It has been argued that such awareness of "what ought to be done to preserve peace" can be seen as a rudimentary *sense of justice* (Pierce & Bekoff 2012; cf. Decety & Yoder 2016).

A rapidly increasing body of research on behavior in social vertebrates that facilitates equity or responds to inequity offers clues about a natural sense of justice as a factor in structural peace. As most of the findings come from captive studies, it will be important to verify and further investigate these behaviors in social animals in their natural habitat wherever possible. Here, I highlight recent studies.

Rats (*Rattus* spp.) are known for their social nature and are widely used in behavioral research (Whishaw & Kolb 2005). In a recent study, food-deprived Sprague–Dawley laboratory rats that were unfamiliar to each other did not compete when a limited amount of food was made available to them; rather, they socialized in a way that resulted in equal access to the available food (Weiss *et al.* 2017). In another study, laboratory rats systematically preferred a food reward that offered an equal quantity of food to them and to another rat to an unequal reward that offered them more food than the other rat in the experimental pair (Oberliessen *et al.* 2016).

As more than 13 million YouTube viewers have seen, aversion to inequity was first demonstrated in capuchin monkeys (*S. apella*) (www.youtube.com/watch?v=meiU6TxysCg). In a now classic study, capuchin monkeys working in pairs on a task that required equal effort refused a food reward (a slice of cucumber) that was less desirable to them than the reward (a grape) that the other monkey in the pair had received (Brosnan & de Waal 2003).

Similar aversion to inequity has been shown in apes and other nonhuman primates (reviewed in Brosnan & de Waal 2014), in wolves and dogs (Essler *et al.* 2017), and in ravens (*Corvus corax*) and carrion crows (*Corvus corone corone*; Wascher & Bugnyar 2013). This so-called first-order inequity aversion is assumed to serve peaceful cooperation as it allows social partners to select each other on the perceived fairness of the outcomes of their cooperative interactions. Cheaters who take more than their fair share from a cooperative interaction are shunned, while those who play fairly are preferred, and peaceful cooperation is optimized (Brosnan & de Waal 2014).

In chimpanzees, but not yet in other nonhuman animals, some individuals have also been shown to reject a reward that was greater than the one offered to a partner. Individuals showing this second-order inequity aversion are believed to anticipate the occurrence of first-order inequity aversion in the partner. The cost associated with refusing the greater reward is believed to be offset by the ultimately greater benefit of stabilizing the valuable relationship with the partner and protecting the actor's investment in it (Brosnan & de Waal 2014).

Much remains to be learned about how different social animals preceive equity in the outcomes of their peaceful cooperation and about how differences in inequity aversion may be linked to individual factors such as dominance, social context, and social temperament ("personality") (Brosnan & Bshary 2016). It seems fair to say, however, that taken together with the data on norm violations and reactions to species-atypical aggression, the findings on inequity aversion provide comparative evidence for structural peace processes in nonhuman animals.

Direct Peace

A Chimpanzee Model of War

In peace ethology's conceptual scheme, direct peace is the social dimension of peace associated with the negation of physical violence, in particular lethal violence such as war. In popular language, appalling aspects of human behavior like war are often labeled with adjectives referring to other animals. We speak, for example, about behavior being "beastly" and about "releasing the dogs of war." Traditional approaches in the social and natural sciences have long looked for clues about war in the behavior of other animals. The guiding ideas were that violence is rampant in nature and that a better understanding of all this violence can help us devise the cultural tools to tame our "natural" instincts.

War can be defined as coordinated attack behavior associated with multiple conspecific deaths. Other than in insects, coordinated attack behavior resulting in multiple conspecific deaths is virtually absent among nonhuman animals. In nonhuman primates, small-scale lethal intergroup conflict has been observed in cebus monkeys, spider monkeys, and chimpanzees, but its occurrence is rare in each of these species, and its function remains unknown (Verbeek 2013, 2017a, 2017b).

Speaking in part based on his experiences as a member of the Dutch resistance and a prisoner of war in the Nazi-occupied Netherlands, Tinbergen suggested that war is caused by the upsetting of the balance between aggression and fear and that this is linked to consequences of cultural evolution. While aspects of our behavior such as territoriality, dominance striving, aggressive emotionality, and preparedness for group defense may be shared with other animals, in and by themselves or even in combination, they do not lead to war. As one of Tinbergen's contemporaries in ethology, Irenäus Eibl-Eibesfeldt, suggested, war, as coordinated attack behavior, requires planning, leadership, destructive weapons, and overcoming empathy by dehumanizing the enemy, all of which can be seen as products or consequences of human culture (Verbeek 2013, 2017a, 2017b).

Anthropologist Richard Wrangham and writer Dale Peterson situate the origins of war in nature. Wrangham and Peterson argue that chimpanzee-like violence preceded and paved the way for human war (Wrangham & Peterson 1996; Wrangham 1999). The brutality of some of the observed lethal intergroup encounters in chimpanzees inspired them to suggest that the killing is driven by an innate appetite for killing conspecifics. From this *chimpanzee model of war* perspective, the lethal intergroup encounters among chimpanzees and the raids and ambushes described as primitive forms of war among early humans are analogues in form and function (Wrangham & Peterson 1996; Wrangham 1999).

Chimpanzees are remarkably astute social actors who, as we saw in this chapter, regularly make peace after fights within their community, console victims of aggression, appease aggressors to reduce social tension, engage in impartial interventions to restore peace in the community, share and help each other, warn fellow group members who are ignorant of danger, and demonstrate a rudimentary sense of justice in experimental settings. And while killing does occur during chimpanzee intergroup encounters, the evidence so far suggests that it is rare. In comparison, evidence for intergroup killing in the closely related bonobos is nonexistent, and, seen from a human standpoint, bonobos appear ready for peace rather than for war (cf. Clay & de Waal 2013).

Work by Fry and others suggests that in the lineage of the human species, war is a recent invention, first occurring just prior to the agricultural revolution 10,000 years ago. Anthropological evidence suggests that peaceful coexistence rather than war characterized the lifestyle of the vast majority of the human genus's existence. Interdisciplinary work by Narvaez and colleagues provides evidence for an "evolved developmental niche for peace," consisting of a complex of child-rearing behaviors that foster peacemakers rather than warriors (Darvaez this volume, Chapter 6; cf. Shepard 1982). Evidence of the form and function of the evolved developmental niche for peace is found in contemporary peaceful societies that continue to practice the small-band hunter-gatherer existence that characterizes our evolutionary history as a species (Darvaez this volume, Chapter 6). Put against the backdrop of the evidence discussed in this chapter, the chimpanzee model appears of limited use for furthering our understanding of war as one of the most appalling forms of violence in our species.

An Elephant System of Peace

In many species, species-typical aggression is subject to behavioral checks, ritualization, and inhibitions (cf. Fry this volume, Chapter 14). Expressed with these natural checks and balances in place, aggression, like peaceful behavior, can enhance fitness, for example through mating competition. Violence, as species-atypical uninhibited aggression, does not enhance fitness. The multiple forms of violence that we see in our own species, directed both at ourselves and at other species, are not an adaptation but an aberration. We are not alone in expressing violence; violence occurs among nonhuman animals, including in places where human activity interferes with the natural order.

Between 1992 and 1997, a group of young orphaned male African elephants (*Loxodonta africana*) that had been introduced to Pilanesberg, South Africa, went on a killing spree. In a remarkable display of violence, they killed more than 40 white rhinoceros (*Ceratotherium simum*) (Slotow *et al.* 2000). The killer males were all showing signs of musth. Musth is a state of heightened aggression that sexually mature male Asian and African elephants temporarily enter and is associated with mating behavior (Slotow *et al.* 2000; Max Wyse *et al.* 2017). A male that engages in musth-driven mate competition with other males faces the possibility of injury or death in the event that the competition escalates into a violent confrontation. Thus, an optimal musth strategy balances the immediate benefit of gaining access to fertile females against the possibility that future benefits will be forgone if the male suffers a musth-related injury. In African elephants, body size and timing of musth have been found to be significant factors in an optimal musth strategy in which violence and injury are averted (Max Wyse *et al.* 2017).

The young killer males in Pilanesberg, South Africa, had been introduced there at an age of less than 10 years old after a cull of their previous population in Kruger Park. They had reached sexual maturity in the absence of adult males. In Amboseli, Kenya, young male elephants were found to be significantly less likely to be in musth when a larger musth male was present. Young males also lose the physical signs of musth minutes or hours after an aggressive interaction with a higher ranking musth male. Larger, older males appear to delay the onset of musth in younger males. To test this idea, six older males were introduced to Pilanesberg's population, which at the time consisted of 85 elephants, including the 17 young males, which were now aged between 15 and 25. The introduction of the adult males ended the killing of rhinoceros by the young males (Slotow *et al.* 2000).

Assessment theory predicts that selection should favor individuals who are able to assess the physical and behavioral traits of rival males and use this knowledge of the costs and benefits of fighting and the probability of winning to adjust their own behavior. In male elephants, a strategy of dropping out of musth when confronted with aggression from higher ranking musth males increases immediate survival, and therefore long-term reproductive fitness. This is a lesson the young males would have learned if they had grown up in the presence of adult males. Their violence was a direct consequence of having been deprived of their natural developmental niche during a crucial time in their development (Slotow *et al.* 2000; cf. Narvaez this volume, Chapter 6).

Elephants are long-lived, large-brained, intelligent mammals that live in a complex social system. Their well-being and survival depend on the optimal functioning of their social system, which revolves around close family relations. Intergroup aggression is subject to behavioral checks that keep it from transgressing into violence. The norms that govern species-typical aggression and the maintenance of peaceful relationships are learned during an extended period of social development. In African elephants, female matriarchs play an important role in making decisions about where to go next to find food, water, and safety (Mutinda *et al.* 2011). These matriarchs are a major source of ecological and social knowledge for the group. Matriarchs play an important role in repairing damage to family relations following aggression. Reconciliation and consolation often involve not only the aggressor and the recipient of aggression but also close family members and the matriarch. By responding vocally to the recipient of aggression and by touching one another, they restore peace (Poole 2011; cf. Asian elephants: Plotnik & de Waal 2014).

Within the context of the peace ethology model (Verbeek & Peters this volume, Chapter 1), the elephant social system is a natural peace system. Regrettably, in Africa, the elephant peace system is under severe threat from the mass deaths and social upheaval resulting from poaching, culling, and habitat loss (Bradshaw *et al.* 2005). Long-term ethological research and field experiments show that the negative effects of anthropogenic disruption on the elephant social system can last for decades (Shannon *et al.* 2013). We are quick to project our own warring ways on the behavior of other animals but have been slow in recognizing their natural peace systems. We can learn from the elephant peace system and must meet it with peace rather than with violence.

Epilogue

> *It is not necessary to "go back" in time to be the kind of creature you are. The genes from the past have come forward to us. I am asking that people change not their genes but their society, in order to harmonize with the inheritance they already have.*

Shepard (1999)

Traditional Western worldviews depicting "us versus them" in a perpetual combat of selfish genes still sustain the kind of science that wittingly or unwittingly perpetuates these worldviews. The metaphor of "our selfish genes versus their selfish genes" still provides a useful logic for the military industrial complex that fuels economic and political activity and funds major aspects of science in the Western world. But buyer beware! The strength and beauty of science are in its ability to upset the proverbial applecart with new knowledge and ideas that challenge everything that came before (Kuhn 1962). I suggest that this is happening throughout the biological sciences in terms of our understanding of natural peace, and the social sciences are following suit. As this chapter and the multidisciplinary work in this volume illustrate, a comprehensive behavioral science of peace is on the rise, supported, in part, by the conceptual framework and methods of the nascent discipline of peace ethology.

Tinbergen (1968) writes that scientific research is one of the finest occupations of our mind and is, with art and religion, one of the uniquely human ways of meeting nature – in fact, the most active way. The research discussed in this chapter meets the "peace of wild things" and suggests that peace has been part of life since long before the human species came along. We tend to be infatuated with our cultural exploits, but we would do well to remember that we are first and foremost a manifestation of nature – a recent one at that – and that important roots of our behavior are grounded in nature. The new knowledge that we gain from studying natural peace in other animals is highly relevant for peace in our own human lives. For too long, research on peace has meant an almost exclusive focus on aggression and violence and on how to negate these behaviors to the detriment of gaining an understanding of peaceful behavior in its own right. The examples of natural peace in nonhuman animals presented in this chapter inspire us to take a good hard look at the natural roots of direct, structural, and sociative peace in our own species.

The new knowledge about natural peace is equally relevant and important with regard to making our peace with the rest of nature. Not to put too fine a point on it, but as the pioneering ecologist Eugene Odum once wrote while calling for greater collaboration between the natural and social sciences, "Since man is a dependent heterotroph, he must learn to live in mutualism with nature; otherwise, like the 'unwise' parasite, he may so exploit his 'host' that he destroys himself" (Odum 1975, p. 142, cited in Cherif 1990).

References

Advanced Consortium on Cooperation, Conflict and Complexity (AC^4). (2015). *The sustainable peace mapping initiative: What is sustainable peace? Expert survey report.* New York: The Earth Institute, Columbia University.

Aires, R.F., Oliveira, G.A., Oliveira, T.F., Ros, A.F.H., & Oliveira, R.F. (2015). Dear enemies elicit lower androgen responses to territorial challenges than unfamiliar intruders in a cichlid fish. *PLoS ONE, 10*(9), e0137705. doi:10.1371/journal.pone.0137705

Andrews, K. (2013). Ape autonomy? Social norms and moral agency in other species. In K. Petrus & M. Wild (Eds.), *Animal minds & animal ethics* (pp. 173–196). Bielefeld, Germany: Transcript Verlag.

Aureli, F., & de Waal, F.B.M. (2000). *Natural conflict resolution.* Berkeley: University of California Press.

Beisner, B.A., & McCowan, B. (2013). Policing in nonhuman primates: Partial interventions serve a prosocial conflict management function in rhesus macaques. *PLoS ONE, 8*(10), e77369. doi:10.1371/journal.pone.0077369

Bekoff, M. (2007). *The emotional lives of animals.* Novato, CA: New World Library.

Bekoff, M., & Pierce, J. (2009). *Wild justice: The moral lives of animals.* Chicago: University of Chicago Press.

Boehm, C. (1992). Segmentary "warfare" and the management of conflict: Comparison of East African chimpanzees and patrilineal-partilocal humans. In A. Hartcourt & F.B.M. de Waal (Eds.), *Us against them: Coalitions and alliances in humans and other animals* (pp. 137–173). Oxford: Oxford University Press.

Boehm, C. (1994). Pacifying interventions at Arnhem Zoo and Gombe. In R.W. Wrangham, W.C. McGrew, F.B.M. de Waal, & P.G. Heltne (Eds.), *Chimpanzee cultures* (pp. 211–226). Cambridge, MA: Harvard University Press.

Bradshaw, G.A., Schore, A.N., Brown, J.L., Poole, J.H., & Moss, C.J. (2005). Elephant breakdown. *Nature, 433,* 807.

Brosnan, S.F., & Bshary, R. (2010). Theme issue: Cooperation and and deception: from evolution to mechanisms. *Philosophical Transactions of the Royal Society B, 365*(1553).

Brosnan, S.F., & Bshary, R. (2016). On potential links between inequity aversion and the structure of interactions for the evolution of cooperation. *Behaviour, 153*(9–11), 1267–1292.

Brosnan, S.F., & de Waal, F.B.M. (2003). Monkeys reject unequal pay. *Nature, 425,* 297–299.

Brosnan, S.F., & de Waal, F.B.M. (2014). Evolution of responses to (un)fairness. *Science, 346*(6207). doi:10.1126/science.1251776

Bshary, R., & Würth, M. (2001). Cleaner fish *Labroides dimidiatus* manipulate client reef fish by providing tactile stimulation. *Philosophical Transactions of the Royal Society B, 268,* 1495–1501.

Burkhardt, R.W. (2005). *Patterns of behavior: Konrad Lorenz, Niko Tinbergen, and the founding of ethology.* Chicago: University of Chicago Press.

Butovskaya, M., Verbeek, P., Ljungberg, T., & Lunardini, A. (2000). A multi-cultural view of peacemaking among young children. In F. Aureli & F.B.M. de Waal (Eds.), *Natural conflict resolution.* Berkeley: University of California Press.

Chase, I.D. (1985). The sequential analysis of aggressive acts during hierarchy formation: An application of the "jigsaw puzzle" approach. *Animal Behaviour, 33,* 86–100.

Cherif, A.H. (1990). Mutualism: The forgotten concept in teaching science. *The American Biology Teacher, 52*(4), 206–208.

Chivian, E., & Bernstein, A. (2008). *Sustaining life: How human life depends on biodiversity.* Oxford: Oxford University Press.

Christie, D.J. (2012). Peace psychology: Definitions, scope, and impact. In D.J. Christie (Ed.), *The encyclopedia of peace psychology.* Chichester: Wiley-Blackwell.

Chuang, M.F., Kam, Y.C., & Bee, M.A. (2017). Territorial olive frogs display lower aggression towards neighbours than strangers based on individual vocal signatures. *Animal Behavior, 123,* 217–228.

Clay, Z., & de Waal, F.B.M. (2013). Bonobos respond to distress in others: Consolation across the age spectrum. *PLoS ONE, 8*(1), e55206. 10.1371/journal.pone.0055206

Clay, Z., Furuichi, T., & de Waal, F. B. M. (2016a). Obstacles and catalysts to peaceful coexistence in chimpanzees and bonobos. *Behaviour, 153*(9–11), 1293–1330.

Clay, Z., Ravaux. L., de Waal, F.B.M., & Zuberbühler, K. (2016b). Bonobos (*Pan paniscus*) vocally protest against violations of social expectations. *Journal of Comparative Psychology, 130*(1), 44–54.

Cohrs, J.P., & Boehnke, K. (2008). Social psychology and peace. *Social Psychology, 39*(1), 4–11.

Coleman, P.T. (2012). Conclusion: The essence of peace? Toward a comprehensive and sustainable model of sustainable peace. In P.T. Coleman & M. Deutsch (Eds.), *Psychological components of sustainable peace.* New York: Springer.

Coleman, P.T., & Deutsch, M. (Ed.). (2012). *Psychological components of sustainable peace.* New York: Springer.

Cools, A.K.A., Van Hout, A.J.M., & Nelissen, M.H.J. (2008). Canine reconciliation and third-party-initiated postconflict affiliation: Do peacemaking social mechanisms in dogs rival those of higher primates? *Ethology, 114*(1), 53–63.

Cordoni, G., Demuru, E., Ceccarelli, E., & Palagi, E. (2016). Play, aggressive conflict, and reconciliation in preschool children: what matters? *Behaviour, 153*(9–11), 1075–1102.

Cordoni, G., & Norscia, I. (2014). Peace-making in marsupials: The first study in the red-necked wallaby (*Macropus rufogriseus*). *PLoS ONE, 9*(1), e86859. 10.1371/journal.pone.0086859

Cordoni, G., & Palagi, E. (2008). Reconciliation in wolves (*Canis lupus*): New evidence for a comparative perspective. *Ethology, 114*, 298–308.

Craik, K.J.W. (1943). *The nature of explanation.* London: Cambridge University Press.

Crook, P. (1994). *Darwinism, war and history.* Cambridge: Cambridge University Press.

Czárán, T., & Aanen, D. (2016). The early evolution of cooperation in humans: On cheating, group identify and group size. *Behaviour, 153*(9–11), 1247–1266.

Daily, G.C. (Ed.). (1997). *Nature's services: Societal dependence on natural ecosystems.* Washington, DC: Island Press.

Dawkins, R. (1976). *The selfish gene.* Oxford: Oxford University Press.

Decety, J., & Yoder, K.J. (2016). The emerging social neuroscience of justice motivation. *Trends in Cognitive Sciences, 21*(1), 6–14.

de Boer, S.F., Caramaschi, D., Natarajan, D., & Koolhaas, J.M. (2009). The vicious cycle towards violence: Focus on the negative feedback mechanisms of brain serotonin neurotransmission. *Frontiers in Behavioral Neuroscience, 3*(52), 1–6. doi:10.3389/neuro.08.052.2009

delBarco-Trillo, J., McPhee, M. E., & Johnston, R. E. (2009). Nonagonistic familiarity decreases aggression in male Turkish hamsters, *Mesocricetus brandti. Animal Behaviour, 77*, 389–393.

Deschner, T., Heistermann, M., Hodges, K., & Boesch, C. (2004). Female sexual swelling size, timing of ovulation, and male behavior in wild West African chimpanzees. *Hormones and Behavior, 46*, 204–215.

de Waal, F.B.M. (1982). *Chimpanzee politics: Power and sex among apes.* Baltimore, MD: Johns Hopkins University Press.

de Waal, F.B.M. (1989). *Peacemaking among primates.* Cambridge, MA: Harvard University Press.

de Waal, F.B.M. (1992). Coalitions as part of reciprocal relations in the Arnhem chimpanzee colony. In A. Harcourt & F.B.M. de Waal (Eds.), *Coalitions and alliances in humans and other animals* (pp. 233–257). Oxford: Oxford University Press.

de Waal, F.B.M. (1996). *Good natured: The origins of right and wrong in humans and other animals.* Cambridge, MA: Harvard University Press.

de Waal, F.B.M. (2000a). The first kiss: Foundations of conflict resolution research in animals. In F. Aureli & F.B.M. de Waal (Eds.), *Natural conflict resolution*. Berkeley: University of California Press.

de Waal, F.B.M. (2000b). Primates: A natural heritage of conflict resolution. *Science, 289*, 586–590.

de Waal, F.B.M. (2006). *Our inner ape*. New York: Riverhead Books.

de Waal, F.B.M. (2014). Natural normativity: The "is" and "ought" of animal behavior. *Behaviour, 151*, 185–204.

de Waal, F.B.M., & Preston, S.D. (2017). Mammalian empathy: behavioural manifestations and neural basis. *Nature Reviews Neuroscience, 18*, 498–509.

de Waal, F.B.M., & van Roosmalen, A. (1979). Reconciliation and consolation among chimpanzees. *Behavioral Ecology & Sociobiology, 5*, 55–66. doi:10.1007/BF00302695

de Waal, F.B.M., & Yoshihara, D. (1983). Reconciliation and redirected affection in rhesus monkeys. *Behaviour, 85*, 224–241.

Diehl, P.F. (2016). Exploring peace: Looking beyond war and negative peace. *International Studies Quarterly, 60*, 1–10.

Douglas, P. H., Hohmann, G., Murtagh, R., Thiessen-Bock, R., & Deschner, T. (2016). Mixed messages: Wild female bonobos show high variability in the timing of ovulation in relation to sexual swelling patterns. *BMC Evolutionary Biology, 16*, 140. doi:10.1186/s12862-016-0691-3

Essler, J., Marshall-Pescini, S., & Range, F. (2017). Domestication does not explain the presence of inequity aversion in dogs. *Current Biology, 27*, 1–5.

Foster, K.R. (2011). The sociobiology of molecular systems. *Nature Reviews Genetics, 12*, 193–203.

Frank, S.A. (2007). All of life is social. *Current Biology, 17*, R648–R650.

Fraser O.N., & Bugnyar, T. (2010). Do ravens show consolation? Responses to distressed others. *PLoS ONE, 5*(5), e10605. doi:10.1371/journal.pone.0010605

Fraser, O.N., & Bugnyar, T. (2011). Ravens reconcile after aggressive conflicts with valuable partners. *PLoS ONE, 6*(3), e18118. doi:10.1371/journal.pone.0018118

Fritze, J.G., Blashki, G.A., Burke, S., & Wiseman, J. (2008). Hope, despair and transformation: Climate change and the promotion of mental health and wellbeing. *International Journal of Mental Health Systems, 2*(13). doi:10.1186/1752-4458-2-13

Fry, D.P. (2006). *The human potential for peace: An anthropological challenge to assumptions about war and violence*. New York: Oxford University Press.

Fry, D.P. (2012). Life without war. *Science, 336*, 879–884.

Fry, D.P. (Ed.). (2013). *War, peace, and human nature. The convergence of evolutionary and cultural views*. New York: Oxford University Press.

Fry, D.P., & Souillac, G. (2017). The original partnership societies: Evolved propensities for equality, prosociality, and peace. *Interdisciplinary Journal of Partnership Studies, 4*(1), art. 4. Retrieved from http://pubs.lib.umn.edu/ijps/vol4/iss1/4

Fuentes, A. (2004). It's not all sex and violence: Integrated anthropology and the role of cooperation and social complexity in human evolution. *American Anthropologist, 106*, 710–718.

Fujisawa, K., Kutsukake, N., & Hasegawa, T. (2006). Peacemaking and consolation in Japanese preschoolers witnessing peer aggression. *Journal of Comparative Psychology, 120*(1), 48–57.

Galtung, J. (1969). Violence, peace, and peace research. *Journal of Peace Research, 6*(3), 167–191.

Galtung, J. (1996). *Peace by peaceful means.* London: Sage Publications.

Galtung, J., & Ikeda, D. (1995). *Choose peace.* London: Pluto Press.

Ghoul, M., Andersen, S.B., & West, S.A. (2017). Sociomics: Using omic approaches to understand social evolution. *Trends in Genetics, 1355.* 10.1016/j.tig.2017.03.009

Gilbert, S.F., Sapp, J., & Tauber, A.I. (2012). A symbiotic view of life: We have never been individuals. *The Quarterly Review of Biology, 87*(4), 325–341.

Gleditsch, N.P., Nordkvelle, J., & Strand, H. (2014). Peace research: Just the study of war? *Journal of Peace Research, 51*(2), 145–158.

Gray, M.W. (2017). Lynn Margulis and the endosymbiont hypothesis: 50 years later. *Molecular Biology of the Cell, 28*(10), 1285–1287.

Heymann, E.W. (2011). Coordination in primate mixed-species groups. In M. Boos, M. Kolbe, P.M. Kappeler, & T. Ellwart (Eds.), *Coordination in human and primate groups.* Berlin: Springer Verlag.

Himmler, S.M., Himmler, B.T., Pellis, V.C., & Pellis, S.M. (2016). Play, variation in play and the development of socially competent rats. *Behaviour, 153*(9–11), 1103–1138.

Hooper, L.V., Wong, M.H., Thelin, A., Hansson, L., Falk, P.G., & Gordon, J.I. (2001). Molecular analysis of commensal host-microbial relationships in the intestine. *Science, 291,* 881–884.

Howell, A.J., Passmore, H., & Buro, K. (2012). Meaning in nature: Meaning in life as a mediator of the relationship between nature connectedness and well-being. *Journal of Happiness Studies, 14*(6), 1681–1696.

Huxley, J. (1955). Foreword. In T.H. Lentz, *Towards a science of peace: Turning point in human destiny* (pp. v–viii). New York: Bookman Associates.

Ikkatai, Y., Watanabe, S., & Izawa, E.I. (2016). Reconciliation and third-party affiliation in pair-bond budgerigars (*Melopsittacus undulates*). *Behaviour, 153*(9–11), 1173–1194.

Ireland, H.M., & Ruxton, G.D. (2017). Zebra stripes: An interspecies signal to facilitate mixed-species herding? *Biological Journal of the Linnean Society, 121*(4), 947–952.

Jaeggi, A.V., Boose, K.J., White, F.J., & Gurven, M. (2016). Obstacles and catalysts of cooperation in humans, bonobos, and chimpanzees: Behavioural reaction norms can help explain variation in sex roles, inequality, war and peace. *Behaviour, 153*(9–11), 1015–1051.

Kahn, P.H., Jr. (1999). *The human relationship with nature: Development and culture.* Cambridge, MA: MIT Press.

Kahn, P.H., Jr. (2002). Children's affiliations with nature: Structure, development, and the problem of environmental generational amnesia. In P.H. Kahn Jr. & S.R. Kellert (Eds.), *Children and nature: Psychological, sociocultural, and evolutionary investigations.* Cambridge, MA: MIT Press.

Kant, I. (1991). Zum ewigen frieden [Perpetual peace]. In H. Reiss (Ed.), *Kant's political writings* (2nd ed., pp. 93–130). Cambridge: Cambridge University Press. (Original work published in 1795)

Kasper, C., Vierbuchen, M., Ernst, U., Fischer, S., Radersma, R., Raulo, A., Cunha Saraiva, F., Wu, M., Mobley, K.M., & Taborsky, B. (2017). Genetics and developmental biology of cooperation. *Molecular Ecology,* 1–14. 10.1111/mec.14208

Kellert, S.R. (1996). *The value of life.* Washington, DC: Island Press.

Kellert, S.R. (2012). *Birthright: People and nature in the modern world.* New Haven, CT: Yale University Press.

Kellert, S.R., & Wilson, E.O. (Eds.). (1993). *The biophelia hypothesis.* Washington, DC: Island Press.

Kelman, H.C. (2010). Peace research: Beginnings. In N. Young (Ed.), *The Oxford international encyclopedia of peace* (Vol. 3, pp. 453–458). New York: Oxford University Press.

Kelman, H.C. (2012). Social psychology and the study of peace: Personal reflections. In L. Tropp (Ed.), *The Oxford handbook of intergroup conflict* (pp. 361–372). New York: Oxford University Press.

Kozo-Polyansky, B.M. (2010). *Symbiogenesis. A new principle of evolution.* Cambridge, MA: Harvard University Press.

Kruk, M. R., & Kruk-de Bruin, M. (2010). *Discussions on context, causes and consequences of conflict.* Leiden: Lorentz Center, Leiden University.

Kruuk, H. (2004). *Niko's nature: The life of Niko Tinbergen and his science of animal behavior.* Oxford: Oxford University Press.

Kuhn, T.S. (1962). *The structure of scientific revolutions.* Chicago: University of Chicago Press.

Kutsukake, N., & Castles, D.L. (2004). Reconciliation and post-conflict third-party affiliation among wild chimpanzees in the Mahale Mountains, Tanzania. *Primates, 45*, 157–165.

Lehtonen, T.K., McCrary, J.K., & Meyer, A. (2010). Territorial aggression can be sensitive to the status of heterospecific intruders. *Behavioural Processes, 84*, 598–601.

Lehtonen, T.K., & Wong, B.B.M. (2017). Males are quicker to adjust aggression towards heterospecific intruders in a cichlid fish. *Animal Behaviour, 124*, 145–151.

Lengen, C., & Kistemann, T. (2012). Sense of place and place identity: Review of neuroscientific evidence. *Health & Place, 18*(5), 1162–1171.

Lentz, T.F. (1955). *Towards a science of peace: Turning point in human destiny.* New York: Bookman Associates Inc.

Lindegaard, M.R., Liebst, L.S., Bernasco, W., Heinskou, M.B., Philpot, R., Levine, M., & Verbeek, P. (2017). Consolation in the aftermath of robberies resembles post-aggression consolation in chimpanzees. *PLoS ONE, 12*(5), e0177725. 10.1371/journal. pone.0177725

López-García, P., Eme, L., & Moreira, D. (2017). Symbiosis in eukaryotic evolution. *Journal of Theoretical Biology, 43*, 20–33.

Mackie, G.O. (1970). Neuroid conduction and the evolution of conducting tissues. *Quarterly Review of Biology, 45*, 319–322.

Margulis, L. (2010). Symbiogenesis: A new principle of evolution rediscovery of Boris Mikhaylovich Kozo-Polyansky (1890–1957). *Paleontological Journal, 44*(12), 1525–1539.

Margulis, L., & Sagan, D. (2001). Marvelous microbes. *Resurgence, 206*, 10–12.

Max Wyse, J., Hardy, I.C.W., Yon, L., & Mesterton-Gibbons, M. (2017). The impact of competition on elephant musth strategies: A game-theoretic model. *Journal of Theoretical Biology, 417*, 109–130.

Mazmanian, S.K., Round, J.L., & Kasper, D.L. (2008). A microbial symbiosis factor prevents intestinal inflammatory disease. *Nature, 453*, 620–625.

Mutinda, H., Poole, J.H., & Moss, C.J. (2011). Decision making and leadership in using the ecosystem. In C.J. Moss, H. Croze, & P.C. Lee (Eds.), *The Amboseli elephants.* Chicago: The University of Chicago Press.

Nadell, C.D., Drescher, K., & Foster, K.R. (2016). Spatial structure, cooperation and competition in biofilms. *Nature Reviews Microbiology, 14*, 589–600. doi:10.1038/ nrmicro2016.84

Nishida, T., & Hosaka, K. (1996). Coalition strategies among adult male chimpanzees of the Mahale Mountains, Tanzania. In W.C. McGrew, L.F. Marchant, & T. Nishida (Eds.), *Great ape societies* (pp. 114–134). Cambridge: Cambridge University Press.

Oberliessen, L., Hernandez-Lallement, J., Schaeble, S., van Wingerden, M., Seinstra, M., & Kalenscher, T. (2016). Inequity aversion in rats, *Rattus norvegicus. Animal Behaviour*, 115–166.

O'Brien, K.J. (2017). *The violence of climate change*. Washington, DC: Georgetown University Press.

Odum, E.P. (1975). *Ecology: The link between the natural and social sciences*. New York: Holt, Rinehart and Winston.

O'Malley, M.A. (2017). From endosymbiosis to holobionts: Evaluating a conceptual legacy. *Journal of Theoretical Biology*, *434*, 1–114.

Palagi, E., Cordoni, G., Demuru, E., & Bekoff, M. (2016). Fair play and its connection with social tolerance, reciprocity and the ethology of peace. *Behaviour*, *153*(9–11), 1195–1216.

Pallante, V., Stanyon, R., & Palagi, E. (2016). Agonistic support towards victims buffers aggression in geladas (*Theropithecus gelada*). *Behaviour*, *153*(9–11), 1217–1246.

Parker, R.J. (1977). Theodore F. Lentz: Retirement for humanity. *Bulletin of the Peace Studies Institute*, *7*(2), 1–6.

Passano, L. (1963). Primitive nervous systems. *Proceedings of the Natural Academy of Sciences*, *50*, 306–313.

Passmore, H. (2014). Eco-existential positive psychology: Experiences in nature, existential anxieties, and well-being. *The Humanistic Psychologist*, *42*, 370–388.

Pierce, J., & Bekoff, M. (2012). Wild justice redux: What we know about social justice in animals and why it matters. *Social Justice Research*, *25*(2), 122–139.

Piersma, T., & van Gils, J.A. (2011). *The flexible phenotype*. New York: Oxford University Press.

Pinheiro, T., Ferrari, S.F., & Lopes, M.A. (2011). Polyspecific associations between squirrel monkeys (*Saimiri sciureus*) and other primates in Eastern Amazonia. *American Journal of Primatology*, *73*, 1145–1151.

Plotnik, J., & de Waal, F.B.M. (2014). Asian elephants (*Elephas maximus*) reassure others in distress. *PeerJ*, *2*, e278; doi:10.7717/peerj.278

Pollin, W. (1962). The behavioral sciences and war and peace. *Science*, *135*, 305–306.

Poole, J.H. (2011). Behavioral contexts of elephant acoustic communication. In C.J. Moss, H. Croze, & P.C. Lee (Eds.), *The Amboseli elephants*. Chicago: The University of Chicago Press.

Preston, S.D., & de Waal, F.B.M. (2002). Empathy: Its ultimate and proximate bases. *Behavioral and Brain Sciences*, *25*(1), 1–20.

Pusey, A., Williams, J., & Goodall, J. (1997). The influence of dominance rank on the reproductive success of female chimpanzees. *Science*, *277*, 828–831.

Rapaport, L.G., Paul, C.E., & Gerard, P. (2016). Hwæt! Adaptive benefits of public displays of generosity and bravery in *Beowulf. Behaviour*, *153*(9–11), 1331–1364.

Rawls, J.F., Samuel, B.S., & Gordon, J.I. (2004). Gnotobiotic zebrafish reveal evolutionarily conserved responses to the gut microbiota. *Proceedings of the National Academy of Sciences of the United States of America*, *101*, 4596–4601.

Reed, E.S. (1996). *Encountering the world: Toward an ecological psychology*. New York: Oxford University Press.

Robinson, G.E., Fernald, R.D., & Clayton, D.F. (2008). Genes and social behavior. *Science, 322*, 896–900.

Robinson, G.E., Grozinger, C.M., & Whitfield, C.W. (2005). Sociogenomics: Social life in molecular terms. *Nature Reviews Genetics, 6.* doi:10.1038/nrg1575

Round, J.L., O'Connell, R.M., & Mazmanian, S.K. (2010). Coordination of tolerogenic immune responses by the commensal microbiota. *Journal of Autoimmunity, 34,* J220–J225.

Romero, T., Onishi, K., & Hasegawa, T. (2016). The role of oxytocin on peaceful associations and sociality in mammals. *Behaviour, 153*(9–11), 1053–1071.

Ruck, M.D., Peterson-Badali, M., & Freeman, M. (2017). *Handbook of children's rights: Global and multidisciplinary perspectives.* New York: Routledge.

Rudolf von Rohr, C., Burkart, J.M., & van Schaik, C.P. (2011). Evolutionary precursors of social norms in chimpanzees: A new approach. *Biology & Philosophy, 26,* 1–30.

Rudolf von Rohr, C.R., Koski, S.E., Burkart, J.M., Caws, C., Fraser, O.N., Ziltener, A., & van Schaik, C.P. (2012). Impartial third-party interventions in captive chimpanzees: A reflection of community concern. *PLoS ONE, 7*(3), e32494. doi:10.1371/journal. pone.0032494

Rudolf von Rohr, C., van Schaik, C.P., Kissling, A., & Burkart, J.M. (2015). Chimpanzees' bystander reactions to infanticide: An evolutionary precursor to social norms? *Human Nature, 26,* 143–160.

Sachs, J.D. (2015). *The age of sustainable development.* New York: Columbia University Press.

Sagan, L. (1967). On the origin of mitosing cells. *Journal of Theoretical Biology, 14*(3), 225–274.

Shannon, G., Slotow, R., Durant, S.M., Sayialel, K.N., Poole, J., Moss, C., & McComb, K. (2013). Effects of social disruption in elephants persist decades after culling. *Frontiers in Zoology, 10*(62), 1–10.

Shepard, P. (1982). *Nature and madness.* Athens: University of Georgia Press.

Shepard, P. (1999). *The only world we've got: A Paul Shepard reader.* San Francisco: Sierra Club Books.

Simonelli, I., Mercer, R., Bennett, S., Clarke, A., Fernandes, G.A.I., Fløtten, K., Maggi, S., Robinson, J.E., Simonelli, F., Vaghri, Z., Webb, E., & Goldhagen, J. (2014). A rights and equity-based "Platform and Action Cycle" to advocate child health and well being by fulfilling the rights of children. *The Canadian Journal of Children's Rights, 1*(1), 199–218.

Slotow, R., Dyk, G.V., Poole, J., Page, B., & Klocke, A. (2000). Older bull elephants control young males. *Nature, 408,* 425–426.

Sridhar, H., Beauchamp, G., & Shanker, K. (2009). Why do birds participate in mixed species foraging flocks? A large-scale synthesis. *Animal Behaviour, 78,* 337–347.

Stensland, E., Angerbjörn, A., & Berggren, P. (2003). Mixed species groups in mammals. *Mammal Review, 33,* 205–223.

Surbeck, M., Boesch, C., Girard-Buttoz, C., Crockford, C., Hohmann, G., & Wittig, R.M. (2017). Comparison of male conflict behavior in chimpanzees (*Pan troglodytes*) and bonobos (*Pan paniscus*), with specific regard to coalition and post-conflict behavior. *American Journal of Primatology, 79,* e22641.

Suzuki, T.N., & Kutsukake, N. (2017). Foraging intention affects whether willow tits call to attract members of mixed-species flocks. *Royal Society Open Science, 4,* 17022. 10.1098/ rsos.170222

Swaisgood, R.R., & Sheppard, J.K. (2010). The culture of conservation biologists: Show me the hope! *BioScience, 60*(8), 626–630.

Tauber, A.I. (2008). The immune system and its ecology. *Philosophy of Science, 75,* 224–245.

Tauber, A.I. (2009). The biological notion of self and non-self. In E.N. Zelta (Ed.), *Stanford encyclopedia of philosophy.* Retrieved from http://plato.stanford.edu/entries/biologyself/

Thompson, J.N. (2006). Mutualistic webs of species. *Nature, 312,* 372–373.

Tinbergen, N. (1968). On war and peace in animals and man. *Science, 160,* 1411–1418.

Townsend, S.W., Slocombe, K.E., Emery Thompson, M., & Zuberbühler, K. (2007). Female-led infanticide in wild chimpanzees. *Current Biology, 17,* R355–R356.

Veenema, H.C., Das, M., & Aureli, F. (1994). Methodological improvements for the study of reconciliation. *Behavioural Processes, 31,* 29–39.

Verbeek, P. (2006). Everyone's monkey: Primate moral roots. In M. Killen & J. Smetana (Eds.), *Handbook of moral development.* Mahwah, NJ: Lawrence Erlbaum Associates.

Verbeek, P. (2008). Peace ethology. *Behaviour, 145,* 1497–1524.

Verbeek, P. (2009). Humanizing conservation. *Science, 325,* 817.

Verbeek, P. (2013). An ethological perspective on war and peace. In D. Fry (Ed.), *War, peace and human nature: The convergence of evolutionary and cultural views.* New York: Oxford University Press.

Verbeek, P. (2016). Speaking of which: The contribution of peace ethology to life science. In I. Norscia & E. Palagi (Eds.), *The missing lemur link. An ancestral step in the evolution of human behavior.* Cambridge: Cambridge University Press.

Verbeek, P. (2017a). The evolution of human violence and aggression: The contribution of peace ethology. In P. Sturmey (Ed.), *The Wiley handbook of violence and aggression.* Chichester: John Wiley & Sons Ltd.

Verbeek, P. (2017b). Animal behavior. In P.I. Joseph (Ed.), *The Sage encyclopedia of war: Social science perspectives* (pp. 69–73). Thousand Oaks, CA: Sage Publications.

Verbeek, P., & de Waal, F.B.M. (1997). Post-conflict behavior of captive brown capuchins in the presence and absence of attractive food. *International Journal of Primatology, 18*(5), 703–725.

Verbeek, P., & de Waal, F.B.M. (2001). Peacemaking among preschool children. *Peace and Conflict: Journal of Peace Psychology, 7*(1), 5–28.

Verbeek, P., & de Waal, F.B.M. (2002). The primate relationship with nature. Biophelia as a general pattern. In P.H. Kahn & S.R. Kellert (Eds.), *Children and nature: Psychological, sociocultural and evolutionary investigations.* Cambridge, MA: MIT Press.

Verbeek, P., & Palagi, E. (2016). Looking at natural peace through Niko Tinbergen's lens. *Behaviour, 153*(9–11), 1005–1011.

Vermeij, G.J. (2009). *Nature: An economic history.* Princeton, NJ: Princeton University Press.

Vindas, M.A., Folkedal, O., Kristiansen, T.S., Stien, L.H., Braastad, B.O., Mayer, I., & Øverli, Ø. (2012). Omission of expected reward agitates Atlantic salmon (*Salmo salar*). *Animal Cognition, 15,* 903–911.

Vollhardt, J.K., & Bilali, R. (2008). Social psychology's contribution to the psychological study of peace. *Social Psychology, 39*(1), 12–25.

Wascher, C.A.F., & Bugnyar, T. (2013). Behavioral responses to inequity in reward distribution and working effort in crows and ravens. *PLoS ONE, 8*(2), e56885. doi:10.1371/journal.pone.0056885

Webb, C.E., & Verbeek, P. (2016). Individual differences in aggressive and peaceful behavior: New insights and future directions. *Behaviour, 153*(9–11), 1139–1172.

Weiss, O., Dorfman, A., Ram, T., Zadicario, P., & Eilam, D. (2017). Rats do not eat alone in public: Food-deprived rats socialize rather than competing for baits. *PLoS ONE, 12*(3), e0173302. doi:10.1371/journal.pone.0173302

Wessells, M.G., McKay, S.A., & Roe, M.D. (2010). Pioneers in peace psychology: Reflections on the series. *Peace and Conflict: Journal of Peace Psychology, 16*, 331–339.

West-Eberhard, M.J. (2003). *Developmental plasticity and evolution.* New York: Oxford University Press.

Whishaw, I.Q., & Kolb, B. (2005). *The behavior of the laboratory rat: A handbook with tests.* Oxford: Oxford University Press.

Wilson, E.O. (1984). *Biophilia: The human bond with other species.* Cambridge, MA: Harvard University Press.

World Wide Fund for Nature (WWF). (2014). *Living planet report: Species and spaces, people and places.* Washington, DC: WWF.

Wrangham, R.W. (1999). The evolution of coalitionary killing. *Yearbook of Physical Anthropology, 42*, 1–30.

Wrangham, R., & Peterson, D. (1996). *Demonic males: Apes and the origins of human violence.* New York: Houghton Mifflin.

Yamamoto, C., Ishibashi, T., Yoshida, A., & Amano, M. (2016). Effect of valuable relationship on reconciliation and initiator of reconciliation in captive bottlenose dolphins (*Tursiops truncates*). *Journal of Ethology, 34*(2), 147–153.

Zylstra, M.J., Knight, A.T., Esler, K.J., & Le Grange, L.L. (2014). Connectedness as a core conservation concern: An interdisciplinary review of theory and a call for practice. *Springer Science Reviews, 2*, 119–143.

Index

Peace Ethology: Behavioral Processes and Systems of Peace, First Edition.
Edited by Peter Verbeek and Benjamin A. Peters.
© 2018 John Wiley & Sons Ltd. Published 2018 by John Wiley & Sons Ltd.